U.K. WASTE MANAGEMENT LAW

AUSTRALIA
LBC Information Services
Sydney

CANADA and USA
Carswell
Toronto

NEW ZEALAND
Brooker's
Auckland

SINGAPORE and MALAYSIA
Thomson Information (S.E. Asia)
Singapore

U.K. WASTE MANAGEMENT LAW

John H. Bates
Barrister

LONDON
SWEET & MAXWELL
1997

Published in 1997 by Sweet & Maxwell Limited of
100 Avenue Road, Swiss Cottage
London NW3 3PF
http://www.smlawpub.co.uk
Phototypeset by LBJ Enterprises Ltd of
Aldermaston and Chilcompton
Printed and bound in Great Britain by
Butler and Tanner Ltd
Frome and London

*No natural forests were destroyed to make this product;
only farmed timber was used and replanted*

A CIP catalogue record for this book is available from the
British Library

ISBN 0421 569 506

PREFACE TO THE SECOND EDITION

The first edition of this work was published in 1992. Since that time there have been major changes in waste management law in the United Kingdom. These result from the implementation of E.C. directives; in particular the amended Framework Waste Directive, the Hazardous Waste Directive and the Transfrontier Shipment of Waste Regulation. The second edition therefore incorporates those directives and United Kingdom implementing legislation.

In addition there have been changes in the way in which waste management is regulated in Great Britain. The Environment Agency and SEPA have been established to provide an integrated approach to pollution control and the waste management aspect of this is considered here. With the Directive on Integrated Pollution Prevention and Control—the IPPC Directive—a more holistic regime will be adopted; at least for those activities covered by the Directive. This book does not deal with the IPPC Directive as its implementation is some way off. However the effect of that Directive is worth noting at this stage.

The title of the book is U.K. Waste Management Law. However the Northern Ireland position is dealt with in outline only. The reason for this is that for some years the government has promised radical reform of the Northern Ireland position so as to bring it in line with waste regulation in Great Britain. This is likely to occur in 1998. Therefore it was felt that an extensive review of the current situation would have little value.

Waste management law is still a rapidly developing area. The draft landfill directive is just one example of this. Further, the subject is developing two distinct aspects—the management of waste produced and the prevention of waste production in the first place. As time goes on this second aspect will become more important. These changes will be dealt with in subsequent editions. This edition states the law at July 1997.

Old Square Chambers,   John H. Bates
1, Verulam Buildings,   July 1997
Grays Inn.

CONTENTS

TABLE OF CASES

TABLE OF STATUTES

TABLE OF STATUTORY INSTRUMENTS

TABLE OF E.C. & INTERNATIONAL TREATIES AND CONVENTIONS

TABLE OF E.C. SECONDARY LEGISLATION

1. E.C. WASTE MANAGEMENT LAW

This chapter looks at the principles of E.C. waste management law as expressed in the policy for waste and the "Framework Waste" directive. Specific provisions as to hazardous waste or the transport of waste will be dealt with where appropriate.

THE BASIS OF COMMUNITY ENVIRONMENTAL LAW

Initially the Treaty of Rome that established the European Economic 1.01 Community made no provision for measures relating to the protection of the environment. Thus early environmental directives were based on the general provisions of articles 2, 100 and 235. Of these the most important is Article 100 which concerns the harmonising of legislation throughout the Community so as to ensure fair competition within the Common Market. While the use of this roundabout way to achieve an environmental law making power has been criticised, it was upheld by the European Court.[1]

The lack of clear and specific environmental protection powers was dealt 1.02 with in Title II of the Single European Act which added new powers to the E.C. Treaties. These powers came into effect on July 1, 1987. Title II added Articles 130R, 130S and 130T to the Treaty. They were subsequently amended by the Maastricht Treaty. Article 130R establishes the basis for community environmental policy; the objectives of which are to preserve, protect and improve the quality of the environment, to contribute towards protecting human health, to ensure a prudent and rational utilisation of natural resources and to promote international co-operation.

By virtue of Article 130R.2   1.03

> "Community policy on the environment shall aim at a high level of protection taking into account the diversity of situations in the various regions of the Community. It shall be based on the precautionary principle and on the principles that preventive

---

[1] Case 91/79, *Re Detergents Directive, E.C. Commission v. Italian Republic* [1980] E.C.R. 1099; [1981] 1 C.M.L.R. 331.

action should be taken, that environmental damage should as a priority be rectified at source and that the polluter should pay. . ."[2]

Article 130R.3 further requires the Community in preparing its action in relation to the environment to take account of available scientific and technical data, environmental conditions in the regions of the E.C., the potential benefits and costs of action or of lack of action and the economic and social development of the Community as a whole and the balanced development of its regions.

1.04   Directives to protect the environment may be adopted under Article 130S while Article 130T allows Member States to adopt stricter environmental standards than those set by a directive, but such measures should only interfere with the harmonisation of trade to the extent necessary to achieve the environmental benefits they are intended to achieve.[3] Further environmental powers are provided by Article 100A which permits the Council to legislate to protect the environment for the purpose of regulating the internal market. These measures may be adopted by a qualified majority but, where this is done, Member States again have the right to adopt stricter measures for the protection of their own environments. The Commission, in making a proposal for a Directive under this Article, must take a high level of environmental protection as the basis for the measure.

ENVIRONMENTAL DIRECTIVES

1.05   Directives are the usual way in which E.C. environmental law is promulgated but Regulations are also proposed. By Article 189 of the Treaty

"a Directive shall be binding as to the result to be achieved upon each Member State to which it is addressed, but shall leave to the Member State the choice of form and method."

An Article in Directives will declare, "This Directive is addressed to the Member States." An Article in a Directive will also require Member States to bring into force the laws, regulations and administrative provisions necessary to comply with this Directive by a certain date and to inform the Commission that they have done so.

1.06   Directives must be implemented in full and within the prescribed time-scale. It is now well established by the European Court that a Member State may not plead provisions, practices or circumstances existing in its internal legal system in order to justify a failure to comply with the obligations and time limits resulting from Community Directives.[4] They can be implemented by statute, regulation or "administrative means" such as circulars or codes of practice provided that the principles of relevant constitutional or

---

[2] Art. 130R.2 as amended by the Maastricht Treaty.
[3] Case 302/86, *E.C. Commission (supported by U.K. intervener) v. Denmark, The Times*, October 3, 1988.
[4] Case 151/81, *E.C. Commission v. Ireland* [1982] E.C.R. III–3573.

administrative law guarantee that national authorities will in fact apply the Directive fully. If the implementing statute, regulation, etc., is incompatible with the Directive, the Directive must prevail.[5]

E.C. WASTE MANAGEMENT POLICY

In September 1989 the E.C. Commission sent the E.C. Council a 1.07 communication on a strategy for waste management. This was endorsed by a subsequent resolution of the Council on Waste Policy.[6] These documents set out the aims of the Community in establishing a common waste policy to protect the environment and to be consistent with the development of the internal market; bearing in mind the special economic characteristics of waste. Waste production in the Community is rising. Much of it is simply dumped, an action that is seen as wasting an economic asset. In addition there is a shortage of waste disposal capacity throughout the community.

The objective in Article 130R.1 of ensuring prudent and rational 1.08 utilisation of natural resources and Article 130R.2 of preventive action and the application of the "polluter pays" policy led the Council to consider that the production of waste should, where possible, be prevented or reduced at source. This involves the development of clean technologies to perfect non-polluting manufacturing processes producing little or no waste. This policy will be promoted through grants under the LIFE Regulation.[7] The second way of preventing waste is to ensure that products placed on the market make the smallest contribution, by their manufacture, use or final disposal, to increasing the amount of waste generated, particularly hazardous waste. This policy will be implemented through a "clean products" campaign which will involve "eco-labelling" to enable consumers to shop in an environmentally friendly manner and possibly a requirement that public procurement policies should specify the use of low waste products. More controversially the Council called on the Commission to bring forward proposals for ecological criteria on the design of products.[8]

The second stage in the waste management policy is to recycle or re-use 1.09 wastes. This will be vigorously promoted through research and development of re-use and recycling techniques—again with grant aid available under the LIFE Regulation. Collection and sorting systems will be optimised, the external costs of re-use and recycling reduced and outlets created for recycled materials. This may involve tax incentives or mandatory deposit systems. In addition specific directives will be introduced to establish schemes for recycling wastes in line with the "Beverage Containers" Directive. All these measures for recycling must be environmentally benign and where necessary regulated through legislation or codes of practice.

[5] R. v. Secretary of State for Transport, ex p. Factortame Ltd & Ors (No. 2) [1990] 3 W.L.R. 818 and Case C–106/89 Marleasing S.A. v. La Commercial Internacional de Alimentacion S.A. [1990] E.C.R. I–4135; [1992] 1 C.M.L.R. 305.
[6] [1990] O.J. C122/2.
[7] Regulation 1973/92 [1993] O.J. L206/1.
[8] Council Resolution, [1990] O.J. C122/2, para. 4.

1.10   Wastes that cannot be recycled or re-used must be disposed of. The first aim in this area is to create an adequate infrastructure for waste disposal throughout the Community based on the principle that each Member State should be self-sufficient in capacity. In particular such a policy would help to reduce movements of waste. Rules on disposal within the Community will aim to reduce exports of waste from one Member State to another. Standards for waste disposal should firstly require pre-treatment to reduce volume and potential harm. After treatment waste should be disposed of in compliance with strict standards. A programme of action on the disposal of particular wastes will be undertaken. Further once waste sites are closed there should be a period of supervision to ensure no harm results from the waste in them. Remedial work to sites that do cause harm should be carried out either under the "polluter pays" principle or through grants made under the LIFE regulation.

1.11   This policy was endorsed by the Community's environmental programme that sets goals for the years 1994–2000.[9]

THE E.C. FRAMEWORK DIRECTIVE ON WASTE

1.12   In 1975 the Council adopted a Directive on waste (75/442).[10] That directive was substantially amended by a further directive adopted in 1991 under Article 130S of the Treaty.[11] This basis for legislation was fiercely opposed by the European Parliament who considered that the correct basis was Article 100A because the purpose of the amendment was to harmonise the laws of the Member States in this area. However the European Court ruled that it was properly based on Article 130S.[12] The directive, as amended, is intended to provide the framework under which the waste management policy will be implemented throughout the Community and is thus known as the "Framework Directive." It is addressed to the Member States[13] who must have implemented its provisions by April 1993.[14]

1.13   One of the main purposes of the Framework Directive is to provide a common definition of waste. This is achieved in Articles 1 and 2 and Annex I. Those provisions are discussed in Chapter 3. However Article 2.2 enables the Council to make "Daughter Directives" in respect of particular categories of waste. These Directives will be concerned with the management of the waste, whether through its production, handling or final disposal.

1.14   Under Article 3 Member States must take appropriate measures to encourage the prevention or reduction of the overall amount of waste

---

[9] [1993] O.J. C138/1.
[10] [1975] O.J. L194/39.
[11] Dir. 91/156 [1991] O.J. L78/32.
[12] Case C–155/91 *Commission v. Council* [1993] E.C.R. I–939.
[13] *ibid.* Art. 21 as amended by Art. 18.2 of the Framework directive.
[14] Framework Waste Dir. 91/156, Art. 2.1.

produced in their countries and to make that waste less harmful. This is to be achieved by the development of clean technologies that use fewer natural resources, product controls to minimise waste and harm resulting from waste and the development of appropriate techniques to remove harmful substances from waste destined for recovery. Where waste is produced it should be recovered through recycling, re-use or reclamation or any other process that will allow the recovery of secondary raw materials. Alternatively waste should be used as a source of energy. Unless the policies concerned are contained in a British Standard the United Kingdom government should notify the Commission of any measures it is taking to achieve these waste reduction and recovery aims.

Member States must take the necessary measures to ensure that waste is 1.15 recovered or disposed of without endangering human health and without using processes or methods that could harm the environment. In particular they must ensure that there is no risk to water, air, the soil or plants and animals, that the recovery or disposal will not cause noise or odour nuisance and that it will not adversely affect the countryside or places of special interest.[15] This last requirement would extend the definition of "pollution of the environment" in section 29(3) of the Environmental Protection Act 1990. It also modifies the duty of a waste regulation authority in section 36(3) of the Act not to grant a waste management licence if it would cause "serious detriment to the amenities of the locality." The definition of "adversely affect" is likely to give the courts some trouble. Under the directive Member States are also required to take the necessary measures to prohibit the abandonment, dumping or uncontrolled disposal of waste; a requirement that is fulfilled by section 33 of the 1990 Act.[16]

The goal of self-sufficiency in Community waste disposal capacity is to be 1.16 achieved under Article 5. Member States are required, in co-operation with each other where necessary, to establish an integrated and adequate network of disposal installations, taking account of the best available technology not involving excessive costs. This should enable the Community as a whole to become self-sufficient in waste disposal. The introduction of the BATNEEC principle here in relation to "waste disposal installations" could raise the argument as to whether landfill, as opposed to incineration, etc., is the "best available technology." The network must also enable waste to be disposed of in one of the nearest appropriate installations, by means of the most appropriate methods and technologies in order to ensure a high level of protection for the environment and public health. This raises the question not only of the type of disposal option (*e.g.*: landfill), but also the method used to landfill—co-disposal, etc.

Article 5 also requires the network to enable the Member States to move 1.17 towards self-sufficiency individually, taking into account geographical circumstances or the need for specialised installations for certain types of waste. However the article does not envisage Member States becoming

---

[15] Framework Waste Dir., Art. 4.
[16] As amended by S.I. 1994 No. 1056, Sched. 4, para. 9(3)–9(5).

completely self-sufficient. They are allowed to co-operate in waste disposal and, under waste transfer rules, to import or export it within the Community.

1.18   Competent authorities are to be established in Member States to be responsible for the implementation of the directive.[17] In Great Britain these will be the Agencies acting as waste regulation authorities under section 30 of the Environmental Protection Act 1990.

1.19   Waste management plans must be drawn up by the competent authorities. "Management" in this context means the collection, transport, recovery and disposal of waste, including the supervision of such operations and the after-care of disposal sites.[18] The plans should, in particular, relate to the type, quantity and origin of waste to be recovered or disposed of, general technical requirements, any special arrangements for particular wastes and suitable disposal sites or installations. Suggestions are made as to the contents of such plans.[19] In the United Kingdom this requirement is met by waste recycling plans and the national waste strategies and the waste local plans under the Planning and Compensation Act 1990. Plans must be notified to the Commission. Member States may take measures to prevent movements of waste which are not in accordance with their waste management plans and, if they do so, should inform the Commission and other Member States of those measures.[20] This provision will in effect be met by the prohibition on unauthorised disposal of waste in the 1990 Act.

1.20   Under Article 8 a duty of care is imposed on "holders" of waste to ensure that their waste is either handled by a private or public waste collector or by an authorised disposal or recovery operator. Alternatively they can deal with it themselves in accordance with Article 11 of the Directive. The term "holder" here means the producer of the waste or the natural or legal person who is in possession of it.[21] This is a comprehensive definition, covering all those categories of person having contact with waste set out in section 34(1) of the Environmental Protection Act 1990 and probably exporters as well.

1.21   Categories of waste disposal operations are listed in Annex IIA of the directive. These categories are:

> D1 tipping above or underground as in a landfill;
> D2 land treatment such as biotreatment of contaminated land;
> D3 deep injection of liquid wastes into wells, etc.;
> D4 surface impoundment as in sludge lagoons;
> D5 specially engineered landfill where the waste is isolated from the environment;
> D6 discharges of solid waste into a watercourse or lake;
> D7 dumping at sea or seabed injection;
> D8 biological treatment;

[17] Framework Waste Dir., Art. 6.
[18] *ibid.* Art. 1(d).
[19] *ibid.* Art. 7.1.
[20] *ibid.* Art. 7.3.
[21] *ibid.* Art. 1(c).

D9 physiochemical treatment of waste, other than that specified elsewhere in the Annex, where the residue is disposed of by an operation categorised under the Annex;

D10 incineration on land;

D11 incineration at sea;

D12 permanent storage such as placement of containers in a mine;

D13 blending or mixture;

D14 repackaging or;

D15 storage before any categorised waste disposal operation is carried out.

However, the fact that an operation is listed in Annex IIA does not mean that it will be an available option. For example incineration at sea is now prohibited under the Oslo "Dumping" Convention.

Under Article 9 these waste disposal operations must be operated under a permit from the competent authority established under Article 6. The permit should ensure that the waste is disposed of without endangering human health or harming the environment.[22] It should specify the types and quantities of waste allowed, technical requirements, security precautions to be taken, the disposal site and any treatment methods. A permit may be granted for a specified period and be renewable. It may be subject to conditions and obligations. However, if it is considered that the intended method of disposal will not protect the environment it may be refused.[23] While these are not mandatory requirements, Community provisions with respect to landfill will be strengthened under the proposed landfill directive. **1.22**

Operations that may lead to waste recovery are listed in Annex IIB to the directive. The categories here are: **1.23**

R1 solvent reclamation or regeneration;

R2 the recycling or reclamation of organic substances that are not used as solvents;

R3 the recycling or reclamation of metals and their compounds;

R4 the recycling or reclamation of other inorganic materials;

R5 the regeneration of acids or bases;

R6 the recovery of components used for pollution abatement;

R7 the recovery of components from catalysts;

R8 oil refining or other reuses of oil;

R9 use principally as a fuel or other means to generate energy;

R10 agricultural uses such as composting—bearing in mind that some agricultural wastes are not "wastes" for the purposes of the directive;

R11 the use of wastes obtained from any of the above mentioned recovery operations;

R12 exchanges of wastes for these purposes;

R13 storage of wastes for recovery is regarded as an operation that may lead to recovery.

---

[22] Framework Waste Dir., Art. 4.
[23] *ibid*. Art. 6.2.

1.24   Installations or establishments carrying out these waste recovery operations must also obtain a permit to do so.[24] However the requirements of such a permit will not be as rigorous as one for a waste disposal operation. The sole purpose of it will be to ensure that the operation does not cause a risk to health or the environment as defined by Article 4. This requirement is implemented by the registration of such establishments or undertakings under Regulation 18 of the Waste Management Licensing Regulations 1994.[25]

1.25   Establishments or undertakings that dispose of waste at the place at which it is produced and those that carry out their own recovery operations may be exempted from the permit requirements of Articles 9 and 10. However, they may be controlled under the directive on hazardous wastes.[26] The exemption will only apply if the relevant authority has adopted guidelines as to how specific types of wastes should be dealt with in these ways and as long as no hazard to human health or harm to the environment ensues. The Commission should be informed of such guidelines. All establishments or undertakings to whom this exemption applies should be registered with the relevant authority.

1.26   Waste collection or transportation undertakings that operate on a professional basis, or which deal in or broker waste for disposal or recovery, and which are not otherwise subject to authorisation, must be registered with the competent authority. A question arises here as to what is meant by "on a professional basis". A charity may not operate for profit but still be organised on a professional basis. In Great Britain the registration of collectors and transporters of waste will be effected through the Control of Pollution (Amendment) Act 1989. Registration of waste dealers and brokers is achieved under Regulation 20 of the Waste Management Licensing Regulations 1994.

1.27   Operations that are subject to authorisation or registration under the Directive must be periodically inspected by the competent authority by virtue of Article 13. Inspection should be made as "appropriate", which presumably relates to the nature and scope of the establishment or installation concerned and its potential for environmental harm or risk to public health.

1.28   Waste disposal and recovery establishments or installations must keep a record of the quantity, nature, origin and, where relevant, the destination, frequency of collection, mode of transport and treatment method in respect of the waste they handle and the operations they carry out.[27] This information should be made available, on request, to the relevant authority. Member States may also require waste producers to comply with these information provisions. In Great Britain records are required to be kept by handlers of waste under regulations made by virtue of section 34(5) of the

---

[24] Framework Waste Dir., Art. 10.
[25] S.I. 1994 No. 1056.
[26] Framework Waste Dir., Art. 11.1.
[27] *ibid.* Art 14.1.

Environmental Protection Act 1990 and paragraph 14 of Schedule 4 to the Waste Management Licensing Regulations 1994.

In accordance with the "polluter pays" principle the costs of waste 1.29 disposal should be borne by the person who transfers the waste to a collector or disposal facility and/or the previous holders or the producer of the product from which the waste came.[28] This can be achieved through charges on waste collection and disposal.

Member States must send triennial reports to the Commission about their 1.30 implementation of the directive.[29] The Commission will produce a consolidated report on this following receipt of the individual communications.[30] In addition Member States must send copies of the principal legislation they have adopted to implement the directive to the Commission.[31] Measures to implement the directive should incorporate a reference to it or be accompanied by such a reference on their official publication.[32]

Articles 17 to 18 establish a committee to assist the Commission in 1.31 ensuring that the provisions of the directive keep pace with scientific and technical advances, particularly by amending its annexes. The procedure of the committee is governed by Article 148 of the Treaty. Any disagreements in the committee will be resolved by the E.C. Council.

The effect of this Directive has been examined by the European Court. In 1.32 *Comitato di Coordiamento per la Difesa della Cava and others v. Regione Lombardia and others*.[33] It was considered that Article 4—which sets out the duty to ensure that waste is not disposed of in an environmentally harmful manner—had no direct effect so as to confer rights on individuals which a national court is required to protect. Article 4 merely indicates a programme to be followed by a state and sets out the objectives which a Member State must observe in their performance of the more specific obligations imposed on them by Articles 5 to 11 of the Directive.

At first sight this may encourage the view that Articles 5 to 11 have 1.33 direct effect in E.C. law. However in *Ministere-Public v. Traen*[34] it was held that Article 10 of the original directive—which was concerned with the supervision of those collecting waste—did not lay down any particular requirement restricting the freedom of Member States regarding the way in which they organise the supervision of the activities referred to therein. Thus it is likely that in relation to any of the other Articles in the Directive the court would hold that it is what it is said to be, a "framework" directive whose provisions are neither unconditional nor sufficiently precise to have direct effect in E.C. law.

---

[28] Framework Waste Dir., Art. 15.
[29] *ibid*. Art. 12 as substituted by Dir. 91/692, Art. 5.
[30] *ibid*. Art. 16.
[31] *ibid*. Art. 20 and see Art. 18.2 of the Framework directive and Art. 2.2. of that directive.
[32] *ibid*. Art. 20.1.
[33] (1996) 8 J.E.L. 313.
[34] [1987] E.C.R. 2141; [1988] 3 C.M.L.R. 511.

2. WASTE CONTROL IN THE UNITED KINGDOM

THE OBJECTIVES OF CONTROL

2.01   Under the Control of Pollution Act 1974 the primary purpose of waste licensing was to prevent pollution of water or danger to public health. It was also to ensure that there should be no serious detriment to the locality in which licensed activities are carried on. These objectives were not defined in the 1974 Act but were drawn from its provisions by the court in *Attorney-General's Reference (No. 2 of 1988)*.[1]

2.02   In Part II of the Environmental Protection Act 1990 the objectives of waste management control are again not specified. However "relevant objectives" are set out in relation to the disposal or recovery of waste in Schedule 4 to the Waste Management Licensing Regulation 1994.[2] These are to ensure that waste is recovered or disposed of without endangering human health and without using processes or methods which could harm the environment. In particular waste should be recovered or disposed of without risk to water, air, soil, plants or animals, without causing a nuisance through noise or odours and without adversely affecting the countryside or places of special interest.[3] These objectives impose no duty on a disposer of waste. It is not an offence for a licence holder to fail to comply with them; unless they have been incorporated as a condition in her licence. However it is the duty of the licensing authority to discharge its functions in accordance with these objectives.[4] Further, licensing exemptions under Regulation 17 of and Schedule 3 to the 1994 Regulations can only be granted to an "establishment or undertaking"[5] if the type and quantity of waste, and the method of disposal or recovery by the activity concerned is consistent with the need to attain these objectives.[6]

2.03   Part II of the Environmental Protection Act 1990 is concerned with the prevention of pollution of the environment and harm to human health. This

---

[1] [1989] 3 W.L.R. 397.
[2] S.I. 1994 No. 1056, Sched. 4, para. 4.
[3] *ibid.* Sched. 4, para. 4(1)(a).
[4] *ibid.* Sched. 4, paras 2(1) & 3.
[5] See para. 16.10.
[6] S.I. 1994 No. 1056, reg. 17(4).

10

can be deduced from the provisions of section 33(1)(c), while section 35(3) adds an additional purpose of preventing serious detriment to the amenities of the locality in which a proposed licensed site will be situated.[7] These objectives in the Act now have to be read in the light of the "relevant objectives" imposed by the Framework Waste Directive and Schedule 4 of the 1994 Regulations.

Prevention of "pollution of the environment" has the wide meaning given **2.04** in section 29(1)–29(5) of the 1990 Act. The "environment", for these purposes, consists of all, or any, of the media of land, water and the air. "Land" includes land covered by waters where the land is above the low-water mark of ordinary spring tides[8]; but in any event "water" is a wide enough term to include the sea. "Pollution" of this "environment" here means pollution of the land, water or air due to the release or escape—into any environmental medium—from the land on which controlled waste is treated kept or deposited; or from fixed plant by means of which such waste is so dealt with. It will also cover pollution from a mobile plant.[9] Such pollution can arise from substances (any natural or artificial substances, whether solid, liquid, gas or vapour[10]) or articles constituting or resulting from the waste (such as rainwater contaminated by passing through the waste or landfill gases), and capable, by reason of the quantity or concentrations involved, of causing harm to man or any other living organisms supported by the environment.[11]

Harmless pollution is therefore not prohibited. If a small amount of clean **2.05** soil from a site falls into a stream and causes a brief discolouration of the waters, this would not be "pollution" as no "harm" would, or could, ensue. "Harm" here means harm to the health of living organisms or other interference with the ecological systems of which they form part. In the case of man such "harm" includes offence to any of his senses (such as smell) or harm to his property. "Harmless" where it occurs in the Act, has a corresponding meaning.[12] An example of "harm" could be the deposit of a small amount of bricks on grass. A few days later the grass can be shown to have discoloured. This would be "harm" for the purposes of the Act. Indeed it would not be necessary to show actual discolouration. The deposit merely has to be capable of causing adverse effects to be labelled "polluting."

The main objective of waste management controls is to prevent pollution **2.06** of the environment, as defined, or adverse effects on the countryside or places of special interest. However, for waste disposal, there is an additional objective namely the establishment of an integrated and adequate network of waste disposal installations, taking account of the best available technology not involving excessive costs.[13] The aim of the network is to make the

---

[7] See also E.P.A. 1990, ss.37(2)(a), 38(1)(b) & (6), 42(1)(a), 54(2)(a) & (10).
[8] E.P.A. 1990, s.29(8).
[9] *ibid.* s.29(4).
[10] *ibid.* s.29(11).
[11] *ibid.* s.29(3).
[12] *ibid.* s.29(5).
[13] S.I. 1994 No. 1056, Sched. 4, para. 4(2)(a).

E.C. self-sufficient in waste disposal, with Member States moving also towards that goal. However this has to take account of geographical circumstances and the need for specialised installations, such as chemical waste incinerators, for certain types of waste. The network should also enable waste to be disposed of in one of the nearest appropriate installations, by means of the most appropriate methods and technologies that will ensure a high level of protection for the environment and public health.[14] This is designed to curb movements of waste between states or regions but to allow facilities like chemical incinerators to remain economically viable.

2.07   Waste planning is important in the control process. Authorities should, as an objective, implement the National Waste Strategies and any waste disposal plan or development plan so far as it deals with waste and the United Kingdom offshore waste management plan.[15] The objectives of those plans and the National Waste Strategy will be set out in Chapter 13.

THE SECRETARY OF STATE

2.08   In England the Secretary of State responsible for waste management and litter control under the Environmental Protection Act 1990 is the Secretary of State for the Environment. The respective Secretaries for Wales and Scotland will exercise control in those countries. In this chapter, and throughout the book unless the context shows otherwise, the "Secretary of State" means all three Secretaries acting within their separate jurisdictions. In general powers will be exercised in accordance with the provisions of the Act but where necessary regulations may be made to modify waste management provisions so as to enable the United Kingdom government to comply with its obligations under Community law or international treaty.[16]

2.09   The Environment Agency and the Scottish Environmental Protection Agency (SEPA) are concerned with day to day control of waste management activities. In England and Wales the Secretary of State for the Environment will supervise the activities of the Agency. He will appoint the majority of its members and its chairman.[17] He will approve the appointment of its chief executive and the terms and conditions of its staff.[18] He will determine its financial duties in respect of all of its functions jointly with the Minister for Agriculture, but will deal with any surpluses held by it and with its accounts himself.[19] In Scotland, the Secretary of State for Scotland will exercise similar powers over SEPA.[20]

2.10   By virtue of section 38 of the Environment Act 1995 any Minister may authorise a relevant Agency to exercise on his behalf any of his functions

---

[14] S.I. 1994 No. 1056, Sched. 4, para. 4(2)(b).
[15] *ibid.* Sched. 4, para. 4(1)(b) and (1) as amended by S.I. 1996 No. 593, Sched. 2, para. 10(5)(a).
[16] E.P.A. s.156.
[17] E.A. 1995, s.1(2) & (3).
[18] *ibid.* Sched. 1, paras 4(2), (3)(a).
[19] *ibid.* s.44.
[20] *ibid.* s.20 & Sched. 6.

that can be appropriately exercised by that Agency. However, this cannot extend to any legislative functions or powers to fix fees or charges.[21] The terms under which the relevant Agency or its employees exercise any Ministerial function will be set out in an agreement between the parties. Any such agreement will not prevent a Minister from exercising any function to which the agreement relates.[22]

The Secretary of State may give directions of a general or specific nature 2.11 with respect to the carrying out of the relevant Agency of any of its functions.[23] Such directions can include those necessary to enable the government to give effect to Community obligations or to international agreements to which the United Kingdom is a party.[24] Except in an emergency the Agency must be consulted before any direction is issued to it.[25] The relevant Agency must comply with any directions issued to it, and with any requests for information from it made by the Secretary of State under section 51 of the Environment Act 1995.

Appeals against the decisions of an Agency will be determined by the 2.12 Secretary of State; although he may delegate this function under section 114 of the Environment Act 1995. The Secretary of State may also hold inquiries of hearings in connection with any of the Agency's functions or in respect of his functions that concern an Agency.[26] The provisions of section 250(2) to 250(5) of the Local Government Act 1972 will apply to such inquiries.[27] In particular, section 250(4) allows the recovery of costs incurred by the relevant Minister in holding an inquiry from a party to it. Costs between parties will be dealt with under section 250(5); probably on a similar basis to that set out in Circular 8/93 in relation to planning inquiries.

The compass of waste management will be determined in regulations 2.13 made by the Secretary of State that define controlled wastes, exempt some wastes or activities from the licensing regime, set out conditions to be included in licences, apply waste rules to radioactive substances or specify when waste is considered as having been treated for the purposes of the Act.[28] In addition special wastes will be defined in accordance with regulations he makes and non-controlled wastes may be brought under control by other regulations.[29] Information about the environmental effects of certain substances may be sought through regulations and orders made under section 142 of the Act. Through a code of practice and regulations he will substantially define the duty of care for waste imposed by section 34 of the 1990 Act.

In the collection and disposal of waste the Secretary of State in England 2.14 and Wales will make the national waste strategy—in Scotland this is done

---

[21] E.A. 1995, s.38(2).
[22] *ibid.* s.38(7).
[23] *ibid.* s.40(1).
[24] *ibid.* s.40(2).
[25] *ibid.* s.40(6).
[26] *ibid.* s.53(1).
[27] *ibid.* s.53(2).
[28] E.P.A. 1990, ss.75(8), 33(3) & (4), 35(6), 78 and 29(7).
[29] *ibid.* ss.62(1) & 63(1).

by SEPA. The Secretary will review waste recycling plans before they are finalised and may require them to be completed in a certain time.[30] He will regulate as to when charges for the collection of household waste may be made and determine the rules under which recycling credits are payable.[31] He may issue notices or directions requiring waste to be accepted, treated, disposed or delivered in the manner set out in them.[32] The role of the Secretary in litter control is set out at paragraph 6.03.

2.15   Guidance notes will be issued by the Secretary to help regulation authorities in licensing sites, supervising closed sites and determining whether someone is a "fit and proper" person to hold a licence.[33] He will have powers to direct an Agency to include particular conditions in a licence, to vary a licence, revoke or suspend it or to exercise their supervisory powers in a specified way in relation to a particular licence.[34] To assist him in exercising these functions special inspectors may be authorised under section 108(1) of the Environment Act 1995. The Secretary of State will approve an Agency's schedule of fees and charges for licensing matters.[35] Regulations or directions made by him will prescribe the contents of licence registers, while he may exclude information on grounds of national security or, on appeal, confidentiality.[36] In addition he will make regulations to control the import or export of wastes.[37]

THE ENVIRONMENT AGENCIES (WASTE REGULATION AUTHORITIES)

2.16   The Environment Act 1995 created the Environment Agency for England and Wales and, for Scotland, the Scottish Environment Protection Agency (SEPA). These two Agencies have taken over responsibility for the functions of the former waste regulation authorities.[38] However the legislation, and in particular part II of the Environmental Protection Act 1990, still refers to waste regulation authorities. Such references are now to be taken as references to the respective Agency and reference to their areas is to be taken as a reference to the area over which the respective agency exercises functions or the function in question.[39]

2.17   Transfers of the property, rights and liabilities from the waste regulation authorities to the respective Agencies is dealt with by schemes made under section 3 of the 1995 Act for the Environment Agency and section 22 for SEPA. Schemes made under both sections are also subject to the provisions

---

[30] E.P.A. 1990, ss.49(4) & (7).
[31] *ibid.* ss.44(3) and 52(5) & (8).
[32] *ibid.* ss.57 and 58 (Scotland).
[33] *ibid.* ss.35(8), 61, (11) and 74(5).
[34] *ibid.* ss.35(7), 37(3), 38(7) & 42(8)—in Scotland similar powers are granted in s.54 with respect to sites of waste disposal authorities.
[35] *ibid.* E.A. 1995, s.42.
[36] *ibid.* ss.64(8), 65 and 66(1) & see (7).
[37] *ibid.* ss.140 and 141.
[38] E.A. 1995, s.2(1)(b).
[39] E.P.A. 1990, s.30(1) as substituted by E.A. 1995, Sched. 22, para. 62(2).

of Schedule 2 to the Act. The transfers are also the subject of D.o.E. Circular 15/95 or Scottish Office Circular 19/95.

The Agency has established a Pollution, Prevention and Control Directo- 2.18 rate which, through its water and waste division, is responsible for national waste regulatory policy and the National Waste Strategy. Although there is a regional structure for the Agency, most of its waste management functions will be conducted at area—usually county—level. Thus it is the area manager who will be responsible for issuing licences, etc.

The way in which the Agency exercises its enforcement powers is 2.19 governed by "The Environment Agency Enforcement Code of Practice."[40] This Code is made under section 5 of and Schedule 1 to the Deregulation and Contracting Out Act 1994. Its main principles with respect to the enforcement of environmental protection law are proportionality in applying the law and securing compliance, consistency of approach, targeting of enforcement action and transparency about how it operates and what those regulated may expect.

The Agencies control the carriage and disposal of waste in their areas. As 2.20 far as waste disposal is concerned waste management sites will be licensed by the Agencies who will supervise the operation of the licence until they accept its surrender. In addition they may determine that someone is not a "fit and proper person" to hold a licence.[41] They will also be concerned with exemptions from licensing and the registration of most exempt activities under the Waste Management Licensing Regulations 1994. The Agencies will charge fees for processing applications for, or in respect of, licences and make subsistence charges in accordance with a scheme drawn up under section 41 of the 1995 Act. An Agency will also supervise closed landfill sites and may require the removal of waste unlawfully deposited. Inspectors may be appointed to enable an Agency to carry out its duties under the Act.

Registers containing details of waste management licences and other 2.21 information must be maintained by an agency under section 64 of the Act. It will be initially for the Agency to determine whether information should be excluded from the register on the grounds of commercial confidentiality.[42] In addition by virtue of section 71(2) an Agency may obtain information from anyone to assist them in exercising their functions.

Under the Control of Pollution (Amendment) Act 1989[43] an Agency will 2.22 also be responsible for the registration of carriers of waste and the enforcement of that legislation. It will register brokers of waste by virtues of Schedule 5 to the Waste Management Licensing Regulations 1994.[44] In addition it will be responsible for the control of transfers of special waste within its area and the transfrontier movement of waste under E.C. or international controls.

---

[40] May 1996.
[41] E.P.A. 1990, s.74.
[42] *ibid.* s.66(1).
[43] As amended by the E.P.A. 1990, Sched. 15, para. 31(2).
[44] S.I. 1994 No. 1056.

2.23   As part of their general pollution control functions the Agencies will have some responsibility for contaminated land under Part IIA of the Environmental Protection Act 1990. They will be the enforcing authorities in respect of sites designated as "special sites" under section 78C of the Act. They can therefore require remediation of such sites under section 78E, or adopt earlier notices served in respect of such a site. In addition under section 78U the Agencies must from time to time prepare and publish a report on the state of contaminated land in their areas. They may also issue guidance to local authorities on how to deal with a particular site.

2.24   The Agencies' powers under Part II and IIA of the Environmental Protection Act 1990 are part of their "pollution control powers" for the purposes of the Environment Act 1995.[45] These powers are exercisable in order to prevent, minimise, or remedy or mitigate the effects of pollution of the environment.[46] They must follow developments in technology and techniques to these ends.[47] They also have the general environmental and recreational duties set out in sections 7 and 32 of the Act and must comply with any codes of practice issued to them in respect of those duties under sections 9 or 36. In addition they have particular duties in respect of Sites of Special Interest and, in Scotland, Natural Heritage Areas by virtue of sections 8 or 35 of the Act.

2.25   In considering the exercise of any of their powers—as opposed to non-discretionary duties—the Agencies must take into account the costs which are likely to be incurred and the benefits which are likely to accrue as a result.[48] For these purposes costs include costs to any person and costs to the environment.[49] However, the duty to make a cost-benefit analysis is not intended to require the Agencies to make a full analysis or follow its results slavishly. Costs and benefits must be considered broadly and in the round.[50] Guidance has been issued by the Secretary of State as to what is required and is contained in Chapter 5 of the Explanatory Document attached to "The Environment Agency and Sustainable Development."[51] In particular an Agency does not have to make such an analysis if doing so, or to such an extent, would be unreasonable in view of the nature or purpose of the power or in the circumstances of the particular case.[52] Thus a defendant in criminal proceedings is unlikely to be able to argue successfully that the costs and benefits of taking those proceedings should have been considered. Further, the duty to conduct a cost-benefit analysis does not affect an Agency's duty to carry out its other duties under any enactment or to fulfil its statutory obligations.[53]

2.26   In England and Wales regional Environment Protection Advisory Committees have been established under section 12 of the 1995 Act. The Agency

---

[45] E.A. 1995, ss.5(5)(e) & 33(5)(e).
[46] *ibid.* ss.5(1) & 33(1).
[47] *ibid.* ss.5(4) & 33(4).
[48] *ibid.* s.39(1).
[49] *ibid.* s.56(1).
[50] *Hansard*, H.L. Vol. 560, col. 1384.
[51] D.o.E. June 1996.
[52] E.A. 1995, s.39(1).
[53] *ibid.* s.39(2).

must consult these Committees as to the general way it carries out its functions in the areas concerned and consider representations made to it by the Committee about the way it operates.

WASTE DISPOSAL AUTHORITIES

In England a waste disposal authority will normally be a county council. 2.27 However in Greater London they will be the London Waste Authorities constituted by the Waste Regulation and Disposal (Authorities) Order 1985[54] and the London Boroughs or the Common Council of the City of London. In Greater Manchester, Wigan district Council and the Greater Manchester Waste Disposal Authority. In Merseyside, the Merseyside Waste Disposal Authority and in any other metropolitan county in England, the district council. In Wales the disposal authority will be a county or county borough council while in Scotland they will be constituted councils.[55]

The roles of waste disposal authorities in England and Wales are different 2.28 from those in Scotland. In England and Wales there was a transition from authority run disposal operations so that council services will be controlled through Local Authority Waste Disposal Companies. This transfer of functions has been effected under the provisions of section 32 of and Schedule 2 to the Environmental Protection Act 1990. The functions of a disposal authority are to arrange for the disposal of waste collected in their area by collection authorities and to provide refuse amenity sites. These functions must be carried out through waste disposal contractors.[56] An authority will retain some powers to enable them to direct a collection authority to deliver waste to specified places, to provide for the recycling of waste and to have an input to waste recycling plans.

In Scotland authorities will run their own disposal operations, providing 2.29 sites and equipment for waste recycling or disposal. Under section 56 of the 1990 Act authorities also have power to recycle waste. While these authorities are not subject to the privatisation regime that applies in England and Wales, they must comply with E.C. public procurement rules.

WASTE COLLECTION AUTHORITIES

In England waste collection authorities are district councils and, in 2.30 Greater London, London borough councils or the Common Council of the City of London.[57] In Wales they are county or county borough councils,[58] while in Scotland they are constituted councils.[59]

---

[54] E.P.A. 1990, s.30(4), & S.I. 1985 No. 1884, Sched. 1, Parts I, II, III, IV & V.

[55] *ibid.* s.30(2) as amended by Local Government (Wales) Act 1994, Sched. 9, para. 17(2) and Local Government, etc., (Scotland) Act 1994, Sched. 13, para. 167(3).

[56] *ibid.* s.51(1).

[57] and Temple authorities.

[58] E.P.A. 1990, s.30(3)(bb) as added by Local Government (Wales) Act 1994, Sched. 9, para. 17.

[59] *ibid.* s.30(3)(c) as amended by Local Government, etc., (Scotland) Act 1994, Sched. 13, para. 167(3).

2.31   It is the duty of each waste collection authority to collect most household waste in their area and they may also make arrangements to collect commercial and industrial waste from premises if so requested.[60] In addition they may serve notices under section 59 of the Act to require the removal of unlawfully deposited waste or remove it themselves. They will also empty privies and cesspools. They may direct the use of particular types of receptacles for the waste they collect and supply receptacles for commercial and industrial waste. In general they should deliver their waste to the disposal authority for the area, unless it is to be recycled by the collection authority. The authority will make a recycling plan for waste collected and, in England and Wales, may buy or otherwise acquire waste to recycle it and dispose of the recycled products.[61] Waste collection authorities will maintain a register containing details of waste disposal operations taking place in their areas.[62]

POWERS OF INSPECTORS

2.32   The Secretary of State and the Agencies are enforcing authorities for the purposes of section 108 of the Environment Act 1995.[63] They may appoint persons who appear to them as suitable to be inspectors under section 108(1) of the Act. An appointment must be made in writing. An inspector will not be liable in any civil or criminal proceedings for anything done in the purported performance of the powers granted him by the Act if the court is satisfied that he did what he did in good faith and that there were reasonable grounds for doing it.[64] The powers of an inspector under section 108 may also be exercised by someone specifically authorised to do so by an Agency for the purpose of complying with a requirement by the Secretary of State to prepare a report or assessment on a serious incident, or a situation that may give rise to one, under sections 5(3) or 33(3) of the Act.[65] It will be an offence under section 110(3) for a person to falsely pretend to be an inspector; an offence which is punishable on summary conviction by a fine not exceeding level 5 on the standard scale.[66]

2.33   An Agency inspector may exercise his powers to discharge any of the functions conferred on his authorising authority by, *inter alia*, Part II of the Environmental Protection Act 1990 or the Control of Pollution (Amendment) Act 1989 or to determine whether, and if so in what manner such a function should be discharged. Further he may use them to determine whether any provision of Part II or the 1989 Act or regulations made under it are being complied with.[67] There is a duty to inspect establishments or undertakings that carry out the disposal or recovery of waste, its collection or

---

[60] E.P.A. 1990, s.45(1) & (2).
[61] *ibid.* s.55(3) & (4).
[62] *ibid.* s.64(4)–64(6).
[63] E.A. 1995, s.108(15).
[64] *ibid.* Sched. 18, para. 6(4).
[65] *ibid.* s.108(2) & (3).
[66] *ibid.* s.110(5).
[67] *ibid.* s.108(1) & (15).

18

transport (on a professional basis) and that act as waste brokers.[68] Inspectors appointed by the Secretary of State can investigate any matter in relation to the Secretary's statutory functions relating to the control of pollution. The powers are exercisable in relation to "premises" which includes any land, vehicle, vessel or mobile plant.[69] Land will include any buildings and structures on it[70] so that an inspector could look at skips and containers under these powers. In order to obtain any relevant information, these powers include powers to carry out experimental borings or other works on the premises and to install, keep or maintain monitoring or other apparatus there.[71]

Under section 108(4)(a) an inspector may enter premises which he has 2.34 reason to believe it is necessary for him to enter. "Reason to believe" here will mean a

> "reasonable and bona fide belief in the existence of such a state of things as would amount to a justification of the course pursued. . . It is not essential in any case that facts should be established . . . as evidence (for) a jury . . ."[72]

Entry may be made at any reasonable time—for example the times that business premises are[73]—open unless there is a situation that the inspector considers there is an immediate risk of serious environmental pollution or harm to human health, or that circumstances exist that are likely to endanger life or health, when entry can be made at any time and, if necessary, by force.[74] In going on to premises the inspector may take anyone else authorised by his enforcing authority or, if he has reasonable cause to believe that there will be serious opposition to his entry, a constable. He will also be able to take any equipment or materials required for his inspection.[75]

Except in an emergency, entry to residential premises or to premises on 2.35 which it is proposed to take heavy equipment should only be effected after seven days' notice has been given to the person who appears to be the occupier of those premises and either with the actual occupier's consent or under a warrant granted by virtue of Schedule 18 to the Act.[76] A warrant is also required, except in emergency, to enter premises to which entry has been refused, or it is reasonable for the inspector to consider that entry will be refused, and to which it is reasonably apprehended that force will be necessary to gain entry.[77]

For a warrant to be issued the requirements of paragraph 2 of Schedule 18 2.36 must be satisfied. The justice of the peace, or, in Scotland, the sheriff or the justice of the peace, must be satisfied on sworn information in writing, that

---

[68] S.I. 1994 No. 1056, Sched. 4, para. 13.
[69] E.A. 1995, s.108(15).
[70] Interpretation Act 1978, s.5 & Sched. 1.
[71] E.A. 1995, s.108(5).
[72] *Hicks v. Faulkner* (1878) 8 Q.B.D. 167 at 173.
[73] *Davis v. Winstanley* (1930) 144 L.T. 433.
[74] E.A. 1995, s.108(4)(a) and see s.108(15) "emergency".
[75] *ibid.* s.108(4)(b).
[76] *ibid.* s.108(6).
[77] *ibid.* s.108(7).

there are reasonably grounds for the Agency or Secretary of State to exercise powers under section 108 in relation to the premises and that either, such exercise has been refused or a refusal is reasonably apprehended, or that the premises are unoccupied, or that the occupier is temporarily absent and the matter is urgent, or that to give notice of the intended entry would defeat the object of the exercise. In addition where entry is sought to residential premises or with heavy equipment, the seven days' notice required by section 108(6) must have expired. If the justice or sheriff is satisfied a case for a warrant is made out he may issue it to the enforcing authority who may designate someone to exercise section 108 powers in relation to the relevant premises in accordance with the warrant and if need be by force. The warrant continues to have effect until the purposes for which it was issued have been fulfilled.

2.37   The inspector must produce evidence of his authorisation before he exercises any powers under section 108.[78] "Produce" here means "make available to be inspected".[79] If the premises are unoccupied or temporarily unoccupied, he must, on leaving, leave them as effectually secured against trespassers as he found them.[80]

2.38   Once on the premises the inspector may make such examination and investigation as may be necessary in the circumstances.[81] For this purpose he may direct that the premises or part of them or anything in them shall be left undisturbed for as long as it is necessary to enable him to conclude his inspection.[82] He may take measurements and photographs of the premises, and make such recordings as he considers necessary for the purpose of the investigation.[83]

2.39   Samples of any articles or substances found on the premises may be taken as well as of the air, water or land in or on them or in their vicinity.[84] Regulations may provide for the procedure to be followed in taking and dealing with such samples[85]; although no such regulations have yet been made.

2.40   Where an article or substance is found, in or on premises he has power to enter, that appears to the inspector to have caused or to be likely to cause pollution or harm to human health he may have it dismantled or subjected to any process or test, as long as this does not cause any unnecessary damage or destruction.[86] Before exercising this power he should consult anyone he thinks appropriate as to the dangers that may be involved[87] and if asked by someone present who is responsible for the premises, conduct the process or

---

[78] E.A. 1995, Sched. 18, para. 3.
[79] *R. v. Longman* [1988] 1 W.L.R. 619 at 627B.
[80] E.A. 1995, Sched. 18, para. 5.
[81] *ibid.* s.108(4)(c).
[82] *ibid.* s.108(4)(d).
[83] *ibid.* s.108(4)(e).
[84] *ibid.* s.108(4)(f).
[85] *ibid.* s.108(9).
[86] *ibid.* s.108(4)(g).
[87] *ibid.* s.108(11).

dismantling in his presence.[88] The inspector may take possession of the article or substance and keep it for as long as it is necessary to either examine it, or to ensure that it is not tampered with before the examination is completed or to ensure that it is available for use as evidence in any proceedings under the relevant statute or regulations.[89]

A person who an inspector has reasonable cause to believe is able to give him information in relation to his examination or investigation may be required to answer any questions the inspector thinks fit to ask and to sign a declaration as to the truth of those answers. The person may be questioned in the absence of anyone else other than someone nominated by him to be present or as allowed by the inspector.[90] It will be an offence for someone such as an employer to prevent an employee from answering an inspector's questions.[91] No answer given by someone in response to questions asked under this power will be admissible in evidence against him in any proceedings in England and Wales or, in Scotland, in any criminal proceedings.[92] Otherwise, however, information obtained during the course of an inspection will be admissible against anyone, as will results of monitoring or other apparatus installed on the premises.[93] **2.41**

An inspector may require the production of records that are required to be kept under Part II of the Environmental Protection Act 1990 or the 1989 Act or of any other records it is necessary for him to see for the purposes of an examination or investigation. These can be in documentary form or extracts from computer entries. He will be able to take copies of such material.[94] However this power will not enable him to inspect records that are subject to legal professional privilege.[95] In addition an inspector may require anyone to afford him such facilities and assistance with respect to anything within that person's control or over which he has responsibilities that are necessary to enable him to exercise his powers of inspection.[96] **2.42**

It will be an offence for anyone to fail, without reasonable excuse, to comply with any requirement imposed by an inspector under these provisions or to refuse to provide facilities or assistance or permit any reasonably required inspection. In addition no one may prevent another person from appearing before or from answering any question an inspector may put to him under section 108(4)(j).[97] It will also be an offence to intentionally[98] obstruct[99] an inspector in the exercise or performance of any of the powers contained in section 108.[1] A person will be liable on summary **2.43**

---

[88] E.A. 1995, s.108(10).
[89] *ibid*. s.108(4)(h).
[90] *ibid*. s.108(4)(j) (*n.b.* there is no para. 108(4)(i)).
[91] *ibid*. s.110(2)(c).
[92] *ibid*. s.108(12).
[93] *ibid*. Sched. 18, para. 4.
[94] *ibid*. s.108(4)(k).
[95] *ibid*. s.108(13).
[96] *ibid*. s.108(4)(l).
[97] *ibid*. s.110(2).
[98] See *Arrowsmith v. Jenkins* [1963] 2 Q.B. 561.
[99] See *Lewis v. Cox* [1984] 3 W.L.R. 875.
[1] E.A. 1995, s.110(1).

conviction for any such offence to a fine not exceeding level 5 on the standard scale.[2]

2.44　If, while he is carrying out an inspection, an inspector finds an article or substance that he has reasonable cause to believe that, in the circumstances in which he finds it, is a cause of imminent danger of serious pollution or harm to human health, he may, under section 109(1) of the Environment Act 1995 seize it and cause it to be rendered harmless even if this involves destroying it. As soon as may be after he has exercised this power he should write and sign a report detailing the circumstances in which he exercised it Copies of this report should be given to a responsible person at the premises concerned and the owner of the substance or article; but if the owner cannot be found after reasonable inquiry this provision will be satisfied by serving his copy on the responsible person.[3] It will be an offence to intentionally obstruct an inspector exercising his powers under section 109, punishable on summary conviction by a fine not exceeding the statutory maximum and on indictment with up to two years imprisonment or a fine or both.[4]

2.45　Where entry has been made under section 108(4)(a) and equipment has been brought on site or powers to carry out works exercised under section 108(5), the agency or Secretary of State may be liable for any loss or damage caused by the exercise of those power or by a failure to secure unoccupied premises.[5] However, compensation is not payable for loss or damage that is attributable to the fault of the person sustaining it, or that is compensatable under other pollution control enactments.[6] Disputes over the entitlement to compensation or as to its amount are settled in England and Wales by a single arbitrator or, in Scotland, an arbiter appointed by agreement between the parties or, in default, by the Secretary of State.[7]

APPLICATION TO THE CROWN

2.46　By virtue of section 159(1) of the Environmental Protection Act 1990 the Act and regulations made under it will bind the Crown; but not Her Majesty in her private capacity.[8] However, the Crown itself cannot be held criminally liable for an offence under the Act. Rather, where an agency or a principal litter authority consider there has been a contravention of a relevant provision by the Crown, they must apply to the High Court or the Court of Session for a declaration that the act or omission alleged to constitute the contravention is unlawful.[9] An individual civil servant can be made liable for a breach for which he is responsible.[10] Some specified crown

---

[2] E.A. 1995, s.110(4)(b) & (5).
[3] *ibid.* s.109(2).
[4] *ibid.* s.110(1) & (4)(a).
[5] *ibid.* Sched. 18, para. 6(1).
[6] *ibid.* Sched. 18, para. 6(2).
[7] *ibid.* Sched. 18, para. 6(3).
[8] E.P.A. 1990, s.159(5).
[9] *ibid.* s.159(2) & (7).
[10] *ibid.* s.159(3).

premises may be exempted by a certificate issued by the Secretary of State on national security, from the powers of entry available to inspectors.[11]

WASTE MANAGEMENT IN NORTHERN IRELAND

Pollution controls in Northern Ireland are operated or supervised by the 2.47 Department of the Environment for Northern Ireland through its Environmental Protection Division. With some exceptions, waste management controls are contained in the Pollution Control and Local Government Order (Northern Ireland) 1978.[12] Those controls are similar to the provisions of Part I of the Control of Pollution Act 1974.[13]

The 26 district councils in Northem Ireland are primarily responsible for 2.48 waste management control. Under Article 3 of the Order it is their duty to ensure that arrangements made by them or others for the disposal of waste are adequate for the purpose of disposing of all controlled waste that becomes, or is likely to become, situated in their districts. They must also arrange for the collection of waste,[14] provide civic amenity sites for household waste[15] and take steps to deal with litter.[16] They will license waste disposal facilities and deal with unauthorised deposits of controlled waste. Their powers of entry and inspection under the provisions of section 98 of the Local Government (Northern Ireland) Act 1972 are extended by the Order to enforce these waste management controls.[17] Not all the provisions of the 1978 Order are in force. Thus, while the Order is concerned with the collection of waste, that function will still be performed under sections 52 and 53 of the Public Health (Ireland) Act 1878 and section 48 of the Public Health Acts Amendment Act 1907.

In 1990 the House of Commons Environment Committee reported on 2.49 Environmental Issues in Northern Ireland.[18] The Committee found that there was room for improvement in waste management standards and recommended that a system of regional waste licensing should be established and that the provisions in the Environmental Protection Act 1990 relating to aftercare of sites and integrated pollution control be enacted in Northern Ireland.

A consultation paper on the reform of Northern Ireland's waste manage- 2.50 ment regime had not been issued at the time of writing this book.

---

[11] E.P.A. 1990, s.159(4).
[12] S.I. 1978 No. 1049 (N.I. 19), as amended by Local Government (Miscellaneous Provisions) (Northern Ireland) Order 1985 (S.I. 1985 No. 1208 (N.I. 15)), Sched. 3, para. 17.
[13] The 1978 Order is supplemented by the Waste Collection and Disposal Regulations (Northern Ireland) 1992.
[14] *ibid.* Art. 14.
[15] *ibid.* Art. 22.
[16] *ibid.* Arts. 25 & 27.
[17] *ibid.* Art 74.
[18] Session 1990–1991, H.C. 39.

3. "DIRECTIVE" AND "CONTROLLED WASTE"

3.01   The provisions of Part I of the Control of Pollution Act 1974 or Part II of the Environmental Protection Act 1990 were initially applied to "controlled waste". If the matter was not "waste" or if the waste was not "controlled" waste then those provisions were not applicable. This has changed with the entry into force of the Framework Directive. Any reference to "waste" in Part II of the 1990 Act is now to be taken as a reference to "Directive waste".[1]

3.02   "Directive waste" is defined in Regulation 1(3) of the Waste Management Licensing Regulations 1994[2] to mean

> "any substance or object in the categories set out in Part II of Schedule 4 which the producer or the person in possession of its discards or intends or is required to discard."

Part II of Schedule 4 repeats Annex I to the Directive.[3] However, this definition does not encompass anything excluded by Article 2 of the Directive. Under Regulation 7A of the Controlled Waste Regulations 1992[4] waste which is not "Directive waste" will also not be treated as household, industrial or commercial waste for the purposes of Part II of the Environmental Protection Act 1990.

3.03   For these purposes "discard" has the same meaning as in the Directive. This has been construed in the light of the French edition wording "*se defait*" or "get rid of". "Producer" means anyone whose activities produce Directive waste or who carries out preprocessing, mixing or other operations resulting in a change in its character or composition.[5]

3.04   Not all waste is "Directive waste". Article 2 of the Framework Waste Directive.[6] excludes gaseous, effluents emitted into the atmosphere from its scope. It also excludes, where they are already covered by other legislation, radioactive waste, waste from mining and quarrying operations, animal carcasses and agricultural wastes consisting of faecal matter and other

---

[1] S.I. 1994 No. 1056, Sched. 4, para. 9(2).
[2] S.I. 1994 No. 1056.
[3] See para. 3.06.
[4] S.I. 1992 No. 588 as amended by S.I. 1994 No. 1056, reg. 24.
[5] S.I. 1994 No. 1056, reg. 1(3) "Directive Waste."
[6] Dir. 91/156 [1991] O.J. L78/32.

natural, non-dangerous substances used in farming, waste waters with the exception of waste in liquid form and decommissioned explosives. Wastes that do not fall under the controls of waste management legislation are discussed in Chapter 4 of this book.

This basic definition of "directive waste" overtakes that of controlled **3.05** waste but does not subsume it. The categorisation of controlled wastes into "industrial, commercial and household wastes" in section 75 of the 1990 Act will still be relevant in relation to the collection of wastes, the licensing of sites to take those categories and charges for collection.

1. "WASTE" AND "DIRECTIVE WASTE"

In Article 1 of the Framework E.C. Directive on Waste,[7] "waste" is **3.06** defined to mean

> "any substance or object in the categories set out in Annex I which the holder discards or intends or is required to discard."

Annex I sets out a number of specific categories of ways in which substances can be discarded. However, these categories are open-ended. The first category is production or consumption residues not otherwise specified below, while the last category is any material or substances or products which are not contained in the above categories. Thus the other categories are examples of how matter can become waste but are not definitive. The other categories are;

- Q2 off-specification products,
- Q3 products whose date for appropriate use has expired,
- Q4 materials spilled, lost or having undergone other mishap, including any materials, equipment, etc., contaminated as a result of the mishap,
- Q5 materials contaminated or soiled as a result of planned actions such as residues from cleaning operations, packaging materials, containers, etc.,
- Q6 unusable parts such as reject batteries,
- Q7 substances which no longer perform satisfactorily such as exhausted tempering salts,
- Q8 residues of industrial processes,
- Q9 residues from pollution abatement processes,
- Q10 machining or finishing residues like sawdust,
- Q11 residues from raw materials extraction and processing,
- Q12 adulterated materials such as oil contaminated with PCBs,
- Q13 any materials, substances or products whose use has been banned by law,
- Q14 products for which the holder has no further use and,
- Q15 contaminated materials, substances or, products resulting from remedial action with respect to land. This categorisation is also

[7] Dir. 91/156 [1991] O.J. L78/32.

found in Schedule 2B to the 1990 Act which was added by paragraph 95 of Schedule 22 of the Environment Act 1995.

3.07   Section 30(1) of the Control of Pollution Act 1974 provided a less wide-ranging definition of "waste." It stated that "waste" includes any substance which constitutes a scrap material or an effluent or other unwanted surplus substance arising from the application from any process. Excavated soil can be "waste" by virtue of this provision, although whether it is or is not will be a question of fact.[8] The statutory definition continued by including any substance or article which requires to be disposed of as being broken, worn out, contaminated or otherwise spoiled. Further, it went on to add that for the purposes of Part I of the Act anything which is discarded or otherwise dealt with as if it were waste shall be presumed to be waste unless the contrary is proved. These definitions were re-enacted by section 75(2) & 75(3) of the Environmental Protection Act 1990 as originally enacted.

3.08   The provisions of section 30(1) have been discussed a number of times by the courts. In *Long v. Brooke*[9] the Crown Court held that

> "Although one man's waste may be another man's valuable material, on its true construction the Act defines waste from the point of view of the person discarding the material . . . and the fact that as occupier of the land he had a use for the discarded material did not prevent its constituting waste within the meaning of section 30."

This view was upheld by the High Court in *Berridge Incinerators Ltd v. Nottinghamshire County Council and Anor*[10] where the judge stated,

> "It is, of course, a truism that one man's waste is another man's raw material. The fact that a price is paid by the collector of material to its originator is, no doubt, relevant; but I do not regard it as crucial. If I have an old fireplace to dispose of to a passing rag and bone man, its character as waste is not affected by whether or not I can persuade the latter to pay me 50 pence for it. In my judgment, the correct approach is to regard the material from the point of view of the person who produces it. Is it something which it produced as a product, or even as a by-product of his business, or is it something to be disposed of as useless? I notice that this was the approach adopted (in) *Long v. Brooke* and I respectfully agree with it."

3.09   This interpretation has been reinforced by the judgment of the European Court in a reference concerning *Criminal proceedings against E. Zanetti and Others. Cases 206/88 and 359/88.* There it was held that the concept of waste within the meaning of the Waste and Hazardous Waste Directives was not to be understood as excluding substances and objects which are capable of economic utilisation.

3.10   These decisions raise the question as to when waste stops being "wastes". This was discussed in *Kent County Council v. Queensborough Rolling Mill Co.*[11] There, discarded material from a former pottery site was sorted into different categories some of which were used to infill areas on another site that was prone to subsidence. It was argued that because the discarded material had

---

[8] *Charles Neil Ashcroft v. Michael McErlain Ltd*, unreported, January 30, 1985.
[9] [1980] Crim. L.R. 109.
[10] Q.B.D. unreported, April 14, 1987.
[11] (1990) 154 J.P. 530.

been put to a useful purpose this meant it was no longer "waste." That argument was rejected on the grounds that the usefulness of the deposit as infill did not change the character of the material. Nor did the fact that the material was separated from other material before deposit deprive it of its identity as waste. Different considerations might apply if material is recycled or reconstituted before the deposit complained of. In *R. v. Rotherham Metropolitan Borough Council, ex p. Rankin*,[12] spent solvents that were to be recycled were regarded as both trade waste and raw materials; although that case was a planning matter rather than one decided under waste management legislation. The conclusion here would seem to be that what is originally a "waste" remains a waste until it undergoes some physical operation that changes its character. Mere settlement and removal of resulting sludge may not be enough. This view is reinforced by section 29(6) of the Environmental Protection Act 1990 which states that

> waste is "treated" when it is subjected to any process, including making it re-usable or reclaiming substances from it."

The entry into force of the 1994 Regulations has led to a different 3.11 definition of "waste" being adopted, namely "Directive waste".[13] This is given statutory effect by the amendments to section 75 of the 1990 Act contained in paragraph 88 of Schedule 22 to the Environment Act 1995. The terms adopted almost mirror those of the 1994 Regulations but the new section 75(2) interposes the "holder" of waste as the person who discards it, etc.; although the "holder" is effectively the waste's producer as defined in the Regulations.[14] In addition the former subsection 75(3) ceases to have affect.[15] Annex I to the Directive forms the new Schedule 2B to the 1990 Act.[16]

Despite this the courts may still have recourse to the cases decided under 3.12 the earlier provision to assist in the interpretation of the new. Clearly such cases should now be treated with caution. A better picture of what is now meant by "waste" is contained in Annex 2 to D.o.E. Circular 11/94. However, this too has to be treated carefully. What is meant in law by the term "waste" is a matter for the courts who may or may not follow what is said in the Circular. This is emphasised in paragraph 2.12 of the Annex.

Annex 2 to the Circular looks at three aspects of the situation—(1) what 3.13 is waste?, (2) when is a substance or object discarded?, (3) and can waste cease to be waste? It does so by introducing the concept of substances or articles which fall out of the commercial cycle or out of the chain of utility.[17] This is a concept drawn from the preambles to the Directive but not expressly dealt with in them. In addition, to distinguish the waste collection and disposal cycle from mainstream activities, the concept is refined to deal

---

[12] [1990] J.P.L. 503.
[13] For the meaning of the "Directive" here see E.P.A. 1990, s.75(11) & (12) as added by E.A. 1995, Sched. 22, para. 88(4).
[14] Act 1995, Sched. 22, para. 88(2).
[15] *ibid.* para. 88(3).
[16] E.P.A. 1990, s.75(10)–(12) inserted by E.A. 1995, Sched. 22, para. 88(4).
[17] D.o.E. Circular 11/94, Annex 2, para. 2.14.

with substances, etc., falling out of the *normal* commercial cycle.[18] It is these substances or articles that are "waste" in the view of the authors of the Circular.

3.14   To determine whether a substance or article is waste the Circular suggests the following question should be asked—"Has the substance or object been discarded so that it is no longer part of the normal commercial cycle or chain of utility?[19] If the answer is "no" then the matter is not waste. If it is being consigned to a disposal or recovery operation as set out in Annex II of the Directive then the answer is likely to be "yes" except in those cases of recovery operations that form part of the normal commercial cycle.[20]

3.15   In the case of a consignment to a recovery operation the question should be, "Can the matter be used in its present form (albeit after repair) or in the same way as any other raw material without being subjected to a specialised recovery operation?" Here a "yes" answer leads to the conclusion that the matter is not waste. On the other hand if the answer to the question "Can the substance or object be used only after it has been subjected to a specialised recovery operation?" is "yes" the matter is likely to be waste. The term "specialised recovery operation" is not in the legislation but is used in the Circular to mean one whose sole purpose is the recovery of waste—*i.e.* the deposit of bottles in a bottle bank—as opposed to one that may be part of a normal commercial operation—*i.e.* use of a substance as fuel. The difference is that in the latter case the substance still has value to the holder and thus makes it likely he will look after it.[21]

3.16   The difficulty with this interpretation is that even if the material is not waste under its terms, it may still be something that the producer has discarded or got rid of. The Circular attempts to come to grips with this in paragraphs 2.32–2.42. The key to understanding this aspect is that it is the producer of the waste whose intentions are important. If he intended to get rid of the material—even to a recovery operation within the normal commercial cycle—then it is likely to be considered as waste by the courts. That would affirm the decisions set out in paragraph 3.07.

3.17   If the material is a product or a by-product of any process then it will not be waste. It is a useful material and so has not fallen out of the normal commercial cycle.[22] However, a distinction has to be drawn between something that the producers discards—even if he sells it—which is waste and a by-product of a process that is not. To a certain extent this was dealt with in *Charles Neil Ashcroft v. Michael McErlain Ltd.*[23] There the court in effect stated that whether a product is surplus to the producer's requirements is in any given case a question of fact. Thus whether something is a product or by-product depends on the intention of the producer. This is borne out by the Oxford English Dictionary definition of "by-product"—

[18] D.o.E. Circular 11/94, Annex 2, para. 2.15.
[19] *ibid.* para. 2.20.
[20] *ibid.* paras 2.23–2.26.
[21] *ibid.* paras 2.27–2.31.
[22] *ibid.* paras 2.32 & 2.39–42.
[23] Divisional Court, unreported, 1985.

"A secondary product; a substance of more or less value obtained in the course of a specific process, though not its primary object."

Cases where there is a dispute as to whether a substance or object is waste 3.18 will turn on the intention of the producer. The use of the term "holder" in paragraph 2.34 and in the new section 75(2) of the Environmental Protection Act 1990 may be confusing. The Directive is clear in that it is the producer's intention that matters. This leads to the conclusion that an intention to re-use the substance by a person into whose hands the waste comes cannot change its status as waste. This is not the view of the Circular which concludes that if the substance can be used without being subjected to a specialised recovery operation it may be re-classified so as not to be waste.[24] As the Directive does not deal with the question of when a substances ceases to be waste this may be right. However, until the courts have ruled on the matter any re-classification should be done cautiously.

Otherwise waste will only cease to be waste after it has undergone a 3.19 specialised recovery operation so that it is once more a useful product— either as a raw material or finished goods. For example, a tin can discarded by the person who ate its contents will remain waste throughout the process of collection, crushing, storing and preliminary treatment. It only ceases to be waste after it emerges from the furnace in which it has been made into new raw material.

2. CONTROLLED WASTE

This section concentrates on "controlled wastes" as defined in section 3.20 75(4) of the Environmental Protection Act 1990. This provides the basic meaning as being "household, industrial and commercial waste or any such waste." More detailed explanation is given by subsections 75(5)—75(7). The basic definition in the Act can be expanded by regulations made under it[25] and, for Great Britain, these regulations are the Controlled Wastes Regulations 1992[26] The Regulations are explained in D.o.E. Circular 14/92.[27] They deal with the definition of "controlled wastes" for the purposes of Part II of the 1990 Act.[28] However regulations made under these provisions cannot deal with mine or quarrying wastes or agricultural wastes.[29]

HOUSEHOLD WASTE

"Household waste" is defined in section 75(5) of the Environmental 3.21 Protection Act 1990 to include waste from "domestic property"—a building or self-contained part of a building which is used wholly for the purposes of

---

[24] D.o.E. Circular para. 2.50.
[25] E.P.A. 1990, s.75(8).
[26] S.I. 1992 No. 588 as amended by S.I. 1994 No. 1056, reg. 24 and S.I. 1995 No. 288.
[27] W.O. Circular 30/92.
[28] *ibid.* paras 19–25.
[29] E.P.A. 1990, s.75(8).

living accommodation—or from a caravan which is usually, and for the time being, situated on a caravan site, as well as from residential homes, premises forming part of educational establishments and hospitals or nursing homes. This definition would not enable waste from a science laboratory at a school or hospital to be classified as "household" waste. In determining the classification of waste from those premises regard must be had to its nature.[30] Thus it is submitted that it is only waste from those parts of such premises that are used for residential accommodation that can be described as "household."

3.22 Under Regulation 2(1) and Schedule 1 of the Controlled Waste Regulations 1992[31] certain types of waste are to be treated, in England and Wales, as "household" waste for the purposes of the 1990 Act. These are wastes from:

1. Waste from a hereditament or premises exempted from local non-domestic rating by virtue of—

    (a) In England and Wales, paragraph 11 of Schedule 5 to the Local Government Finance Act 1988 (places of religious worship, etc.)
    (b) In Scotland, section 22 of the Valuation and Rating (Scotland) Act 1956 (churches, etc.)

2. Waste from premises occupied by a charity and wholly or mainly used for charitable purposes.
3. Waste from any land belonging to or used in connection with domestic property, a caravan or a residential home.
4. Waste from a private garage which either has a floor area of 25 square metres or less or is used wholly or mainly for the accommodation of a private motor vehicle.[32]
5. Waste from private storage premises used wholly or mainly for the storage of articles of domestic use.
6. Waste from a moored vessel used wholly for the purposes of living accommodation.
7. Waste from a camp site.
8. Waste from a prison or other penal institution.
9. Waste from a hall or other premises used wholly or mainly for public meetings.
10. Waste from a royal palace.
11. Waste arising from the discharge by a local authority of its duty under section 89(2)—highway cleansing.

A camp site, for the purpose of these Regulations means land on which tents are pitched for the purposes of human habitation and land the use of which is incidental to the land where the tents are. A charity means any body of persons or trust established for charitable purposes.[33]

[30] *Westminster Corp. v. Gordon Hotels Ltd* [1906] 2 K.B. 39.
[31] S.I. 1992 No. 588.
[32] See *ibid.* Sched. 4, para. 4.
[33] *ibid.* reg. 1(1).

However even though the general waste from these premises will be 3.23
regarded as "household waste", waste mineral or synthetic oil or grease,
asbestos or clinical wastes[34] emanating from them are not to be treated as
such for the purposes of section 33(2) of the 1990 Act.[35] Section 33(2)
enables a householder to dispose of his waste within the curtilage of the
dwelling. These exceptions are designed to provide an additional safeguard
against the unsatisfactory disposal by householders of potentially hazardous
wastes.[36] Clinical waste is defined in Regulation 1(1).

For the purposes of the duty of care certain wastes are to be treated as 3.24
household wastes produced on domestic property. These are wastes arising
from works of construction or demolition, including preparatory works, and
septic tank sludge.[37] This means that the occupier of the property is not
subject to the duty in respect of those wastes. A builder or collector of septic
tank sludge, however, remains liable to the duty.[38]

Charges for collection of certain types of household wastes can be made by 3.25
virtue of section 45(3) of the Act. The types of waste concerned are set out
in Schedule 2 to the 1992 Regulations. These are dealt with in Chapter 8 on
"Waste Collection."

INDUSTRIAL WASTE

Industrial waste consists of waste from any factory within the meaning of 3.26
the Factories Act 1961. A factory is defined in section 175 of the Factories
Act. In addition to waste from factories, waste from premises used for or in
connection with, public transport services, gas, water, electricity or sewerage
services or postal or telecommunication services, will also be "industrial
waste" under section 75(6) of the Environmental Protection Act 1990.

Regulation 5(1) of, and Schedule 3 to, the Controlled Waste Regulations 3.27
1992[39] set out descriptions of waste that are to be treated as industrial waste
for the purposes of the 1990 Act. Eighteen categories of such waste are set
out.

1. Waste from premises used for maintaining aircraft, vehicles or
   vessels, not being waste from a private garage to which paragraph
   4 of Schedule 1 applies.
2. Waste from a laboratory
3. Waste from a workshop or similar premises not being a factory
   within the meaning of section 175 of the Factories Act because the
   people working there are not employees or because the work is not
   carried on by way of trade or for purposes of gain. (Here
   "workshop" does not include premises at which the main activities
   are computer operations or photocopying.)

[34] See para. 3.29.
[35] S.I. 1992 No. 588, reg. 3.
[36] D.o.E. Circular 14/94, Annex I, paras 1.20–22.
[37] S.I. 1992 No. 588, reg. 2(2).
[38] See *Waste Management: the Duty of Care, A Code of Practice*, Annex A, para. A.7.
[39] S.I. 1992 No. 588.

4. Waste from premises occupied by a scientific research association approved by the Secretary of State under section 508 of the Income and Corporation Taxes Act 1988.

5. Waste from dredging operations.

6. Waste from tunnelling or from any other excavation.

7. Sewage not falling within a description in Regulation 7 which—

   (a) is treated, kept or disposed of in or on land, other than by means of a privy, cesspool or septic tank;

   (b) is treated, kept or disposed of by means of mobile plant;

   (c) has been removed from a privy or cesspool.[40]

8. Clinical waste[41] other than—

   (a) clinical waste from a domestic property, caravan, residential home, or from a moored vessel used wholly for the purposes of living accommodation;

   (b) waste collected under section 22(3) of the Control of Pollution Act 1974;

   (c) waste collected under sections 89, 92(9) or 93 (of the 1990 Act) (litter).

9. Waste arising from any aircraft, vehicle or vessel which is not occupied for domestic purposes.

10. Waste which has previously formed part of any aircraft, vehicle or vessel and which is not household waste.

11. Waste removed from land on which it has been previously deposited and any soil with which such waste has been in contact, other than—

   (a) waste collected under section 22(3) of the Control of Pollution Act 1974;

   (b) waste collected under sections 89, 92(9) or 93 (of the 1990 Act) (litter).

12. Leachate from a deposit of waste.

13. Poisonous or noxious waste arising from any of the following processes undertaken on premises used for a trade or business—

   (a) mixing or selling paints;

   (b) sign writing;

   (c) laundering or dry cleaning;

   (d) developing photographic film or making photographic prints;

   (e) selling petrol, diesel fuel, paraffin, kerosene, heating oil or similar substances; or

   (f) selling pesticides, herbicides or fungicides.

14. Waste from premises used for the purposes of breeding, boarding, stabling or exhibiting animals.

---

[40] See para. 4.19.
[41] See para. 3.30.

15. Waste oil, waste solvent or scrap metal, other than—

   (a) waste from a domestic property, caravan or residential home;
   (b) waste falling within paragraphs 3 to 6 of Schedule 1.

   Here "waste oil" means mineral or synthetic oil which is contaminated, spoiled or otherwise unfit for its original purpose and "waste solvent" means solvent which is in similar condition.

16. Waste arising from the discharge by the Secretary of State of his duty under section 89(2) of the 1990 Act highway cleansing.

17. Waste imported into Great Britain.

18.[42] Tank washings or garbage landed in Great Britain. Here "tank washings" has the same meaning as in paragraph 36 of Schedule 3 to the Waste Management Licensing Regulations 1994 and "garbage" has the same meaning as in Regulation 1(2) of the Merchant Shipping (Reception Facilities for Garbage) Regulations 1988.

Sewage or sewage sludge or septic tank sludge that is deposited on land **3.28** within the curtilage of a sewage treatment works as an integral part of the operation of those works is neither industrial nor commercial waste.[43] However if it is dealt with by mobile plant it will not gain the benefit of this exemption. Sludge here is the residue produced at a sewage treatment works that treats domestic or urban waste waters and from plants treating waste waters of a similar composition.[44] Sludge that is deposited directly onto land for agricultural purposes in accordance with the Sludge (Use in Agriculture) Regulations 1989 also falls outside this classification.[45] Agricultural purposes are defined as the growing of commercial food crops including the growing of such crops for stock rearing purposes.[46] Sludge from septic tanks, but not from cesspools, that is used on agricultural land in accordance with the 1989 Regulations is also neither industrial nor commercial waste.[47]

Animal by-products which are collected and transported in accordance **3.29** with Schedule 2 to the Animal By-Products Order 1992[48] will not be treated as either commercial or industrial waste for the purposes of the duty of care on waste imposed by section 34 of the 1990 Act.[49] For these purposes animal by-products means any carcass or part of any animal or product of animal origin not intended for direct human consumption. This does not include animal excreta or catering waste or meat cooked at a knacker's yard for use as food for animals whose flesh is not intended for human consumption.[50]

Clinical waste is defined in Regulation 1(1) of the 1992 Regulations as **3.30** including any waste which consists wholly or partly of human or animal

---

[42] As amended by S.I. 1996 No. 972, reg. 24.
[43] S.I. 1992 No. 588, reg. 7(4).
[44] *ibid.* reg. 1(1) and S.I. 1989 No. 1263, reg. 2(1).
[45] *ibid.* reg. 7(1)(b).
[46] S.I. 1989 No. 1263, reg. 2(1).
[47] S.I. 1992 No. 588, reg. 7(1)(c).
[48] S.I. 1992 No. 3303.
[49] S.I. 1992 No. 588, reg. 7(3) as added by S.I. 1994 No. 1056, reg. 24.
[50] *ibid.* reg. 7(4) and S.I. 1992 No. 3303, reg. 3(1).

tissue, blood or other body fluids, excretions, drugs or other pharmaceutical products, swabs, or dressings or syringes, needles or other sharp instruments, being waste which unless rendered safe may prove hazardous to any person coming into contact with it. It also includes any other waste arising from medical, nursing, dental, veterinary, pharmaceutical or similar practice, investigation, treatment, care, teaching or research, or the collection of blood for transfusion, being waste which may cause infection to any person coming into contact with it.

COMMERCIAL WASTE

3.31   Under section 75(7) of the Environmental Protection Act 1990 "commercial waste" consists of waste from premises used wholly or mainly for the purposes of a trade or business or for the purposes of sport, recreation or entertainment. However, household and industrial wastes fall outside this category, as do agricultural and mining and quarrying wastes and any other wastes prescribed for these purposes. A "trade" is an "organised seeking after profits"[51] while a "business" is almost anything which is an occupation[52] and thus is wider in scope than "trade."

3.32   In Great Britain certain wastes are to be treated as "commercial" by virtue of Regulation 6 of and Schedule 4 to, the Controlled Waste Regulations 1992.[53] These are wastes;

1. from an office or showroom;
2. from a hotel within the meaning of section 1(3) of the Hotel Proprietors Act 1956, or, in Scotland, section 139(1) of the Licensing (Scotland) Act 1976;
3. from any part of a composite hereditament—as defined by section 64(9) of the Local Government Finance Act 1988—or, in Scotland, of part residential subjects—as defined in section 99(1) of the Local Government Finance Act 1992—which is used for the purposes of a trade or business;
4. from a private garage which either has a floor area exceeding 25 square metres or is not used wholly or mainly for the accommodation of a private motor vehicle;
5. from premises occupied by a club, society, or any association in which activities are conducted for the benefit of the members;
6. from premises occupied by a court, government department, local authority, a company or individual appointed by or under any enactment to discharge any public functions, a body incorporated by Royal Charter; unless waste from the particular premises is "household" or "industrial" under another provision of the 1992 Regulations;
7. from a tent pitched on land other than a camp site;

[51] *Aviation & Shipping Co. v. Murray* [1961] 1 W.L.R. 974.
[52] *Rolls v. Miller* (1894) 27 Ch.D. 71 at 88.
[53] S.I. 1992 No. 588.

8. from a market or fair;

9. waste collected under section 22(3)—street cleaning—of the Control of Pollution Act 1974, or section 25(2) of the Local Government and Planning (Scotland) Act 1982.

THE EUROPEAN WASTE CATALOGUE

Under Article 1(a) of the Framework Waste Directive the Commission is 3.33 required to draw up a list of wastes belonging to the categories listed in Annex 1 of that directive. This has been implemented by the Commission Decision establishing such a list[54] which is commonly referred to as the European Waste Catalogue. It applies to all wastes, irrespective of whether they are destined for disposal or recovery. However, the inclusion of a material in the list does not mean that it is a waste in all circumstances. It must first fall under the definition of "waste" as set out in Article 1(a), subject to the exceptions in Article 2.1(b).

The purpose of the list is to provide a common terminology throughout 3.34 the Community. Under the list wastes are allocated six figure numbers and divided into industrial section—for example, waste from the leather and textile industry has the number 040000. Within each grouping there are subdivisions—*i.e.* waste from the leather industry (040100). The types of waste from that sub-division are then classified—*i.e.* liming wastes (040102).

The United Kingdom is currently drawing up a national waste classifica- 3.35 tion scheme which will be based on the European List. Classification numbers under the EWC are required on consignment notes under the Special Waste Regulations 1996 and on notes for transfrontier shipments.[55] They should also be entered under the description of waste required by transfer notes for the purposes of the duty of care.[56]

---

[54] (94/3 E.C.) [1994] O.J. L5/15.
[55] S.I. 1996 No. 972, Sched. 1.
[56] See *Waste Management, The Duty of Care, A Code of Practice*, March 1996, para. 1.8.

4. NON-CONTROLLED WASTES OR MATERIALS

4.01   Certain wastes are excluded from the definition of "waste" in the Framework Directive and are thus not "Directive waste" for the purposes of the 1990 Act or the Controlled Waste Regulations 1992. Waste which is not "Directive waste" will not be household, industrial or commercial waste for the purposes of Part II of the 1990 Act.[1] Those wastes excluded by the Directive are gaseous effluents emitted into the atmosphere and, where they are already covered by other legislation, radioactive waste, mining and quarrying wastes, animal carcasses, faecal matters and other natural non-dangerous substances used in farming, waste waters (sewage effluent)—but not liquid effluent—and decommissioned explosives.

1. AGRICULTURAL WASTES

4.02   Wastes from premises used for agriculture, as defined in section 109(3) of the Agriculture Act 1947 (in Scotland the Agriculture (Scotland) Act 1948), are not commercial waste for the purposes of section 75(7) of the Environmental Protection Act 1990 nor may they be specified as controlled waste in any regulations made under section 75(8).[2] Substances regulated under Part IV of the Agriculture Act 1970 or Part III of the Food and Environment Protection Act 1985 cannot be made the subject of regulations requiring information about them under section 142 of the 1990 Act.[3] Wastes from premises used for the purposes of breeding, boarding, stabling or exhibiting animals is industrial waste for the purposes of the Act.[4]

4.03   Under the Control of Pesticides Regulations 1986[5] a pesticide may only be used if the person using it takes all reasonable precautions to protect the health of human beings, creatures and plants, to safeguard the environment and, in particular, to avoid the pollution of water.[6] A disposal of pesticide

---

[1] S.I. 1992 No. 588, reg. 7A(1) as added by S.I. 1994 No. 1056.
[2] But see E.P.A. 1990, s.63(1), at para. 4.22. For N.I. see S.I. 1978 No. 1049 (N.I. 19), Art. 36(2)(c) & (3).
[3] E.P.A. 1990, s.142(2)(b) & (7).
[4] S.I. 1992 No. 588, reg. 5(1) & Sched. 3, para. 14.
[5] S.I. 1986 No. 1510, as amended by S.I. 1997 No. 188.
[6] *ibid.* reg. 6(1)(c) & Sched. 3, para. 1.

wastes contrary to this requirement would, it is submitted, be a breach of Regulation 4(5)(b). Such a breach may be an offence under section 16(12)(a) of the Food and Environment Protection Act 1985 under which a person is liable, on summary conviction, to a fine not exceeding the statutory maximum or on indictment to a fine.[7] The enforcement officer of a local authority or the Minister[8] may require a person he considers to have committed an offence under section 16(12)(a) to remedy the situation.[9]

Many agricultural wastes are no longer excluded from the definition of 4.04 "waste" in the Framework Directive. Regulations are to be introduced to bring such wastes under control.

Animal carcasses are dealt with under the Community Directive laying 4.05 down veterinary rules for the disposal and processing of animal waste and other matters.[10] The Directive was adopted under Article 43 of the E.E.C. Treaty and its main purpose is to deal with waste from slaughterhouses and to impose controls on the production of pet food. Thus "animal waste" is defined by it as

> "carcasses or parts of animals or fish, or products of animal origin not intended for direct human consumption, with the exception of animal excreta and catering waste."

Animal carcasses that are to be disposed of on a farm will in general be 4.06 excepted from the more stringent requirements of the Directive as, even if they are classified as "high risk" it could be argued that the quantity of the waste involved and the distance to be covered does not justify collecting them. In such cases the carcasses must be burnt or buried. If they are to be buried the burial must be deep enough to prevent carnivorous animals from digging them up. They must be interred in suitable ground so as to prevent contamination of the water table or any environmental nuisance. Before burial the carcass should be sprayed as necessary with a suitable disinfectant authorised by the relevant competent authority.[11] The Directive is implemented in Great Britain by the Animal By-Products Order 1992[12] and is enforced by local authorities.

The carcasses of animals that have been slaughtered subsequent to a 4.07 ministerial direction must be disposed of in accordance with section 34 of the Animal Health Act 1981. It will be an offence under section 35(4)(a) of the Act to put, or cause to be put, the carcass of an animal that has died of a disease or been slaughtered as diseased or suspected to be so, into any inland water or into the sea within 4.8 kilometres of the shore. The offence is punishable on summary conviction in the manner set out in section 75 of the Act.[13]

---

[7] Food and Environment Protection Act 1985, ss.21(3) & (4) and 22.

[8] *ibid.* s.19, as amended by the Pesticides (Fees and Enforcement) Act 1989, s.2.

[9] *ibid.* s.19(5).

[10] Dir. 90/667 [1990] O.J. L365/51.

[11] Dir. 90/667, Arts 3.2 & 5.1.

[12] S.I. 1992 No. 3303.

[13] See also Diseases of Animals Order (Northern Ireland) 1981, S.I. 1981 No. 1115 (N.I. 22), Art. 52(1)(d).

2. EXPLOSIVE WASTES

4.08   The definition of "waste" in section 75(2) of the Environmental Protection Act 1990 excludes "a substance which is an explosive within the meaning of the Explosives Act 1875.[14] In section 3 of the 1875 Act explosives are defined as,

> "gunpowder, nitro-glycerine, dynamite, gun-cotton, blasting powders, fulminate of mercury or of other metals, coloured fires and every other substance, whether similar to those above mentioned or not, used or manufactured with a view to produce a practical effect by explosion or pyrotechnic effect."

The term includes fog-signals, fireworks, fuzes, rockets, percussion caps, detonators and every adaptation and preparation of an explosive as defined.[15] This definition may have been extended by orders in council made under section 106 of the Act. These wastes are also excluded from regulations made under section 142 of the 1990 Act requiring information about substances.[16]

4.09   Explosive wastes can be disposed of or destroyed by burning, detonation, dissolution or dilution by a solvent, chemical destruction or drowning.[17] If they have to be disassembled or broken down prior to disposal they will be being "manufactured" for the purposes of the Explosives Act.[18] Such an operation may only be carried out in a factory licensed by the Health and Safety Executive under section 4 of that Act. Once an explosive has been subjected to treatment that takes it outside the definition in the 1875 Act it is submitted that the resulting matter is "waste" for the purposes of the Environmental Protection Act.

3. MINING AND QUARRYING WASTES

4.10   Wastes from mining or quarrying operations are not controlled wastes for the purposes of the Environmental Protection Act nor may regulations under section 75 so designate them.[19] Wastes from tunnelling or from any other excavation are industrial waste and so controlled.[20] However, a waste management licence is not required where such waste is being used for the purposes of construction currently being carried out, or to be carried out, on the land on which it is deposited, provided the waste is not stored for more than three months and the occupier of the relevant land consents to its deposit there.[21] The deposit of excavated material from peatworking does not

---

[14] See also S.I. 1978 No. 1049 (N.I. 19), Art. 36(1).
[15] Explosives Act 1875, s.3(2).
[16] E.P.A. 1990, s.142(2)(a) & (7).
[17] See *Disposal of Explosives Waste . . .* , Health & Safety Executive, HS/G 36 (1987).
[18] Explosives Act 1875, s.105.
[19] E.P.A. 1990, s.75(4)–(8), but see s.63(1) at para. 4.22.
[20] S.I. 1992 No. 588, reg. 5 & Sched. 3, para. 6.
[21] S.I. 1994 No. 1056, Sched. 3, para. 19(1) & (2).

require a licence if done at the site of operations[22] nor does the deposit of such material from mineral exploration activities. But this last exemption will only apply if the development from which the waste arises is authorised by the General Permitted Development Orders of 1992 or 1995 and complies with any limitation or condition specified in that order.[23]

Deposits of mineral wastes are controlled under the Town and Country   4.11
Planning Act 1990. Conditions imposed on a planning permission for mineral extraction may stipulate where waste should be tipped on the site and regulate the appearance of tips. Conditions will also be set, after consultation with the relevant Agency, to protect groundwater from pollution by mining and quarrying waste.[24] Restoration and aftercare of a mineral working site may be required by conditions imposed under Schedule 5 to the Town and Country Planning Act 1990,[25] or Schedule 3 the Town and Country Planning (Scotland) Act 1997.

Provisions as to the safety of tips are set out in the Mines and Quarries   4.12
(Tips) Act 1969. Under Part II of that Act a county or London borough council (in Scotland a constituted council and in Wales a county or county borough council)[26] may require information as to the stability of a tip in their area from any person they consider may be able to assist them.[27] They may also investigate the tip to ensure that it is safe.[28] If, as a result of such information or investigations, they consider the tip to be unstable as defined by section 36(2) of the Act[29] they may serve a notice requiring the owner to take remedial action.[30] An authority can also carry out remedial works themselves. In an emergency these works can be done without prior notification to the owner.[31] In Northern Ireland these matters are dealt with by Article 9 of the Quarries Order (Northern Ireland) 1983.[32]

A number of mining and quarrying wastes no longer fall within the   4.13
exemption for waste in the Framework Directive. Regulations to bring such wastes under control have yet to be issued.

4. RADIOACTIVE WASTES

The regime for radioactive wastes is set out in the Radioactive Substances   4.14
Act 1993. Radioactive waste is defined as waste consisting wholly or partly of a substance or article which, if it were not waste would be radioactive

[22] S.I. 1994 No. 1056, Sched. 3, para. 33.
[23] *ibid.* Sched. 3, para. 35.
[24] Mineral Planning Guidance Note 2, paras 93–97 & 101.
[25] As amended by Planning and Compensation Act 1991, s.18 and Sched. 1.
[26] Mines and Quarries (Tips) Act 1969, s.11 as amended Local Government, etc. (Scotland) Act 1994, Sched. 13, para. 81 and Local Government (Wales) Act 1994, Sched. 16, para. 34.
[27] *ibid.* s.12.
[28] *ibid.* s.13.
[29] See *Lanark County Council v. Doonin* [1974] S.L.T.(Sh.Ct.) 13.
[30] Mines and Quarries (Tips) Act 1969, s.14.
[31] *ibid.* s.17(1) & (3).
[32] S.I. 1983 No. 150 (N.I. 4).

material, or a substance or article that has become contaminated by radioactive material or waste.[33] Radioactive material is defined in section 1 as a substance, or article made wholly or partly from such a substance, that is not waste and which either naturally contains an element like radium, that is specified in Schedule 1 to the 1993 Act, in a greater amount of microcuries per gramme than specified for it in that Schedule, or that is radioactive wholly or partially as a result of nuclear fission, or other artificial methods of inducing radioactivity, or that has become contaminated by radioactivity from another source, unless its radioactivity is below the levels set for that substance in regulations.

4.15   Waste includes scrap material, an effluent or other unwanted surplus substances, such as gases, arising from the application of any process and also includes any substance or article that must be disposed of as being broken, worn out, contaminated or otherwise spoilt.[34] Moreover, any substance or article that is treated as waste by an undertaking[35] is deemed for the purposes of the 1993 Act to be waste until the contrary is proved.[36] Thus "waste" in this context has a different meaning from that in the Directive as it can include gases, however, this may have little practical importance.

4.16   Section 78 of the Environmental Protection Act 1990 provides that unless otherwise prescribed in regulations made by the Secretary of State, nothing in Part II of the 1990 Act will apply to radioactive wastes as defined by the 1993 Act.[37] If regulations are made under this section, and none have been made so far, they can provide for the provisions of Part II, the 1993 Act and any other legislation to have effect with such modifications as the Secretary of State considers appropriate. Provision is made by Regulation 3 of the Special Waste Regulations 1996 to enable the Secretary of State to deal with the chemical components of radioactive wastes as special waste where necessary.

4.17   Under the Radioactive Substances Act 1993 the disposal of radioactive wastes is subject to the control of the relevant Agencies. They will authorise the disposal of wastes in accordance with the provisions of sections 13 and 16 of the Act. The Agencies will also authorise accumulations of waste under section 14.[38] They will follow the procedures set out in a "Review of Radioactive Waste Management Policy"[39] in authorising disposals or accumulations. In particular the Agencies, before granting an authorisations must be satisfied that any practice giving rise to radiation exposures is justified.[40]

4.18   Records of waste disposal sites and of disposals may be required under section 20. If the appropriate Agency considers that a person authorised

[33] Radioactive Substances Act 1993, s.2.
[34] R.S.A. 1993, s.47(1).
[35] ibid. s.47(1).
[36] ibid. s.47(4).
[37] See also S.I. 1978 No. 1049 (N.I. 19), Art. 36(4).
[38] S.I. 1996 No. 972.
[39] July 1995, Cm. 2919.
[40] R. v. Secretary of State for the Environment, ex p. Greenpeace [1994] Env. L.R. 401.

under sections 13 or 14 is failing, or is likely to fail, to comply with any condition or limitation in the authorisation it may serve an enforcement notice under section 21 of the Act. Further, if it considers that the continuation of a disposal or accumulation operation will involve an imminent risk to the environment or human health, it may serve a prohibition notice under section 22. Authorisations may be varied or revoked under section 17 while rights of appeal are granted by section 26.

5. SEWAGE

Sewage is only "controlled waste" as far as regulations made under section 4.19 75(8) of the Environmental Protection Act 1990 provide. Regulation 7(1) of the Controlled Waste Regulations 1992[41] specifically excludes three types of sewage waste from being industrial or commercial waste. These are sewage, sludge and septic tank sludge which is treated, kept or disposed of (otherwise than by means of mobile plant) in the curtilage of a sewage treatment works as an integral part of the operation of those works, sludge supplied in accordance with the Sludge (Use in Agriculture) Regulations 1989[42] and septic tank sludge that is used on agricultural land within the meaning of those Regulations. On the other hand sewage that falls outside this exemption which is treated, kept or disposed of in or on land otherwise than by means of a privy, cesspool or septic tank or is treated, kept or disposed of by means of mobile plant or that has been removed from a privy or cesspool will be industrial waste.[43] The key distinction here has to be drawn between waste from a septic tank which can benefit from the exemption and waste from a cesspool which cannot. "Septic tank sludge" is defined in the 1989 Regulations[44] as residual sludge from septic tanks and other similar installations for the treatment of sewage. Cesspools do not treat sewage.

Septic tank sludge that is not dealt with as an integral part of the 4.20 operations of a sewage treatment works or deposited on agricultural land under the 1989 Regulations will be industrial waste.[45] If it emanates from a household then the duty of care imposed by section 34 of the 1990 Act does not apply to the house's occupier in respect of it[46]; but a contractor removing it would still be bound by the duty. Sewage from a sanitary convenience on a passenger train or, if buried, from a removable receptacle from a toilet on premises other than a private dwelling will not require a waste management licence as long as it does not give rise to an environmental hazard and as long as the quantity does not exceed the limits specified in the exemption.[47]

---

[41] S.I. 1992 No. 588 as amended by S.I. 1995 No. 288, reg. 2(2).
[42] S.I. 1989 No. 1263.
[43] ibid. Sched. 3, para. 7.
[44] S.I. 1989 No. 1263, reg. 2(1) and see S.I. 1992 No. 588, reg. 1(2).
[45] S.I. 1992 No. 588, reg. 5(2)(a).
[46] ibid. reg. 2(2)(b) & E.P.A. 1990, s.34(2).
[47] S.I. 1994 No. 1056, Sched. 3, paras 31 and 32.

4.21   Requirements relating to the agricultural use of sewage or septic tank sludge are set out in the Sludge (Use in Agriculture) Regulations 1989.[48] Where such sludge for use in agriculture under those Regulations is deposited in a holding lagoon or container on land used for agriculture no waste management licence will be required if the lagoon or container is designed or adapted so that, as far as is practicable, the sludge cannot escape from it and the public cannot have access to it.[49]

6. THE DEPOSIT OF NON-CONTROLLED WASTE

4.22   Agricultural and mining or quarrying wastes are not controlled wastes by virtue of section 75(4)–(8) of the 1990 Act. However, they may become controlled by regulations made by the Secretary of State under section 63(1). Before making them he should consult with such persons as he considers appropriate. Regulations would specify an area within which certain agricultural waste such as pesticides or certain mining or quarrying waste should be treated as controlled waste, or a type of controlled waste, for the purposes of particular provisions of Part II of the Act. The regulations may make necessary modifications to those provisions and to other enactments as the Secretary considers appropriate.

4.23   While the deposit of non-controlled waste does not require a licence an offence may be committed where particularly noxious waste is disposed of. The offence will be committed by a person who deposits, or knowingly causes or knowingly permits[50] the deposit of any waste other than controlled waste. If he does this in a case where, if the waste were special waste (in other words if it has the characteristics of special waste) and if any waste management licence were not in force in respect of the place and type of waste deposited he would be guilty of an offence under section 33 of the Act, he will also be guilty of an offence under section 63(2) and liable to be punished in accordance with section 33(9).[51] As this offence is concerned with "waste" the deposit of explosives would not be caught by it.

4.24   The defendant would have to be guilty of an offence under section 33 for the offence under section 63(2) to be committed. Thus if he has a defence under section 33(7) he will not be guilty here. Further, if the act he is charged with was done under and in accordance with any consent, licence or approval or authority granted under any enactment, no offence will have been committed.[52] Deposit under planning permission is excluded from this exemption but disposal of matter in accordance with an order under section 35 of the Animal Health Act 1981 for example, would afford a defence. In addition submitted that steps taken in relation to mineral waste as a result of

---

[48] S.I. 1989 No. 1263 as amended by S.I. 1990 No. 880. For N.I. see S.R. 1990 No. 245.
[49] S.I. 1994 No. 1056, Sched. 3, paras 7(7) & 8(3).
[50] See para. 15.05.
[51] s.63(2) as replaced by E.A. 1995, Sched. 22, para. 81.
[52] E.P.A. 1990, s.63(3).

orders by a mineral planning authority made under Schedule 9 of the Town and Country Planning Act 1990[53] would also come within this exemption. In Northern Ireland a similar offence to that under section 63 is contained in Article 18 of the Pollution Control and Local Government Order (Northern Ireland) 1978.[54]

A waste collection authority may collect non-controlled waste at the 4.25 request of an occupier of premises in its area under section 45(2) of the Act and supply him with receptacles for it under section 47(1).[55] Where wastes other than controlled wastes are to be kept, treated or disposed of at a licensed site, conditions in the site licence may regulate the way in which that waste is to be dealt with.[56]

---

[53] As amended by the Planning and Compensation Act 1991, Sched. 1. Scotland; T.&C.P. (Scotland) Act 1997, Sched. 8.
[54] S.I. 1978 No. 1049 (N.I. 19).
[55] E.P.A. 1990, s.63(4).
[56] *ibid.* s.35(5).

5. THE DUTY OF CARE

5.01   The concept of a duty of care for waste was first introduced in the Eleventh Report of the Royal Commission on Environmental Pollution[1] which recommended that responsibility for ensuring that wastes are properly handled and disposed of should lie with the person who produces them. However, they suggested that, within the framework of such a duty, a waste producer may assign responsibility to a person who he has good reason to believe is competent to handle the waste safely. The intention was to create a policy of security in the waste stream through a continuity of care exercised by the producer of the waste, the transporter or handler of that waste and those responsible for its ultimate disposal.[2] This basic approach is also contained in Article 8 of the Framework Waste Directive.[3]

5.02   Section 34 of the Environmental Protection Act 1990 enacts this recommendation and sets out the framework of the duty. The substantive provisions of this section were brought into force from April 1, 1992, although powers to make regulations and codes of practice became effective earlier.[4] The duty is explained in Department of the Environment Circular 19/91.[5] While the provisions of section 34 extend to Scotland, they have not yet been applied to Northern Ireland.

5.03   By virtue of section 34(7) the Secretary of State must, after consulting interested parties, prepare and issue a code of practice which will provide practical guidance on how those subject to the duty of care should discharge it. Different codes may be issued to cover different areas.[6] Codes prepared under this provision must be laid before both Houses of Parliament.[7] A code that has been issued may be revised from time to time by the Secretary of State who can revoke, amend or add to its provisions.[8] *Waste Management, The Duty of Care: A Code of Practice*[9] has been issued under these provisions and covers all of Great Britain. The current edition is the second edition, which revokes the first.

[1] Cm. 9675 (1985) paras 3.4–3.8.
[2] *ibid.* paras 8.12–8.15 & 8.25.
[3] Dir. 75/442 as amended by 91/156 [1991] O.J. L78/32.
[4] S.I. 1991 No. 2829.
[5] W.O. Circular 63/91; S.O. Circular 25/91.
[6] E.P.A. 1990, s.34(11).
[7] *ibid.* s.34(9).
[8] *ibid.* s.34(8).
[9] March 1996 H.M.S.O. ISBN 0 11 753210 X.

A code of practice that has been issued under section 34(7) is admissible **5.04** in evidence and if any provision of such a code appears to the court to be relevant to any question arising in the proceedings it shall be taken into account in determining that question.[10] However, someone may not have taken the steps recommended in the code but still have complied with the duty of care by other means. As is said in the introduction to the code it: "cannot cover every contingency; the legal obligation is to comply with the duty of care itself rather than with the code." However, the use of the code is not confined to the provisions of section 34. It could, for example, be used in a civil claim for negligence arising out of the handling of waste, or in criminal proceedings under section 33 of the 1990 Act.

THE SCOPE OF THE DUTY

Section 34(1) imposes the duty of care on anyone who imports, produces, **5.05** carries, keeps, treats or disposes of controlled waste or who, as a broker, has control of such waste. None of these terms are defined in the legislation; although they may not cause much difficulty. However, in the circular and the code of practice the shorthand term "holder" or "waste holder" is adopted. While this is convenient, and will be used later in this chapter, it must be remembered that there is no such person as a "holder" of waste under section 34. A charge against someone as a "holder" of waste would be invalid. Thus, it will be necessary to determine the capacity in which the person on whom it is sought to impose the duty acted.

The duty under section 34 requires the "holder" of the waste "to take all **5.06** such measures applicable to him in that capacity as are reasonable in the circumstances" to prevent anyone else committing an offence under section 33 of the 1990 Act, to prevent the escape of the waste from his control or that of any other person and, when the waste is transferred, to secure that it is only transferred to an authorised person or to a person for "authorised transport purposes" and that it is accompanied by proper documentation.[11] These duties are examined in more detail below.

WASTE "HOLDERS"

An importer will be someone who brings the waste to the place where it **5.07** first enters Great Britain. This may pose problems in respect of waste landed on shore from a ship and sent to a reception facility. If the waste is the "garbage" from a ship then those operating the ship would be "producers" rather than importers. If, however, the waste is residue from a cargo they may be "importers." In either event the operators would be subject to the duty. As far as the code of practice is concerned, importers should act in most respects as if they were the producers of the waste they import. They

[10] E.P.A. 1990, s.34(10).
[11] *ibid.* s.34(1).

bear the primary responsibility for ensuring the adequacy of the accompanying description, the packaging of the waste on its entry to Great Britain and the fitness of the destination to deal with the waste.[12]

5.08   A producer of waste is not, it is submitted, the person who manufactured an article or substance that has become waste but the person in whose hands it is when it changes to waste. Where a factory supplies a drum of chemicals to a wholesaler, who breaks it down into smaller drums and sells it directly to a printer who uses most of the chemical and discards the container with the residue, it is the printer who "produces" the waste. However, if the chemical was placed into a dirty container by the wholesaler, which so contaminated the chemical as to make it useless, then the waste will have been "produced" by the wholesaler. There may be potentially more than one "producer" in relation to a particular waste. In construction and demolition operations the "producer" could be the developer, the main contractor or the sub-contractor who actually carries out the work. The Department of the Environment considers that here it would be the sub-contractor who would be subject to the duty on the basis that the "producer" should be the person undertaking the works that give rise to the waste.[13]

5.09   Under the code of practice it is said that waste producers are solely responsible for the care of their waste while they hold it.[14] They bear the main responsibility for the description of waste which leaves them being correct and, if they select the final destination, of ensuring that the disposal is lawful. In addition, they should ensure that the waste is properly packaged for transport. A producer will still be responsible for the waste after it leaves his premises to the extent that if there is a suspicion that it is not being properly dealt with he should investigate the matter and take steps to end the situation.[15]

5.10   A carrier of waste will be the person transporting it within Great Britain. This will generally be the operator of the vehicle or other means of carriage rather than the driver. The code of practice suggests that the carrier is responsible for the adequacy of packaging while the waste is under his control. While he does not have to provide a new description of it he should give it at least a quick visual inspection to see that it appears to match the description. If he mixes, treats or repackages it or if the waste deteriorates or decomposes in his care then a new description may be required.[16]

5.11   The "keeper" of waste will be someone storing it or in charge of it for the time being. A person treating waste is someone subjecting it to any process, including making it re-usable or reclaiming substances from it.[17] A "disposer" of waste will be someone who carries out most of the operations listed in Annex IIA of the E.C. amended Waste Directive[18]; although some

---

[12] Code of Practice, para. B.8.
[13] D.o.E. Circular 19/91, paras 16 & 17.
[14] Code of Practice, Annex B, para. B.3.
[15] *ibid.* para. B.6.
[16] *ibid.* paras B.8–B.11.
[17] E.P.A. 1990, s.29(6).
[18] Dir. 91/156 [1991] O.J. L78/32.

of those are more appropriate to keeping or treating waste. For example, operation D4 of that Annex, the placement of liquid or sludge discards into pits, ponds or lagoons, etc., could be keeping or disposing, depending on the intention of the person having control of the waste. In the code of practice these categories of holders are termed "waste managers." They should check the descriptions of waste they receive. Sample checks on the composition of waste received would be good practice. They should also follow up evidence of previous misconduct.[19]

The code of practice defines a broker as a person who arranges the **5.12** transfer of waste, which he does not himself hold, to such an extent that he controls what happens to it. In *Milford v. Hughes*[20] a broker was defined as someone who contrives, makes and concludes bargains and contracts between merchants and tradesmen for reward. While a broker may have control of the waste, he will rarely have physical custody of it so that there will normally be someone else involved as its keeper. In the code of practice it is said that where the broker controls what happens to the waste he is taking responsibility for the legality of the arrangement.[21]

In addition to these specific duties on individual categories of holder, the **5.13** code of practice also sets out general duties. No-one should accept waste from a source that seems to be in breach of the duty or that is badly packaged or mis-described.[22] Further the transferee should co-operate with the transferor to enable him to comply with the duty.[23]

The duty imposed by section 34(1) does not apply to an occupier of **5.14** domestic property as respects the household waste produced on that property.[24] However, if the occupier has a workshop on the premises the waste from that workshop must be dealt with in accordance with the duty, as must waste brought onto them from other premises. Further, waste produced by a builder carrying out work on the property will be subject to the duty.[25]

THE DUTIES

The holder must take steps to prevent any contravention of section 33 of **5.15** the 1990 Act by any other person in respect of the controlled waste for which he is responsible.[26] Section 33 concerns the unlawful deposit, or keeping, or treatment of controlled waste.[27] Thus the holder should ensure that the waste is either exempt from licensing controls or will be taken to a disposal or intermediate facility that is authorised to take that type of waste.

---

[19] Code of Practice, para. 3.13. and Annex B, paras. B.13 & B.14.
[20] (1846) 16 M. & W. 174 at 177.
[21] Code of Practice, para. 3.19 and Annex B, para. B.12.
[22] *ibid.* paras 4.2–4.6.
[23] *ibid.* paras 4.7–4.10.
[24] E.P.A. 1990, s.34(2).
[25] Code of Practice, Annex A, para. A.9.
[26] E.P.A. 1990, s.34(1)(a).
[27] See paras 15.05–15.09.

This, therefore, involves the holder in knowing the proper description of his waste, whether or not it is controlled, whether the facility is licensed and the conditions of that licence. The code of practice suggests that the holder should check the licensing situation before sending any waste by looking at the actual licence and asking the operator whether his waste is covered by it.[28] If the operator says no licence is necessary because of an exemption he must specify the exemption concerned. This procedure should be followed by all holders, not just producers, who are sending waste to a facility; although a carrier taking waste under contract between the producer and the facility need not make such checks.[29]

5.16 The most contentious part of this duty may be in making checks after the waste has been transferred. While generally a producer is not responsible for the waste after it has left his premises circumstances may arise when he would fail in his duty by either not reporting suspicions of unlawful handling to the relevant waste regulation authority or terminating his current arrangements. Section 5 to the code of practice deals with this situation and in particular suggests that contracts should provide for termination if a breach of the duty occurs and is not rectified.[30]

5.17 The second duty is to prevent the escape of the waste from his control or that of anyone else.[31] This involves ensuring that it is properly packaged, handled and secured to prevent spillage or leakage. The duty here is to secure that the waste will be packaged so as to survive transport to its final destination, unless it is to be immediately mixed with other waste. It should be stored to prevent damage by vandals or others and, if necessary, to prevent the mixing of incompatible types of wastes. Waste containers should be adequate for their purpose and drums or similar closed containers labelled with a note of their contents. Skips with loose materials should be covered to prevent waste escaping. Waste left for collection outside premises should be packaged so as to resist wind and rain and animal disturbance.[32]

5.18 The third duty is to ensure that when the waste is transferred it is only handed over to an authorised person or a person for authorised transport purposes.[33] Authorised persons for these purposes are a waste collection authority; any holder of a waste management or disposal licence, anyone exempted from the provision of section 33(1) of the Act by regulations made under section 33(3); a carrier registered under section 2 of the Control of Pollution (Amendment) Act 1989 or exempted from registration under Regulation 2 of the "Carriers" Regulations[34] and, in Scotland, a waste disposal authority.[35] "Authorised transport purposes" are the transport of waste on premises between different parts of those premises, the import of

[28] Code of Practice, para. 3.15.
[29] *ibid.* para. 3.18.
[30] *ibid.* para. 5.8.
[31] E.P.A. 1990, s.34(1)(b).
[32] Code of Practice, paras 2.1–2.7.
[33] E.P.A. 1990, s.34(1)(c)(i).
[34] See para. 9.04.
[35] E.P.A. 1990, s.34(3).

waste to its first landing in Great Britain and the export of waste beyond Great Britain.[36] "Transport" here includes transport by road, rail, air, sea or inland waterway but does not include moving the waste from one place to another by means of a pipe or other apparatus joining those places.[37] The list of authorised persons may be added to by regulations made under section 34(3A) of the Act.[38]

On transfer of the waste the holder must ensure that it is accompanied by **5.19** an adequate written description to enable others to avoid a contravention of section 33 and to prevent it from escaping.[39] Guidance on written descriptions is contained in section 1 of the code of practice which states that it should provide enough information to enable subsequent holders to avoid mismanaging the waste. For some wastes a simple description of the type or premises or business from which they originate together with the name of the substance or substances and the waste classification number will be enough—for example, paper waste from an office. A more detailed description will be needed for other wastes including the process producing it and, where industrial waste is mixed or its source unknown, a chemical and physical analysis. In addition, the description should always include any information of special problems that might affect its handling. These include the use of special containers to store the waste safely, whether it can or cannot be mixed with other wastes, whether there might be problems with its processing, incineration or landfill or where a significant quantity or an unusual substance is present in a load of otherwise familiar waste—for example, a large amount of photocopier waste in a load of office waste.[40]

The duty only requires a holder "to take all such measures applicable to **5.20** him in that capacity as are reasonable in the circumstances." The code of practice suggests that the capacity of the holder is who he is—producer, carrier, etc.,—and how much control he has over what happens to the waste.[41] He must take such measures as are reasonable in the circumstances. A court looking at this provision would have regard to the defendant's knowledge and foresight at all material times and is unlikely to require him to take measures against unknown and unexpected risks.[42] The relevant circumstances here will include the nature of the waste, the dangers it presents in handling and treatment, how it is dealt with and what the holder might reasonably be expected to know or foresee.[43] The duty of one holder may extend to the way in which another holder of the waste deals with it if he knows of, or might reasonably foresee, a breach of the duty by the other.[44] The effect of this on particular holders is considered at paragraphs 5.07 to 5.12.

---

[36] E.P.A. 1990, s.34(4).
[37] *ibid.* & Control of Pollution (Amendment) Act 1989, s.9(1).
[38] As added by E.A. 1995, Sched. 22, para. 65.
[39] *ibid.* s.34(1)(c)(ii).
[40] Code of Practice, paras 1.7–1.17.
[41] *ibid.* Annex A, para. A.13.
[42] *Austin Rover Group v. Inspector of Factories* [1989] 3 W.L.R. 520.
[43] Code of Practice, para. A.12.
[44] *ibid.* para. A.14.

5.21   The courts may be assisted in determining the scope of the duty by the decision in *Seaboard Offshore Limited v. Secretary of State for Transport*[45]; a case on section 31 of the Merchant Shipping Act 1988 but with relevance here. In that case the House of Lords held that a company could not be in breach of its duties under that section because of the acts or omissions of employees.[46] The only people for whose acts the company is responsible are its corporate officers. This is because of the wording of the section "taking of all steps as are reasonable for him to take", a wording similar to that for the duty of care—"take all such measures applicable to him in that capacity." Thus a company could be liable if it fails to institute system or ensure it is operated properly, but not for the failure of an employee to put the system into practice. The House of Lords went on to say that liability under section 31 was for a failure to take steps which by an objective standard are held to be reasonable steps to take (in the interests of the safe operation of a ship). For the purposes of the duty of care the liability would be a failure to take measures which by an objective standard are reasonable measures to comply with the duty.

5.22   The code recognises that a waste holder may not always have the knowledge or expertise to discharge his duty of care. Section 6 of it therefore sets out sources of expert help and guidance. However, it stresses that the duty of care is still that of the holder and cannot be transferred to the expert.

5.23   The Duty of Care and Scrap Metal is the title to section 7 of the Code of Practice. While it mainly is concerned with documentation, it does emphasise the need to take account of the distinctive features of scrap metal and the circumstances in which it is recovered. It repeats the advice in D.o.E. Circular 6/95 that regard should be had, in considering enforcement, to the benefits of metal recycling sites. It also points out that the code is not intended to place dealers in the position where they have to seek excessive verification from suppliers.[47] Nor may it be an offence for a scrap metal dealer to receive waste from an unregistered carrier.[48]

DOCUMENTATION

5.24   The Environmental Protection (Duty of Care) Regulations 1991[49] have been made under section 34(5) of the 1990 Act to impose requirements on waste holders to ensure that when the description of the waste is transferred it is accompanied by a transfer note completed and signed on behalf of the transferor and the transferee.[50] For these purposes the holder who provides the description is the transferor and the one that receives it is the transferee.[51] This does not mean that every transfer of waste must be

---

[45] [1994] 2 All E.R. 99.
[46] But see Annex A, para. A.8 of the Code of Practice (*contra*).
[47] Code of Practice, para. 7.18.
[48] *ibid.* paras 7.19 & 7.20.
[49] S.I. 1991 No. 2839.
[50] *ibid.* reg. 2(1).
[51] *ibid.* reg. 1(2).

accompanied by a note. Where there are regular collections of the same type of waste involving the same parties a single note can cover multiple consignments; although this should be limited to those taking place in the same year.[52] In addition consignment notes under the Special Waste Regulations 1996 will take the place of transfer notes under the 1991 Regulations.[53]

Section 34(4A) of the 1990 Act—added by section 33 of the Deregulation 5.25 and Contracting Out Act 1994—makes it clear that, for the purposes of the transfer duty, a transfer of waste in stages shall be treated as taking place when the first stage of the transfer takes place. Where a series of transfers of waste of the same description are made between the same parties the operation is to be treated as a single transfer that takes place when the first of the transfers in the series takes place. This allows the use of a single transfer note.

The transfer note must identify the waste to which it relates, stating the 5.26 quantity, whether during transfer it is loose or in a container, if it is in a container the type it is in and the time and place of transfer.[54] It must also show the name and address of the transferor and transferee, state whether the transferor is the producer or importer of the waste, and if the transfer is made for authorised transport purposes, state which of those purposes is applicable.[55] Finally, it should also describe the transferor and transferee by reference to the table to Regulation 2. This lists the "authorised persons" designated in section 34(3) of the Act and requires the number of any relevant licence or registration, together with the name of the authorising body, to be stated.[56] While these are the prescribed requirements under section 34(5), the description of the waste and the transfer note can be combined into one document where this is convenient.[57] A suggested standard form of transfer note is set out in the Annex to Circular 19/91 and in Annex C to the code of practice. Where the waste is special waste, the consignment note used for that waste or from its transfrontier shipment may be adapted to form the necessary transfer note.[58]

The transferor and the transferee must each keep the written description 5.27 and the transfer note, or copies of them, for two yeas after the transfer of the waste to which they relate.[59] They could be kept in any form, such as on paper or on computer.

Only waste regulation authorities (the agencies) have a right of access to 5.28 these records. Under Regulation 4 of the Duty of Care Regulations they may serve a written notice on anyone, whether in or outside their area, requiring the production of a specified or described document. If that document is one

---

[52] D.o.E. Circular 19/91, para. 21 and Interpretation Act 1978, s.6(c).
[53] S.I. 1991 No. 2839, reg. 2 as amended by S.I. 1996 No. 972, reg. 23.
[54] *ibid.* reg. 2(2)(a).
[55] *ibid.* reg. 2(2), 89(b)–(d).
[56] *ibid.* reg.2(2)(e).
[57] D.o.E. Circular 19/91, para. 20.
[58] *ibid.* para. 27.
[59] S.I. 1991 No. 2839, reg. 3.

that he is required to have by virtue of Regulation 3 then the person served must furnish the authority with a copy of it, at the office they specify, within the period set out in the notice. However, at least seven days must be allowed. Service will be governed by section 160 of the Environmental Protection Act 1990 and section 7 of the Interpretation Act 1978. Failure to comply with such a notice will be an offence under section 34(6) of the 1990 Act. Holders can ask another holder for access to his records. If he refuses it may give rise to a suspected breach of duty about which the agencies should be informed.[60]

OFFENCES

5.29   Section 34(6) provides that any person who fails to comply with the duty of care or with a requirement of the regulations made under section 34(6) will be liable on summary conviction to a fine not exceeding the statutory maximum and on conviction on indictment to a fine. Any such conviction must be entered on the waste regulation authority's public register.[61] In addition, section 34(6) is a prescribed offence for the purposes of the Control of Pollution (Amendment) Act 1989[62] and the fit and proper persons provision of the Environmental Protection Act 1990[63] so that a conviction can lead to the defendant being regarded as unfit to be registered to carry waste or to hold a waste management licence.

5.30   There is no restriction on who can bring proceedings for such an offence but neither is any authority given the function of enforcing the duty of care. Nevertheless, it is envisaged that waste regulation authorities (the agencies) will enforce section 34 as part of their role of waste regulation; although they will be unable to use their powers to obtain information under section 71 of the Act for this purpose.[64]

5.31   The offence of failing to comply with the duty of care involves failure to take all such measures applicable to the relevant holder in his capacity as such, as are reasonable in the circumstances, to prevent a specified breach, for example an escape of the waste. Here, it is submitted, it will be for the prosecution to prove the breach of duty and that it applied to the relevant holder. It must then go on to show that there were reasonable measures the holder could have taken to prevent the breach. However, the defence will have to show that the holder took all such measures as were reasonable in the circumstances to prevent it.[65]

5.32   Where such a defence is raised the question of whether the breach was reasonably foreseeable will arise. In this context, section 8 of the Criminal Justice Act 1967 provides that a court or jury, in determining whether a

---

[60] Code of Practice, Annex C, para. C.6.
[61] S.I. 1994 No. 1056, reg. 10(1)(f).
[62] S.I. 1991 No. 1624, reg. 1(1) and Sched. 1.
[63] S.I. 1994 No. 1056, reg. 3(k).
[64] D.o.E. Circular 19/91, paras 34 & 35.
[65] *Austin Rover Group v. Inspector of Factories* [1989] 3 W.L.R. 520 at 534E–534G.

person has committed an offence, shall not be bound in law to infer that he intended or foresaw a result of his actions by reason only of its being a natural and probable result of those actions. However, a court or jury shall decide whether he did intend or foresee that result by reference to all the evidence, drawing such inference from the evidence as appears proper in the circumstances. This does not mean that guilt is dependent on proof of foresight but merely directs the way a court is to determine whether the breach was foreseeable.[66]

The circumstances of a defendant may play a role in what he should **5.33** possibly foresee or do to prevent a breach. A small business employing one or two people may be in a different position from a multi-national company. In the context of section 4(2) of the Health and Safety at Work Act 1974 it was considered that in the phrase, "to take such measures as it is reasonable for a person in his position to take", "reasonable" related to the person and not the measures.[67] Thus it is that the question of foresight has to be addressed from the defendant's eyes; to what it is reasonable for him to foresee. On this basis, different standards may well be imposed on different types of holder.

As far as the offence of failing to comply with the documentation **5.34** regulations is concerned no such questions arise. The duty is an absolute one so that failure to follow the transfer note procedure is enough. No defences, such as a reasonable excuse, are provided in the legislation. While it is not specifically stated it is also probably an offence to provide a transfer note that gives false or incomplete information.[68]

---

[66] See *D.P.P. v. Majewski* (1976) 62 Cr.App.R. 282 *per* Lord Edmund-Davies at 288.
[67] *Austin Rover Group v. Inspector of Factories* [1989] 3 W.L.R. 520 at 534C.
[68] See D.o.E. Circular 19/91, para. 36.

6. LITTER

6.01   "Litter" is defined in the Oxford English Dictionary, in this context, as,

"Odds and ends, fragments and leavings lying about, rubbish; a state of confusion or untidiness; a disorderly accumulation of things lying about."

"Litter" is not defined in the Environmental Protection Act 1990 but in the Pollution Control and Local Government Order (Northern Ireland) 1978 (The 1978 Order) it is stated to mean,

"any refuse, filth, garbage or any other nauseous, offensive or unsightly waste; or any waste which is likely to become nauseous, offensive or unsightly."[1]

Whether something is or is not litter is a question of fact.[2] In *Vaughan v. Briggs*[3] a derelict motor car was held to constitute litter while in Australia a pool of motor oil and water drained from a car sump was held to be waste and therefore litter for the purposes of the Australian statute concerned.[4] Generally "litter" in the 1990 Act should be given its natural meaning of miscellaneous rubbish left lying about.[5]

6.02   Part IV of the Environmental Protection Act 1990 is concerned with "litter." A number of provisions of that part control the deposit of "litter or refuse." Under section 86(14) of the 1990 Act the Secretary of State may by order provide that the droppings of specified animals shall be treated as refuse in prescribed circumstances. The Litter (Animal Droppings) Order 1991[6] has been made under this power. It provides that dog faeces on specified land that is not heath or woodland or used for the grazing of animals will be litter for the purposes of Part IV. The land specified includes public parks, beaches used for recreation, picnic sites and car parks provided under section 32 of the Road Traffic Regulation Act 1984. Similar provision is made in Northern Ireland by the Litter (Dog Faeces) Order (Northern Ireland) 1995.[7]

---

[1] S.I. 1978 No. 1049 (N.I. 19), Art. 36(1).
[2] *Hill v. Davies* (1903) 88 L.T. 464.
[3] [1960] 2 All E.R. 473.
[4] *Stewart v. Lizars* [1965] V.R. 210 Aus.
[5] *Westminster City Council v. Riding* [1996] Env. L.R. D1.
[6] S.I. 1991 No. 961.
[7] S.R. 1995 No. 235.

1. LITTER CONTROLS

LITTER AUTHORITIES

The Secretary of State for the Environment in England and the Secretaries 6.03
of State for Scotland and Wales will direct the operation of litter controls in
Great Britain. In particular they will regulate the duty to keep land and
highways clear of litter and designate land that may be the subject of litter
control areas and the type of premises and land that may be the subject of
street litter control notices. In addition they may, with the consent of the
Treasury, make grants to any body for anti-litter campaigns.[8] In Northern
Ireland litter is the responsibility of the Department of the Environment
who may also make grants for anti-litter measures.[9]

In England county, district and London borough councils, and other local 6.04
authorities designated by order, are principal litter authorities for the
purposes of Part IV of the Environmental Protection Act 1990.[10] In Wales
they will be county or county borough councils.[11] For the purposes of the
Litter Act 1983 litter authorities are, in addition to the principal authorities,
parish and community councils, a joint board of any two or more of those
authorities and a Park board.[12] Byelaws made under other Acts may allow a
statutory body such as the Environment Agency to deal with the deposit of
litter on land under its control.

In Scotland principal litter authorities for the purposes of Part IV of the 6.05
1990 Act are councils constituted under section 2 of the Local Government,
etc., (Scotland) Act 1994[13] A joint board within the definition of section
235(1) of the Local Government (Scotland) Act 1973 may also be a principal
litter authority.[14] In Northern Ireland district councils exercise powers under
the Litter (Northern Ireland) Order 1994.[15]

Principal litter authorities have the powers granted them by the 1990 Act 6.06
and other legislation. In particular they will have power to establish litter
control areas and issue street litter notices to keep areas or streets prone to
litter tidy. They may enforce these controls through litter abatement notices
or an order of a magistrates court. In addition, they may provide litter bins
in their areas, conduct anti-litter campaigns[16] appoint litter wardens and
may, by virtue of regulation 8(1) of the Controlled Waste Regulations
1992,[17] treat the litter they collect as controlled waste.[18]

[8] Litter Act 1983, s.3.
[9] S.I. 1978 No. 1049 (N.I. 19), Art. 27(3).
[10] As are the Common Council of the City of London and the Council of the Isles of Scilly—
E.P.A. 1990, s.86(2).
[11] E.P.A. 1990, s.86(2) as amended by Local Government (Wales) Act 1994, Sched. 9, para.
17(6).
[12] Litter Act 1983, s.10.
[13] E.P.A. 1990, s.86(3) as amended by Local Government, etc., (Scotland) Act 1994, Sched.
13, para. 167(10).
[14] *ibid.* ss.86(3) & 98(4).
[15] S.I. 1994 No. 1896 (N.I. 10).
[16] E.P.A. 1990, s.87(6).
[17] S.I. 1992 No. 588.
[18] E.P.A. 1990, s.96(1).

6.07   County litter authorities (constituted councils in Scotland) will have to produce county or area litter control plans when section 4 of the Litter Act 1983 is brought into force. For these purposes a National Park authority will act as a litter authority and draw up a plan for its park.[19] These plans must be discussed with other litter authorities in the county (in England) and any appropriate voluntary bodies before they are finalised. The plans must be publicised locally and kept available for free public inspection. They should be revised from time to time.[20] In Northern Ireland plans will be made under Article 27(1) of the 1978 Order.

LITTER BINS

6.08   In England and Wales any litter authority other than a joint or park board[21] may provide litter bins in any street[22] or public place by virtue of section 5(1) of the Litter Act 1983. Highways authorities, district or London borough councils, will also be able to provide them under section 185 of the Highways Act 1980. Where they have such powers a litter authority may also put up and maintain notices to discourage the leaving of litter.[23] Bins and notices must comply with town and country planning legislation and that relating to ancient monuments[24] and may not be put on an open space without the consent of the authority that controls it, nor on land other than a street without the consent of the owner and occupier of that land.[25] Bins or notices in certain streets must also have the consent of the relevant street authority listed in Schedule 1 to the Litter Act 1983. Such a consent must not be withheld unreasonably but may be granted subject to conditions.[26]

6.09   It is an offence under section 5(9) of the Litter Act 1983 for any person to interfere with any litter bin or notice board provided or erected under the 1983 or 1980 Acts. The offence is punishable on summary conviction with a fine not exceeding level 1 on the standard scale. In addition, a court may also order someone convicted of this offence to pay up to £20 in compensation to the authority concerned.[27]

6.10   Litter authorities must make arrangements to empty and cleanse regularly the bins they have provided. They may also empty and cleanse bins provided by others in their area.[28] They must carry out this duty in a way that ensures that no bin or its contents becomes a nuisance or gives reasonable ground for complaint.[29] An authority may sell anything they remove from a bin in

---

[19] See E.A. 1995, Sched. 9, para. 12.
[20] For Scotland see s.4(4) of the 1983 Act as substituted by Local Government, etc. (Scotland) Act 1994, Sched. 13, para. 132(2).
[21] Litter Act 1983, s.6(8).
[22] See Public Health Act 1936, s.343 — applied by Litter Act 1983, s.6(7).
[23] *ibid.* s.5(4).
[24] *ibid.* s.5(8).
[25] *ibid.* s.5(5) — but see s.13 City of London (Various Powers) Act 1971.
[26] *ibid.* Sched. 1, para. 2.
[27] *ibid.* s.5(10).
[28] *ibid.* s.5(21).
[29] *ibid.* s.5(3).

exercise of these powers.[30] Grants may be made by counties to parishes or community councils and by those councils to any person in order to facilitate the provision and emptying of litter bins.[31] The Secretary of State may make orders repealing local Acts or orders that are inconsistent with the provisions of section 5.[32]

In Scotland, by virtue of section 7(1) of the Litter Act 1983, litter bins **6.11** may be provided and maintained by constituted councils[33] in their area and be set up on or adjacent to a public road or on land owned or occupied by the council. Bins can also be put on other land with the consent of everyone who has an interest in that land and subject to such terms as may be agreed between the parties.[34] Any such agreement must be entered in the Land Register for Scotland or, where relevant, recorded in the Register of Sasines. Such agreements will usually be enforceable by the litter authority against the persons who entered into it or their successors in title.[35]

Litter bins set up under these provisions must be emptied and cleansed **6.12** from time to time. Litter in or around them may be removed and disposed of as the authority think fit.[36] In order to remove or dispose of litter the authority may provide such plant and equipment as they need to deal with it.[37]

An authority may delegate their powers and duties under section 7 to any **6.13** other person by agreement.[38] If the agreement concerns rights over land the provisions of section 8(3) to (5) of the 1983 Act will apply to it. An authority may also make grants to any other local authority or voluntary body for the provision and emptying of litter bins and may receive contributions from others for the services it provides under section 7.[39] A council may apply to the Secretary of State for the repeal or amendment of provisions of certain local Acts that are inconsistent with, or have become unnecessary because of, section 7.[40]

In Northern Ireland litter bins may be provided and maintained by a **6.14** district council by virtue of Article 25(4) of the 1978 Order. They should be regularly emptied and cleansed so that they do not become a nuisance or give reasonable grounds for complaint.[41]

---

[30] See Public Health Act 1936, s.343 — applied by Litter Act 1983, s.5(7).
[31] *ibid.* s.6(1) & (2).
[32] *ibid.* s.6(4) & (5) and s.9(4).
[33] *ibid.* s.8(7) as amended Local Government, etc., (Scotland) Act 1994, Sched. 13, para. 132(3).
[34] *ibid.* s.7(4).
[35] *ibid.* s.8(3)–8(5).
[36] *ibid.* s.7(2).
[37] *ibid.* s.7(3).
[38] *ibid.* s.7(5).
[39] *ibid.* s.8(1) & (2).
[40] *ibid.* ss.8(6) & 9(4).
[41] S.I. 1978 No. 1049, Art. 25(5).

COLLECTION OF LITTER

6.15   The collection and removal of litter, including leaves but not derelict cars
or scrap metal, is a cleansing operation for the purposes of section 2(2)(c) of
the Local Government Act 1988.[42] Thus, contracts under which this is done
must have been put out to competitive tender. If the authority carries out
the work itself it must have competed for it in accordance with section 6 of
the 1988 Act. All authorities should have made arrangements in accordance
with Part 1 of the Act by August 1, 1993 at the latest; unless exempted by
an order under section 2(9). However, in Wales these provisions have been
suspended by order until September 30, 1997.[43] In Scotland the Secretary of
State may modify these provisions, from a date specified until December 31,
2001 at the latest, by an order made under section 2(10).[44]

6.16   Not all litter collected will be controlled waste in the definition in section
75 of the Environmental Protection Act 1990. Leaves, for example, would
not fall into any of the categories provided. Therefore, under Regulation 8 of
the Controlled Waste Regulations 1992,[45] litter collected by an authority
under its various powers can be treated as household, industrial or
commercial waste, according to the power under which it has been collected.
In addition litter collected under the provisions of sections 89(1), 92(9) or 93
can be treated as waste collected under section 45 of the Act so that it can
be recycled or otherwise used in accordance with the relevant provisions of
Part II of the Act.[46]

ABANDONED TROLLEYS

6.17   While local Acts may contain provisions relating to abandoned trolleys,
additional measures relating to them have been made under section 99 of
and Schedule 4 to the Environmental Protection Act 1990. The Act gives
local authorities—district councils or London borough councils[47] in England,
county or county borough councils in Wales[48] and, in Scotland, councils
constituted under section 2 of the Local Government, etc., (Scotland) Act
1994[49]—powers to deal with "shopping" and "luggage" trolleys. For these
purposes a "shopping" trolley is one provided by the owner of a shop for
customers to use for carrying goods bought in the shop, while a "luggage"
trolley is one provided by persons with statutory authority to run a railway,
light railway, tramway or road transport undertaking or an authorised

---

[42] Local Government Act 1988, Sched. 1, para. 3(1).
[43] Local Government Act 1988 (Defined Activities) (Exemptions) (Wales) Order 1994, S.I.
1994 No. 339 as amended by S.I. 1997 No. 528.
[44] *ibid.* as added by Local Government, etc., (Scotland) Act 1994, Sched. 13, para. 156(3).
[45] S.I. 1991 No. 588.
[46] *ibid.* reg. 8(c).
[47] And the Common Council of the City of London and the Council of the Isles of Scilly.
[48] E.P.A. 1990, s.99(5)(dd) as added by Local Government (Wales) Act 1994, Sched. 9, para.
17(11)..
[49] E.P.A. 1990, s.99(5) as amended by Local Government, etc., (Scotland) Act 1994, Sched.
13, para. 167(16).

airport operator[50] for travellers to carry their luggage to, from or within the premises used for the undertaking. However, power-assisted trolleys are excluded from the scope of these provisions.[51]

Before these provisions of the Act can have effect the relevant council, 6.18 after consultation with those who appear to them to be likely to be affected by it,[52] must pass a resolution applying them within their area.[53] At least three months must elapse between the date of the resolution and the date of it coming into force.[54] After it has been passed the council should publish a notice about it in at least one newspaper circulating in their area, stating its effect.[55] In addition they must, from time to time, consult those who appear to be affected by it about its operation.[56]

The resolution will allow the council to collect any shopping or luggage 6.19 trolleys found on land in the open air within their area. However, their powers do not extend to all such land. They may not be exercised in respect of shopping trolleys on land owned or occupied by their owners or that are at a place specifically provided for them. Luggage trolleys cannot be removed from land used for the statutory purposes of the undertakings providing them.[57]

Once a resolution has come into force an authorised officer of the relevant 6.20 authority who finds a trolley that appears to him to be abandoned on land in the open air may take it to a designated collection point.[58] However, if the land appears to be occupied the trolley can only be removed with the consent of the occupier or if the authority serves a notice on him stating their intention to remove it and he does not, within fourteen days of service, object in writing to their doing so.[59]

Where a trolley has been removed under these provisions the authority 6.21 must, as soon as is reasonably practicable and in any event within fourteen days of the removal, serve on the apparent owner a notice stating that they have it, the place where it is being kept and that they will dispose of it unless the owner claims it.[60] The trolley must be retained for six weeks and if during that period the owner claims it, it must be handed over to him as long as he pays the charges due to the authority.[61] After the six weeks have elapsed the authority may sell or otherwise dispose of the trolley as long as they have made reasonable efforts to trace the owner.[62]

[50] Within the meaning of Part V of the Airport Act 1986.
[51] E.P.A. 1990, Sched. 4, para. 5.
[52] *ibid.* s.99(3).
[53] *ibid.* s.99(1).
[54] *ibid.*
[55] *ibid.* s.99(2).
[56] *ibid.* s.99(4).
[57] *ibid.* Sched. 4, para. 1(2).
[58] *ibid.* Sched. 4, paras 1(1) and 2(1).
[59] *ibid.* Sched. 4, para. 2(2).
[60] *ibid.* Sched. 4, para. 3(2).
[61] *ibid.* Sched. 4, para. 3(1)(a), (3) & (4).
[62] *ibid.* Sched. 4, para. 3(1)(b) & (5).

6.22   The recovery charges the authority make should be sufficient to recoup the expenses they incur in removing, storing and disposing of trolleys from year to year.[63] "Sufficient" here would mean sufficient to operate the recovery system and no more. Alternatively, the authority may have made an agreement with a supermarket or other user of trolleys in their area under which the store will collect its abandoned trolleys from them. In such a case no charges may be made in respect of those trolleys under Schedule 4 to the Act[64]; although the agreement may specify its own scale of charges.

6.23   Similar powers can be adopted by district councils in Northern Ireland under the provisions of the Litter (Northrn Ireland) Order 1994.[65]

LITTER CONTROL AREAS

6.24   Litter control areas may be designated under the provisions of section 90 of the Environmental Protection Act 1990. Designations may be made by any principal litter authority other than an English county council or joint board.[66] The power to make such an order cannot be delegated under the Local Government Act 1972.[67] In Northern Ireland areas are designated under Article 10 of the Litter (Northern Ireland) Order 1994.[68]

6.25   Only certain descriptions of land that are set out in Article 2(1) of the Litter Control Areas Order 1991,[69] may be designated as litter control areas.[70] Such land includes public car parks, shopping centres, industrial estates, places of entertainment, inland beaches or coastal resorts, aerodromes, marinas or similar recreational boating facilities, motorway service stations, open air markets, camping or caravan sites and picnic areas. In addition, land in the open air under the direct control of certain statutory bodies such as a parish or community council, the Broads Authority, a housing action trust or a health service body may also be designated if the public are entitled or allowed to go onto it. However, land that a body is already under a duty to keep clean of litter under section 89(1)(a) to (f) of the Act cannot also be designated as a litter control area.[71]

6.26   An order designating an area may only be made if the authority consider that because of the presence of litter or refuse on the land it is, and unless it is so designated is likely to continue to be, detrimental to local amenities.[72] Before making such an order the authority must notify anyone who appears to them likely to be affected by it, of their proposal. A person notified must

---

[63] E.P.A. 1990, Sched. 4, para. 4(1).

[64] *ibid.* Sched. 4, para. 4(2).

[65] S.I. 1994 No. 1896 (N.I. 10).

[66] E.P.A. 1990 s.90(3) as amended by Local Government (Wales) Act 1994, Sched, 9, para. 17(7) and Local Government, etc. (Scotland) Act 1994, Sched. 13, para. 167(12).

[67] *ibid.* s.90(5).

[68] S.I. 1994 No. 1896 (N.I. 10).

[69] S.I. 1991 No. 1325 as amended by S.I. 1997 No. 633 and see S.R. 1995 No. 237 for N.I.

[70] E.P.A. 1990, s.90(1) & (2).

[71] S.I. 1991 No. 1325, Art. 2(2).

[72] E.P.A. 1990, s.90(4).

be given 21 days to comment on the proposal and his comments must be considered before the order is made.[73] The order must identify the land to which it applies and must be in the form prescribed in the Schedule to the Litter Control Areas Order; which is slightly different in Scotland.[74]

STREET LITTER CONTROL NOTICES

A principal litter authority order than an English county council or a joint 6.27 board.[75] may issue street litter control notices under sections 93 and 94 of the 1990 Act in order to prevent accumulations of litter or refuse in and around any street or adjacent open land. These notices will impose requirements on the occupiers of premises designated under the Street Litter Control Notices Order 1991[76] to keep the area specified in the notice clear of litter and refuse. In Northern Ireland notices are issued under Article 14 of the 1994 Order and the Street Litter (Control Notices) Order (Northern Ireland) 1995.[77]

Notices may be issued in respect of commercial or retail premises that are 6.28 used for the sale of food or drink for consumption off the premises—such as off-licenses, take-away food shops or sweet shops—or premises where food and drink is consumed in the open air on land that is not part of a street. They may also be served on service stations, places of entertainment and on banks, building societies and other premises that have automatic cash machines on their outside walls.[78]

Notices may only be issued in respect of designated premises that either 6.29 give rise to a recurrent litter problem on streets[79] or land in the open air in their vicinity, or whose frontages are, and are likely to continue to be, unsightly due to litter or refuse, or that produce quantities of litter or refuse as a result of the activities carried out on them that is likely to render streets or land in the open air in their vicinity unsightly.[80]

A notice may be served on the occupier or, if the premises are unoccupied, 6.30 the owner of the premises in the manner required by section 160 of the 1990 Act.[81] Before serving the notice the authority must inform the person on whom it will be served of their intention, give him 21 days within which to make representations about it and take his comments into consideration before reaching a final decision.[82] If the notice is to impose requirements to clear an area off the premises of litter the authority must also take into account their own duties to keep such areas litter free and those of any other relevant local authority in respect of it.[83]

---

[73] E.P.A. 1990, s.90(6).
[74] *ibid.* s.90(7) and S.I. 1991 No. 1325, reg. 3. In N.I. see S.I. 1995 No. 238.
[75] See *ibid.* s.93(1) as amended.
[76] S.I. 1991 No. 1324 as amended by S.I. 1997 No. 632.
[77] S.I. 1994 No. 1896 (N.I. 10) and S.R. 1995 No. 42.
[78] S.I. 1991 No. 1324, Art. 2.
[79] See E.P.A. 1990, s.93(4).
[80] *ibid.* s.93(2).
[81] *ibid.*
[82] *ibid.* s.94(6).
[83] *ibid.* s.94(5).

6.31   A notice must identify the premises to which it relates and state the grounds on which it is issued. Within the limits set by the 1991 Order it must identify the specified area of open land adjoining the frontage, whether on one or two sides, to which it applies. The land to which the notice may apply can be part of the premises, part of a street (other than a carriageway when it is open to vehicles), land of a principal litter authority or that is under the direct control of any other local authority.[84] For most premises the notice can apply to such land within 100 metres of them; but for banks, etc., this limit is reduced to 10 metres.[85] Finally, it should specify the authority's reasonable requirements for ensuring that the specified area is kept clean.[86] To achieve this the authority may, amongst other things, require the occupier or owner of the premises to provide litter bins and empty them regularly, to clear the area of all litter or refuse within a specified period and to continue to do so at specified times or periods. However, they may only require a carriageway to be cleared at a time when it is closed to all vehicular traffic.[87]

6.32   A person in England and Wales on whom a street litter notice is served may appeal against it to a magistrates court.[88] These would be civil rather than criminal proceedings so that the appeal should be made by complaint.[89] By virtue of section 127 of the Magistrates' Courts Act 1980 the complaint should be made within six months of the service of the notice. In Scotland an appeal is made by summary application to the Sheriff. On hearing the appeal the court may quash the notice or may quash, vary or add to any requirement imposed by it.[90]

6.33   Where it appears to the litter authority that a person on whom they served a notice is not complying with any of its requirements they may apply to a magistrates court or, in Scotland to the Sheriff, for an order requiring him to comply with it within a specified period.[91] Again these would be civil proceedings. A person who, without reasonable excuse, fails to comply with the court's order will be guilty of an offence and liable on summary conviction to a fine not exceeding level 4 on the standard scale.[92]

PUBLIC REGISTERS

6.34   Litter authorities must maintain a register containing copies of all orders they have made designating litter control areas and all street litter notices they have issued and of any order of a court varying or adding to the requirements of a notice.[93] The register can be in documentary or other

---

[84] S.I. 1991 No. 1324, Art. 3(1).
[85] *ibid.* Art. 3(2)
[86] E.P.A. 1990, ss.93(3) & 94(4).
[87] *ibid.* s.94(4).
[88] *ibid.* s.94(7).
[89] Magistrates' Courts Act 1980, s.51.
[90] E.P.A. 1990, s.94(4).
[91] *ibid.* s.94(8).
[92] *ibid.* s.94(9), and see ss.157 & 158.
[93] *ibid.* s.95(1) & (2).

form.[94] Copies of orders and notices must be kept on the register for as long as they are in force.[95] The register must be available for free public inspection at all reasonable times and anyone must be allowed to take copies of entries in it for a reasonable charge.[96]

2. LITTER CLEARANCE DUTIES

By virtue of section 89 of the Environmental Protection Act 1990, local 6.35 authorities and certain statutory bodies and landowners are required to keep the public land in the open air under their control clear of litter and refuse as far as it is practicable to do so. "Land in the open air" includes covered land that is open to the air on at least one side.[97] Orders can be made to exclude certain categories of land from the duty.[98] The duty can be enforced by individuals under section 91 of the 1990 Act and by some litter authorities by the service of litter abatement notices under section 92 in respect of some of the categories of land to which the duty applies. The duty only applies to public land or private land in a litter control area. Litter on other private land may be dealt with under a street litter notice or by a notice under section 215 of the Town and Country Planning Act 1990.

PERSONS SUBJECT TO THE DUTY

The duty applies to each local authority in England and Wales in respect 6.36 of any relevant highway, other than a motorway, for which they are responsible.[99] A highway for these purposes is any land over which the public have a right to pass and repass and includes a pavement as well as a roadway.[1] A relevant highway is one that is maintainable at public expense by virtue of section 36 of the Highways Act 1980 or any other enactment.[2] The local authorities responsible for a highway in their area for the purposes of the duty are, in Greater London, London borough councils or the Common Council and outside London, in England, district councils or the Council of the Isles of Scilly and in Wales, county or county borough councils.[3] However, at the request of a highways authority the Secretary of State, after consulting the local authority, may, by an order made under section 86(11) of the 1990 Act, transfer the duty in respect of any highway, class of highway or specified highway from them to the highways authority in order to minimise interference with the passage or safety of traffic along

---

[94] E.P.A. 1990, s.95(5).
[95] *ibid.* s.95(3).
[96] *ibid.* s.95(4).
[97] *ibid.* s.86(13).
[98] *ibid.* s.86(8).
[99] *ibid.* s.89(1)(a).
[1] *ibid.* s.98(5), and Highways Act 1980, s.328.
[2] *ibid.* s.86(9), and Highways Act 1980, s.329(1).
[3] *ibid.* as amended by Local Government (Wales) Act 1994, Sched. 9, para. 17.

the highway. In Scotland constituted councils are responsible, under the Act, for public roads as defined in the Roads (Scotland) Act 1984, except motorways, in their area; although transfer orders can be made in the same way under section 86(11).[4] The Secretary of State for Transport (or Scotland or Wales) must clear litter from any motorway or other highway or public road for which he is responsible[5]; although his responsibility for trunk roads under the Act has passed to local authorities by virtue of section 86(9).

6.37 Each principal litter authority is responsible for clearing their relevant land.[6] Relevant land for these purposes is land in the open air that is under their control and to which the public are entitled to have access with or without payment. Land that the authority control as an education authority is omitted from this requirement as are highways and public roads.[7] Land below the ordinary high water mark is also excluded.[8]

6.38 The appropriate Crown authority, that is the Crown Estates Commissioners, the Minister in charge of the relevant government department or a body occupying or managing land on behalf of the Crown,[9] is under a duty to keep the Crown land for which they are responsible clear of litter.[10] This duty only extends to Crown land in the open air to which the public are entitled to have access free of charge or on payment of a fee. It does not, however, extend to highways or public roads,[11] or to land below the regular high water mark.[12] The enforcement of the duty may be precluded on national security grounds in respect of Crown premises designated by the Secretary of State under section 159(4) of the 1990 Act.

6.39 Certain statutory undertakers may be designated by the Secretary of State as those to whom the duty applies in respect of their relevant land.[13] This only applies to transport undertakers[14]; so that the land of a water or electricity undertaker is not subject to the duty. Undertakers have been designated by the Litter (Statutory Undertakers) (Designation and Relevant Land) Order 1991[15] and are basically all statutory railway, tramway, road transport (other than taxis), canal, dock, harbour, pier or airport operators. The duty applies to land, whether in the open or not, to which the public have access and has been extended, by the 1991 Order, to land to which they do not have access but which is in the direct control of an undertaker and that forms part of a railway, other than a tunnel or goods yard, etc.,

---

[4] E.P.A. 1990, ss.89(1)(a), 86(10) & 98(5).
[5] *ibid.* s.89(1)(b).
[6] *ibid.* s.89(1)(c).
[7] *ibid.* s.86(4).
[8] The Litter (Relevant Land of Principal Litter Authorities and Relevant Crown Land) Order 1991 (S.I. 1991 No. 476).
[9] E.P.A. 1990, s.86(5).
[10] *ibid.* s.89(1)(d).
[11] *ibid.* s.86(5).
[12] S.I. 1991 No. 476.
[13] E.P.A. 1990, s.89(1)(e).
[14] *ibid.* s.98(6).
[15] S.I. 1991 No. 1043, as amended by S.I. 1992 No. 406. For N.I. see S.R. 1994 No. 449 and S.R. 1996 No. 22.

which is within 100 metres of a railway station platform to which the public have access. Land below the regular high water mark is excluded from the duty.

The governing bodies of designated universities, colleges of higher or further education, polytechnics, or state maintained or grant-aided schools in England and Wales are responsible for keeping land under their direct control that is open to the air clear of litter.[16] In Scotland this duty rests on the governing body of any designated university, college of further education, grant-aided college, technology academy or state maintained or aided school or the education authority responsible for its management. All such establishments have been designated by the Litter (Designated Educational Institutions) Order 1991[17] with the exception, in Scotland, of grant-aided special schools. Other educational institutions may be designated for this purpose by the Secretary of State.[18] **6.40**

The occupier of land included in a litter control area is under a duty to keep it clear of litter if it is land, whether open to the air or not, to which the public have access.[19] **6.41**

Under section 89(2) of the 1990 Act it is the duty of the local authority responsible under the Act[20] for a highway or road, and the Secretary of State for Transport as regards a motorway and any other highway or road for which he is responsible, to keep it clean as far as it is practicable to do so. The Secretary of State can make regulations defining matter as litter for these purposes but has not yet done so.[21] Authorities will set standards of road cleansing. In doing so they must have regard to the character and use of the land, highway or road as well as the measures that are practicable in the circumstances for its cleansing.[22] **6.42**

A code of practice[23] has been issued under section 89(7) of the Act to provide general guidance on the discharge of the duties specified in section 89 by establishing what are considered to be reasonable and generally acceptable standards of cleanliness which those under the duty should be capable of meeting. It does this by establishing four standards of cleanliness, grades A to D, and providing photographs of various types of land to show how the grades apply to them. Eleven "zones"—or types of land—are categorised and standards recommended for each zone. The code also sets out guidance as to the practicability of compliance with the duty. Anyone who is subject to a duty under section 89(1) or (2) must have regard to the code currently in force in discharging that duty.[24] **6.43**

In carrying out their street cleansing duties an authority may, by virtue of section 23 of the Control of Pollution Act 1974, stop cars parking during set **6.44**

---

[16] E.P.A. 1990, ss.89(1)(f), 98(2) and 86(7).
[17] S.I. 1991 No. 561. For N.I. see S.R. 1994 No. 338.
[18] E.P.A. 1990, ss.89(1)(f), 98(3) & 86(7).
[19] *ibid.* ss.89(1)(g) & 86(12).
[20] See para. 10.35.
[21] E.P.A. 1990, s.89(3).
[22] *ibid.* s.89(3).
[23] *Code of Practice on Litter and Refuse.*
[24] E.P.A. 1990, s.89(5).

times in the area to be cleaned. Where cleaning is taking place the authority must place traffic signs and barriers in the highway or road so as to warn traffic and prevent accidents. Once the cleaning has finished the signs and barriers must be removed.[25] In placing these signs, etc., the local authority must comply with any directions given them by the relevant highways or roads authority and may not carry out cleansing operations at times forbidden by those authorities. The highways or roads authority may permit the litter authority to exercise their powers under section 14 of the Road Traffic Regulation Act 1984[26] to temporarily prohibit or restrict traffic on a road or may themselves restrict or prohibit traffic so that the litter authority can carry out their duties under section 89.[27]

LITTER ABATEMENT NOTICES

6.45   Section 92 of the 1990 Act requires a principal litter authority, other than an English county council or a joint board, to serve a litter abatement notice to enforce the duties imposed by section 89(1)(d)–89(1)(g) if they are satisfied that relevant land is defaced by litter or refuse or such a situation will recur. A notice can be served on the appropriate Crown authority in respect of relevant Crown land,[28] on a designated statutory undertaker and on a designated educational institute in respect of their relevant land and on the occupier of relevant land in a litter control area (or on the owner if the land is unoccupied).[29] The notice may either require the land to be cleared of litter or refuse within a certain time or prohibit it from becoming defaced by litter or refuse or both.[30]

6.46   The person served with a notice may appeal against it within 21 days of the date on which it was served by complaint to a magistrates court or by summary application to a sheriff's court. The appeal must be allowed if the appellant can show that he is complying with his duty under section 89(1) in respect to the land in question; taking into account any code of practice under section 89(7).[31] Presumably it must also be allowed if he does not control the land or if it is not "relevant land."

6.47   If a person without reasonable excuse fails to comply with the requirements of an abatement notice or contravenes them he will be guilty of an offence and liable on summary conviction to a fine not exceeding level 4 on the standard scale and a further fine of one twentieth of that level for each day the offence continues after conviction. It will be a defence for him to show that he has complied with his duty in respect of the land in question, taking into account any code of practice under section 89(7).[32] In addition,

[25] E.P.A. 1990, s.89(5).
[26] As amended by E.P.A. 1990, Sched. 15, para. 23.
[27] In Scotland see also Roads (Scotland) Act 1984, s.62.
[28] See E.P.A. 1990, s.159(3) & (7).
[29] *ibid.* s.92(1) & (3).
[30] *ibid.* s.92(2).
[31] *ibid.* s.92(4), (5) & (8).
[32] *ibid.* s.92(6)–92(8).

the authority may enter relevant land of a designated educational institute or in a litter control area and clear it of litter or refuse; recovering their necessary costs from the person on whom the abatement notice was served.[33]

PRIVATE LITTER ABATEMENT ACTIONS

A private individual may make a complaint to a magistrates court, or **6.48** summary application to a sheriff, on the grounds that land subject to the duty under section 89, or any motorway, is defaced by litter or refuse.[34] He may make his complaint if he is a person aggrieved by the defacement. The Act is silent on who constitutes a person aggrieved other than to say that a principal litter authority does not.[35] However, the traditional approach by the courts to this phrase that "a person aggrieved" must have some interest to be protected[36] may not apply here as no one other than an authority will have a legal interest in most "relevant land." Instead it is submitted that someone whose house overlooks the land, who stops at a littered picnic site or who travels past a littered roadside should be treated as "aggrieved" for these purposes.

At least five days before bringing the action the complainant should give **6.49** to the person whose duty it is to keep the land or motorway clean, written notice of his intention to go to court and he should specify what it is he is aggrieved about.[37] If the court is satisfied that the land in question is defaced by litter or refuse, or that the road is not clean it may, unless the defendant can show that he has complied with the duty of care under section 89(1) or 89(2) in respect of it (or if a highway is in the state it is in as a result of directions by the highway authority[38]) make a litter abatement order requiring the defendant to clear it within a specified time.[39] In considering a case brought under this section the court should take account of any code of practice issued under section 89(7).[40] Where a court is satisfied that the complaint was justified, and that there were reasonable grounds for bringing it, it may order the defendant to pay the complainant's reasonable costs.[41] In Scotland an appeal on a point of law lies to the Court of Session against the making of an order.[42]

If a person against whom an abatement order has been made fails to **6.50** comply with it without reasonable excuse he will be guilty of an offence and liable on summary conviction to a fine not exceeding level 4 on the standard scale, together with a further fine of up to one twentieth of that level for

---

[33] See E.P.A. 1990, s.92(9) & (10).
[34] *ibid.* s.91(1), (2) & (13).
[35] *ibid.* s.91(3).
[36] *e.g.* Lord Denning in *Att-Gen (Gambia) v. Pierre Sarr N'Jie* [1961] A.C. 617 at 634.
[37] E.P.A. 1990, s.91(4) & (5).
[38] *ibid.* s.91(8).
[39] *ibid.* s.91(6) & (7).
[40] *ibid.* s.91(11).
[41] *ibid.* s.91(12).
[42] *ibid.* s.91(13).

each day on which the offence continues after conviction.[43] It will be a defence for him to show that he has complied with his duty under section 89(1) or 89(2) in respect of the land or highway and the court may take into account a code of practice issued under section 89(7) in determining the matter.[44] However, if the order concerned Crown land the only remedy would be for a principal litter authority to apply to the High Court or Court of Session for an order declaring that the Crown has acted unlawfully; although a responsible civil servant could be liable to prosecution.[45]

3. THE OFFENCE OF LEAVING LITTER

6.51　　The offence of leaving litter has three stages. First, the offender must throw down, drop or otherwise deposit something in a place subject to the duty under section 89 of the Act or a public open place.[46] Second, he must leave it there as it is not intended that an offence should be committed if someone deposits litter then immediately clears it up.[47] Finally, what is thrown down, etc., must cause, contribute to or tend to lead to, the defacement by litter of the place where it was deposited.[48] For the purposes of this section a "public open place" means a place in the open air to which the public have free access or that is covered but open to the air on at least one side and which is available for public use. Thus, the seashore, public parks or railway stations could be public open places for these purposes but not a football stadium, village fete charging an entrance fee or covered shopping precinct.[49]

6.52　　It is a defence here for a person to show that the depositing or leaving of the thing was authorised by law or that it was done with the consent of the owner, occupier or other person or authority having control of the place in or into which it was deposited.[50] A local authority can, as a highway authority, give themselves consent to leave rubbish by the side of the road with the intention of having it collected, although any terms of that consent should be complied with.[51] A prosecution may also fail if it cannot be shown that the litter was left within six months of the commencement of proceedings.[52]

6.53　　Anyone found guilty of this offence will be liable on summary conviction to a fine not exceeding level 4 on the standard scale.[53] In Scotland a conviction may be based on the evidence of one witness.[54]

---

[43] E.P.A. 1990, s.91(9).
[44] *ibid.* s.91(10) & (11).
[45] *ibid.* s.159(2) & (3).
[46] *ibid.* s.87(3).
[47] Lord Parker C.J. in *Vaughan v. Biggs* [1960] 2 All E.R. 473 at 474.
[48] E.P.A. s.87(1).
[49] *ibid.* s.87(4).
[50] *ibid.* s.87(2).
[51] *Camden London Borough Council v. Shinder* (1987) 86 L.G.R. 129.
[52] *Vaughan v. Briggs* [1960] 2 All E.R. 473.
[53] E.P.A. 1990, s.87(5).
[54] *ibid.* s.87(7).

A problem often arises in litter cases because a litter warden or other **6.54** person has no power to obtain the name and address of the alleged offender. A constable will have powers under section 25 of the Police and Criminal Evidence Act 1984 to arrest someone he has reasonable grounds to believe has committed an offence under section 87 of the 1990 Act. Otherwise, unless strong identification evidence is available a prosecution is unlikely to succeed.

FIXED PENALTIES

A district or London borough council in England, a county or county **6.55** borough council in Wales or a constituted council in Scotland or any National Park authority may appoint, in writing, litter wardens to issue fixed penalty notices under section 88 of the 1990 Act. The Secretary of State may by order designate any English county council or joint board to appoint wardens for specified areas outside National Parks.[55] In Northern Ireland fixed penalty notices are issued under Article 6 of the Litter (Northern Ireland) Order 1994.[56]

A warden may issue a fixed penalty notice at any time where he has **6.56** reason to believe that, at that time, a person has committed an offence under section 87 of the Act in the area of his authorising authority.[57] The notice must be in the form prescribed by the Litter (Fixed Penalty Notices) Order 1991.[58] The effect of the notice is to stay the institution of proceedings for the offence for fourteen days from its date. If the penalty is paid within that time the offender cannot be convicted under section 87.[59]

The fixed penalty is, at present, £25 although this can be varied by **6.57** order.[60] It should be sent to the person named in the notice by posting it or otherwise. If it is posted the payment will be deemed to have been made at the time when the letter would be delivered in the ordinary course of the post.[61] An authority receiving a penalty should either, in England and Wales, send it to the Secretary of State, or in Scotland treat it as if it were a fine imposed by a district court.[62] In any dispute as to payment, a certificate by the chief finance officer of an English or Welsh litter authority, or the proper financial officer of a Scots authority, that states that it was or was not received by a specified date will be evidence of the facts stated.[63]

---

[55] E.P.A. 1990, s.88(9) & (10) as amended by Local Government (Wales) Act 1994, Local Government, etc., (Scotland) Act 1994 and E.A. 1995, Scheds 9 (para. 12) and 25.
[56] S.I. 1994 No. 1896 (N.I. 10).
[57] E.P.A. 1990, s.88(1).
[58] S.I. 1991 No. 111. For N.I. see S.R. 1995 No. 17.
[59] E.P.A. 1990 s.88(2)
[60] *ibid.* s.88(6) & (7), and see S.I. 1996 No. 3055.
[61] *ibid.* s.88(3) & (4).
[62] *ibid.* s.88(6).
[63] *ibid.* s.88(8) & (10).

7. RUBBISH CLEARANCE

1. REMOVAL OF REFUSE

7.01   As well as the duties of collection authorities to collect waste that are imposed by the Environmental Protection Act 1990, local authorities will also have other powers to clear rubbish from land in their area. Moreover, the unlawful deposit of waste on land may be an offence under legislation other than the 1990 Act. These provisions are intended to prohibit the deposit of waste on open or public land and to enable rubbish to be cleared. They are in addition to powers to control litter or, under section 59, to require the removal of waste unlawfully deposited.

7.02   The deliberate abandonment of any matter, other than a motor vehicle, on land in the open air or on any other land forming part of a highway is an offence under section 2(1)(b) of the Refuse Disposal (Amenity) Act 1978 and is punishable on summary conviction by a fine not exceeding level 4 on the standard scale or up to three months imprisonment or both. If the prosecution can prove that a defendant left something on land in circumstances, or for such a period, that it may reasonably be assumed that he brought it onto the land in order to abandon it there he will be deemed to have done so unless he can prove otherwise.[1] In Scotland this offence may be dealt with by a sheriff's court or district court that has jurisdiction in the place where it was committed.[2]

7.03   Where it appears to a district or London borough council, a county or county borough council in Wales, or a constituted council in Scotland that anything in their area, other than a motor vehicle, has been abandoned without lawful authority on any land in the open air or on any other land forming part of a highway they may, if they think fit, remove it.[3] However, they may not remove something from land that appears to be occupied unless they first notify the occupier that they intend to remove it and he does not object to their action within a period that has yet to be specified.[4] Powers of entry are provided under section 8 of the 1978 Act to allow an authority to inspect land or to clear it of rubbish. Once matter has been

---

[1] Refuse Disposal (Amenity) Act 1978, s.2(2).
[2] *ibid.* s.2(3), and Criminal Procedure (Scotland) Act 1975, s.462.
[3] *ibid.* s.6(1).
[4] *ibid.* s.6(2).

removed the authority may take it to a site or depot and treat and dispose of it there or sell it.[5] The authority by whom the rubbish was cleared may recover the costs of the operation from the person who dumped it at the place from which it was removed or from anyone convicted of an offence under section 2(1) of the Act in respect of it.[6] Expenses can be recovered as either a simple contract debt or, from an offender, through an order of the court made on conviction.[7]

In England and Wales[8] a similar power to clear rubbish is granted to **7.04** district, London borough and and county councils[9] (or in Wales county and county borough councils), by section 34 of the Public Health Act 1961. This allows them to remove rubbish that is seriously detrimental to the amenity of the neighbourhood from any land in the open air in their area. However, these powers cannot be used to deal with material on a licensed waste disposal site.[10] Before taking action under this section they must give at least 28 days notice to the owner and occupier of the land stating their intention and the rights of a person served with such a notice. These rights allow him to serve a counter-notice within the 28 day period stating that he will clear the rubbish himself.[11] If a counter-notice is served the authority cannot do anything under their original notice unless either he does not take the specified steps within a reasonable time or, having started the work, fails to make reasonable progress towards its completion.[12] Alternatively, the person served may appeal to a magistrates court on the grounds that the authority should not take action under section 34 in respect of his land or that the steps proposed by the notice are unreasonable.[13] If an appeal is brought the effect of the notice is suspended until the appeal is finally determined or withdrawn. A court may order the authority to take no further action, allow them to take such steps as it directs or dismiss the appeal.[14] Appeal from the magistrates' decision lies within the Crown Court.[15]

By virtue of section 215 of the Town and Country Planning Act 1990 (in **7.05** Scotland section 197 of the Town and Country (Planning) (Scotland) Act 1997) if it appears to a local planning authority that the amenity of part of their area, or an adjoining one, is adversely affected by the condition of land in their area they may serve a notice on its owner and occupier requiring them to take specified steps to remedy its condition within a certain time; which must be at least 28 days.[16-17] In Scotland this notice is known as a section 63 notice.[18] This power is more extensive than that under section 34

[5] Refuse Disposal (Amenity) Act 1978, s.6(3), as replaced by s.6(8).
[6] *ibid*. s.6(4).
[7] *ibid*. ss.6(6) & 5(2), (3).
[8] Public Health Act 1961, s.3.
[9] Local Government Act 1972, Sched. 14, para. 37.
[10] Public Health Act 1961, s.34(5).
[11] *ibid*. s.34(2)(a).
[12] *ibid*. s.34(2)(b).
[13] *ibid*. s.34(3).
[14] *ibid*. s.34(4).
[15] Public Health Act 1936, s.301, and P.H.A. 1961, s.1(1).
[16-17] Town and Country Planning Act 1990, s.215(4).
[18] For service see T.&C.P.(S) Act 1997, s.179(2).

of the Public Health Act 1961 as it allows the authority to do more than order the mere clearance of rubbish. It also extends to any land whether in the open or not. It enables an authority to take action to deal with land that is in use as opposed to land that is derelict.[19]

7.06   An appeal against such a notice is made under section 217 of the Act. The grounds of appeal are that the condition of the land does not adversely affect amenity, that its condition results from operations carried out under a valid planning permission, that the requirements of the notice are excessive or that the period allowed for compliance is too short.[20] The appeal is made to the magistrates' court with jurisdiction over the area in which the land is situated.[21] In Scotland appeal is to the Secretary of State[22] on the same grounds but an additional ground of procedural irregularity in the service of the notice is added. Until an appeal is finally determined or withdrawn the notice will have no effect.[23] In determining the appeal any immaterial defect, informality or error in the notice may be corrected.[24] When the appeal is determined the notice may be quashed, varied in favour of the appellant or directions given to give effect to the determination.[25] In England and Wales an appeal from the magistrates' decision lies to the Crown Court.[26]

7.07   If a person on whom a section 215 notice is served fails to take the steps required by it within the specified period—which may be extended by the planning authority[27]—he will be guilty of an offence and liable on summary conviction to a fine not exceeding level 3 on the standard scale.[28] If the owner or occupier has given up his interest in the land before the specified time elapsed and are prosecuted he may, after giving the prosecution three clear days notice of his intention, have the new owner or occupier brought before the court.[29] If no one is in occupation, the owner at the time the notice was served may be made liable.[30] In any such proceedings the former owner or occupier must show that the failure to take the required steps was due, in whole or in part, to the default of the person notified to the prosecution; who may then be convicted of the offence. The original defendant is entitled to be acquitted if he can show that he took all reasonable steps to ensure compliance with the notice.[31] If a person convicted of this offence does not, as soon as is reasonably practicable, do everything in his power to comply with the notice he will be guilty of a further offence and liable on summary conviction to a fine not exceeding £40 a day for each day on which any of the requirements of the notice remain unfulfilled.[32]

[19] *Britt v. Buckinghamshire County Council* [1964] 1 Q.B. 77.
[20] Town and Country Planning Act 1990, s.217(1).
[21] *ibid.* s.217(2).
[22] Town and Country Planning (Scotland) Act 1997, s.180.
[23] T.&C.P. Act 1990, s.217(3), and T.&C.P.(S) Act 1997, s.180(5).
[24] *ibid.* s.217(4), and *ibid.* s.180.
[25] *ibid.* s.217(5), and *ibid.* s.180.
[26] *ibid.* s.218.
[27] *ibid.* s.216(7).
[28] *ibid.* s.216(2).
[29] *ibid.* s.216(3) & (4).
[30] *ibid.* s.216(4).
[31] *ibid.* s.216(5).
[32] *ibid.* s.216(6).

Alternatively, the authority that served the notice can enter the land and **7.08** take the steps required in it. They may recover their reasonable expenses for doing so from the landowner[33] as a simple contract debt.[34] The owner, or the occupier, may recover any expenses he incurs in complying with the notice from anyone who caused or permitted the land to be in the condition that caused it to be served.[35] Under regulation 14 of the Town and Country Planning General Regulations 1992[36] the procedural provisions of sections 276, 289 and 294 of the Public Health Act 1936 have been applied to section 215 notices.[37] In Scotland the remedy for failure to comply with a section 179 notice is for the Council to clear the rubbish themselves and recover their expenses from the owner, lessee or occupier of the land under the provisions of section 135 of the 1997 Act.[38]

A district or London borough council, or in Wales a county or county **7.09** borough council,[39] can take action under section 79 of the Building Act 1984 to deal with rubbish or other material, other than advertisements,[40] resulting from or exposed by, the demolition or collapse of a building or structure that is lying on the site or any adjoining land. If they consider that the material makes the land a serious detriment to the amenity of the neighbourhood they may serve a notice on the landowner requiring him to take such steps as are specified in it to clear the land within a specified time.[41] "Rubbish" here will not include chattels—such as machinery—left as a result of the demolition.[42] The power to serve a notice must not prejudice provisions that concern the conservation of buildings.[43] An appeal against such a notice may be made to the magistrates' court under section 102 of the 1984 Act. Subject to any appeal, if a person served with a clearance notice fails to comply with it in the time given, the council can enter onto the land and remove it themselves and recover the reasonable expenses of the operation from him.[44] In addition he may be liable on summary conviction to a fine not exceeding level 4 on the standard scale and a further fine of up to £2 for each day the rubbish remains on the site after conviction.[45] In Northern Ireland a similar power is given to a district under article 66(2) of the Pollution Control and Local Government (N.I.) Order 1978.[46]

A highways authority in England and Wales may require anyone who has **7.10** deposited anything that constitutes a nuisance on a highway to remove it

---

[33] T.&C.P. Act 1990, s.217(3), and T.&C.P.(S) Act 1997, s.219(1).
[34] *ibid.* s.219(6).
[35] *ibid.* s.219(2).
[36] S.I. 1992 No. 1492.
[37] See also T.&C.P. Act 1990, s.219(3)–219(5).
[38] Town and Country Planning (Scotland) Act 1997, s.179(6).
[39] Building Act 1984, s.126 as amended Local Government (Wales) Act 1994, Sched. 9, para. 15(3).
[40] *ibid.* s.79(4).
[41] *ibid.* ss.79(2) & 99(1).
[42] *McVittie v. Bolton Corporation* [1945] K.B. 281.
[43] Building Act 1984, s.79(5), as added by Housing and Planning Act 1986, Sched. 9.
[44] *ibid.* ss.79(3) & 99(2)(a).
[45] *ibid.* ss.79(3) & 99(2)(b).
[46] S.I. 1978 No. 1049 (N.I. 19).

forthwith by serving them with a notice under section 149 of the Highways Act 1980. If the notice is not complied with the authority may make a complaint to a magistrates court for a removal and disposal order.[47] On hearing the complaint the court may authorise them to remove the nuisance and to dispose of it.[48] But if what was deposited is considered to be a danger to users of the highway, so that it ought to be removed without the delay that would be occasioned by obtaining an order, the authority may remove it immediately.[49] They may recover their expenses from the person who deposited the matter, or who claims to be entitled to it, or may apply to the magistrates for an order allowing them to dispose of it.[50] Where an authority disposes of something under these provisions, then, after meeting their removal expenses, funds received for it should be applied to highway maintenance.[51] If they do not realise enough from the disposal to meet their removal expenses they may recover the remainder from the person who deposited the matter.[52] Any sums due under section 149 can be recovered as a civil debt or as money due under statute.[53] In Northern Ireland similar provision is made by article 49 of The Roads (Northern Ireland) Order 1980.[54] In Scotland objects that have fallen onto a road so as to cause an obstruction can be removed by a roads authority by virtue of section 89 of the Roads (Scotland) Act 1984. The deposit, without the consent of a roads authority, on a road of anything that causes an obstruction is an offence under section 59 of the Act.

7.11   If refuse falls onto a highway maintainable at public expense from land, the highways authority may serve a notice on the owner or occupier of that land requiring the execution of works that will prevent it from doing so in such a quantity that will obstruct the street or choke any sewer or gully in it.[55] Anyone aggrieved by a requirement of such a notice may appeal against it to a magistrates court.[56] Otherwise if it is not complied with within the specified period the person on whom it was served will be guilty of an offence and liable on summary conviction to a fine not exceeding level 3 on the standard scale. If the offence is continued after conviction a penalty of up to £1 may be imposed for each day it continues.[57] In Northern Ireland similar provision is made by the 1980 Order.[58] In Scotland it will be an offence under section 99 of the Roads (Scotland) Act 1984 if the owner and occupier of land fail to prevent filth, dirt or other offensive matter from flowing onto a road. It is doubtful whether this section would encompass solid waste falling onto a road.

---

[47] Highways Act 1980, s.149(1).
[48] *ibid.* s.149(4)(a).
[49] *ibid.* s.149(2).
[50] *ibid.* s.149(3).
[51] *ibid.* s.149(4)(b).
[52] *ibid.* s.149(5).
[53] *ibid.* s.305(5).
[54] S.I. 1980 No. 1085 (N.I. 11).
[55] Highways Act 1980, s.151(1).
[56] *ibid.* s.151(2).
[57] *ibid.* s.151(3).
[58] S.I. 1980 No. 1085 (N.I. 11), Art. 32.

An accumulation or deposit of waste that is prejudicial to public health or 7.12
a nuisance can be treated as a statutory nuisance under the provisions of Part
III of the Environmental Protection Act in Great Britain.[59] To constitute a
statutory nuisance the accumulation must give rise to an actionable nuisance
or be prejudicial to health, because of its smell or because it attracts flies or
other pests.[60] Builder's rubble or other inert rubbish that does not have such
an effect cannot be a statutory nuisance even though it may be unsightly
and constitute a risk to those who walk over it.[61] However, "contaminated
land" that comes under the regime of Part IIA of the 1990 Act cannot be
dealt with as a statutory nuisance.[62] In Northern Ireland statutory nuisances
are dealt with under sections 110 to 116 of the Public Health (Ireland) Act
1878.

2. ABANDONED VEHICLES

There are two statutory provisions that are concerned with abandoned 7.13
vehicles: the Refuse Disposal (Amenity) Act 1978 which gives local
authorities powers to deal with vehicles that have been dumped in their area,
and sections 99 to 103 of the Road Traffic Regulation Act 1984 which is
more concerned with the removal of vehicles that are illegally parked,
causing an obstruction or that have broken down; although powers to deal
with abandoned vehicles are also contained in those sections. The powers
under the 1984 Act are exercised by the police and local authorities. The
exercise of either of these powers by an authority is not subject to the
competitive tendering provisions of Part I of the Local Government Act
1988.[63]

Any person who, without lawful authority abandons a motor vehicle (or a 7.14
part of one that was removed while dismantling it) on any land in the open
air, or on any other land forming part of a highway, will be guilty of an
offence under section 2(1)(a) of the Refuse Disposal (Amenity) Act 1978 and
liable, on summary conviction, to a fine not exceeding level 4 on the
standard scale or up to three months imprisonment or both. For these
purposes someone who leaves anything on any land in such circumstances, or
for such a period, that he may reasonably be assumed to have abandoned it
will be deemed to have abandoned it unless he can show otherwise.[64]
"Abandon" in this context means to leave the vehicle completely and finally
with no intention to retrieve it.[65] This provision applies to a vehicle that is
abandoned on any land in the open air, such as a field or on the roadside and
to one left on a highway that is not in the open air such as a road tunnel.

---

[59] E.P.A. 1990, s.79(1)(e), applied to Scotland by E.A. 1955, s.107 & Sched. 17.
[60] *Bland v. Yates* (1914) 58 S.J. 612.
[61] *Coventry City Council v. Cartwright* [1975] 2 All E.R. 99.
[62] E.P.A. 1990, s.79(1A) & (1B) as added by E.A. 1995, Sched. 22, para. 89(3).
[63] Local Government Act 1988, Sched. 1, para. 3(2)—"litter."
[64] Refuse Disposal (Amenity) Act 1978, s.2(2).
[65] *Ellerman's Wilson Line Ltd v. Webster* [1952] 1 Lloyd's Rep. 179.

For the purposes of the Act, a motor vehicle means a mechanically propelled vehicle intended or adapted for use on roads, whether or not it can be so used, and includes a trailer for such a vehicle, any chassis or body of a vehicle or trailer or anything attached to one.[66] A wider definition is given under the Road Traffic Regulation Act 1984 which treats "vehicles" for its purposes as including any vehicle, whether or not a "motor" vehicle.[67]

REMOVAL OF VEHICLES

7.15   Local authorities—in England, district and London Borough councils or the Common Council of the City of London, in Wales, county or county borough councils, and in Scotland, constituted councils—have a duty under section 3(1) of the 1978 Act to remove vehicles that appear to them to have been abandoned without lawful authority on any land in the open air or on any other land that forms part of a highway. Police and local authorities have a power to remove vehicles falling within the scope of the 1984 Act by virtue of Part II of the Removal and Disposal of Vehicles Regulations 1986.[68] These regulations have been made under both the 1978 Act and sections 99 and 101 of the Road Traffic Regulation Act 1984. The powers in Part II of the Regulations concerning the removal of illegally parked vehicles are not of relevance here. However, regulation 5(1) clarifies a council's powers of removal by extending them to cases where it appears to them that a vehicle has been abandoned either because it has broken down or because it has been left in a position, condition or in circumstances that lead them to this view. However, these powers may not be used in respect of vehicles on the Severn Bridge.[69] In order to exercise their powers under the 1978 Act an officer of the authority may enter any land in order to ascertain if a vehicle on it should be removed.[70]

7.16   An authority's duties under section 3 do not extend to abandoned vehicles on a highway that is not a carriageway if they consider that the cost of getting it to the nearest convenient carriageway would be too much.[71] They do not have to give any notice before removing a vehicle that is actually on a highway unless they consider that its condition is such that it ought to be destroyed. Then they must, seven days before removing it,[72] fix a notice to it stating that it will be removed and destroyed at the end of the seven day period.[73]

7.17   Where a vehicle is on occupied land the authority must notify the occupier of their intention to remove it and may not do so if he objects to

[66] Refuse Disposal (Amenity) Act 1978, s.11(1).
[67] Road Traffic Regulation Act 1984, s.99(5).
[68] S.I. 1986 No. 183.
[69] *ibid.* reg. 7.
[70] Refuse Disposal (Amenity) Act 1978, s.8.
[71] *ibid.* s.3(3).
[72] S.I. 1986 No. 183, reg. 10.
[73] Refuse Disposal (Amenity) Act 1978, s.3(5), and Road Traffic Regulation Act 1984, s.99(4).

their proposed action[74] within fifteen days from the date of being served with the notice.[75] The notice should be in the form set out in Schedule 2 to the Regulations and addressed to the occupier of the land. It may be served on him by delivering it to someone who appears to be the occupier, by leaving it at his last known residential address, posting it to him or, if the occupier is a company, by delivering it to the secretary or clerk of the company at their registered or principal office. Alternatively, a notice should be marked "Important—This Communication affects your property" and delivered to the land or affixed to it.[76] Any objections must be in writing and served on the local authority by post or delivery to their offices.[77] These procedures only apply to removals under the 1984 Act; notices under the 1978 Act are fixed to the vehicle.

A vehicle may be towed or driven away in the exercise of these powers or 7.18 any other methods or steps as are thought necessary may be used or taken to effect a vehicle's removal.[78] Vehicles are removed under the 1978 Act by waste collection authorities. They must deliver them to the relevant waste disposal authority under arrangements agreed between them, any dispute being resolved by arbitration.[79] It will be the duty of the disposal authority to take such steps as are reasonably necessary for the safe custody of vehicles they receive[80] unless a vehicle is to be destroyed. Similar duties are imposed in respect of vehicles removed under the Road Traffic Regulation Act 1984 by section 100 of that Act. A vehicle that is thought to be abandoned and is removed by a police officer in accordance with regulation 4 of the 1986 Regulations should be delivered to a local authority; usually also the waste disposal authority.[81] An authority in Scotland will not be liable in a civil action for any failure to carry out their duty of safe storage under the 1978 Act, but the Secretary of State may order them to pay compensation or take other steps if he finds that they have not kept a vehicle in safe custody.[82] Such an order may be enforced in the High Court of Justiciary.[83] No power to award compensation exists in England and Wales or for a breach of the duty under section 100(4) of the 1984 Act so that recourse to the civil courts would be necessary.

DISPOSAL OF VEHICLES

Abandoned vehicles that have been removed by the police or a local 7.19 authority may be disposed of in accordance with the provisions of section 4 of the Refuse Disposal (Amenity) Act 1978 or section 101 of the Road

---

[74] Refuse Disposal (Amenity) Act 1978, s.3(2), and Road Traffic Regulation Act 1984, s.99(4).
[75] S.I. 1986 No. 183, reg. 9(2).
[76] *ibid.* reg. 8.
[77] *ibid.* reg. 9(1).
[78] *ibid.* reg. 6.
[79] Refuse Disposal (Amenity) Act 1978, s.3(6) & 3(7).
[80] *ibid.* s.3(8), and see Road Traffic Regulation Act 1984, s.100(1), 100(2) & 100(4).
[81] Road Traffic Regulation Act 1984, s.100(3).
[82] *ibid.* ss.3(9) & 1(5), as amended by E.P.A. 1990, Sched. 14, Part II.
[83] *ibid.* ss.3(9) & 1(6), as amended by *ibid.*

Traffic Regulation Act 1984 in the manner set out in Part III of the Removal and Disposal of Vehicles Regulations 1986. Where a notice was fixed to a vehicle before it was removed stating that it would be disposed of on removal, and that vehicle has no current excise licence, the authority may dispose of it at any time after it has been removed.[84] If there is a current excise licence in respect of it they must wait until it has expired before disposal.[85]

7.20 In any other case an attempt must be made to find the vehicle's owner.[86] If it has G.B. registration plates, vehicle licensing records should be checked and, if a possible owner is discovered, he should be notified that the authority have it and intend to dispose of it. The owner will then have twenty one days to collect it. In addition a check should be made with the local police force and H.P. Information Ltd.[87] If the vehicle was registered in Northern Ireland, the local police force of the area in which it was found, the Secretary of State for Transport and H.P. Information Ltd should be notified but, if it was registered in Eire only the first two bodies need be informed.[88] If it was registered elsewhere only the police force of the area in which it was abandoned and H.P. Information Ltd must be notified.[89] Where there is no way of finding out where the vehicle was registered and it was removed by the police they should contact the local authority; while if an authority removed it the police should be informed.[90]

7.21 Once the police or local authority have found someone whom they believe to be the owner of the vehicle, either before or after taking those steps, they should serve him with a notice stating that they have removed the vehicle and that unless he claims it within twenty one days they will dispose of it.[91] These notices should, if possible, specify the registration number and make of the vehicle,[92] and be served in the manner provided by regulation 13 of the 1986 Regulations. If the attempt to find an owner fails, or if he does not remove it from the authority's custody within twenty one days from the time he was served with the notice about it (fourteen days in the case of a subsequent notice[93]), they may proceed with its disposal once its vehicle excise licence has expired.[94]

7.22 If, before a vehicle is disposed of, it is claimed by someone who can satisfactorily prove to the authority that he owns it, and who pays the charges (£105 for the removal and £12 for each day on which it has been

---

[84] Refuse Disposal (Amenity) Act 1978, ss.4(1)(a) & 11(1), Road Traffic Regulation Act 1984, s.101(3)(a) & 101(8).

[85] ibid. s.4(1)(b), and ibid. s.101(3)(b).

[86] ibid. s.4(1)(c), and ibid. s.101(3)(c).

[87] S.I. 1986 No. 12(1)(a) & No. 14.

[88] ibid. reg. 12(1)(b) & (c).

[89] ibid. reg. 12(1)(d).

[90] ibid. reg. 12(1)(e).

[91] ibid. regs. 12(2) & 14.

[92] ibid. reg. 11.

[93] ibid. reg. 14.

[94] Refuse Disposal (Amenity) Act 1978, s.4(1)(c), and Road Traffic Regulation Act 1984, s.101(3)(c).

stored) prescribed by the Removal, Storage and Disposal of Vehicles (Prescribed Sums and Charges) Regulations 1989[95] the authority must allow him to take it from them.[96] In such a case he has seven days from the time the authority accept his claim, or, if later, until they dispose of it, to take it away.[97]

An abandoned vehicle may be disposed of by selling it for scrap or in any 7.23 other way. If it was sold but, within a year of the sale, someone convinces the authority that he owned it at the time of sale he must be paid the proceeds, less the authority's removal, storage and disposal charges.[98] Where the authority consider that more than one person owned it they may select one of them and treat him as its owner for the purposes of releasing it to him or for paying him the balance of the proceeds.[99] Once a vehicle has been finally disposed of the bodies specified in regulation 15 of the 1986 Regulations must be told about it if it was removed under section 101 of the 1984 Act.[1] Where a vehicle is removed under the provisions of section 3(1) of the 1978 Act or the 1986 Regulations an authority that removed it or stored it or disposed of it may recover the charges for doing so in accordance with the Prescribed Sums and Charges Regulations 1989, as amended, from the person who was responsible for that vehicle.[2] The person who was responsible for it will be its owner at the time it was abandoned unless he can show that he was not concerned in, and did not know of, its being abandoned. Otherwise the person responsible will be someone who left it at the place from which it was removed or anyone convicted of an offence under section 2(1) of the 1978 Act in respect of its abandonment.[3] In this latter case the court convicting the offender may, on the application of the authority, order him to pay the relevant charges in respect of their dealings with the vehicle in addition to any other order they may make.[4]

More than one authority may recover charges in respect of a vehicle. For 7.24 example, if it was removed by the police and stored and disposed of by a county council both authorities could recover the relevant amounts. However, in England, if a vehicle is removed or stored by a district or London Borough Council or a metropolitan district council it will be deemed to have been in the custody of the relevant waste disposal authority for the purposes of the 1978 and 1984 Acts[5]; in Wales, county and county borough councils are the disposal authorities. Sums recoverable under these provisions

---

[95] S.I. 1989 No. 744, as amended by S.I.s 1991 No. 336 and 1993 No. 550.
[96] Refuse Disposal (Amenity) Act 1978, s.4(5), and Road Traffic Regulation Act 1984, s.101(4).
[97] S.I. 1986 No. 183, reg. 16.
[98] Refuse Disposal (Amenity) Act 1978, s.4(6), and Road Traffic Regulation Act 1984, s.101(5).
[99] *ibid.* s.4(7), and *ibid.* s.101(6).
[1] Road Traffic Regulation Act 1984, s.101(7).
[2] Refuse Disposal (Amenity) Act 1978, s.5(1) & (4), and Road Traffic Regulation Act 1984, s.102(1)(b), (2) & (8).
[3] *ibid.* s.5(4), and *ibid.* s.102(8).
[4] *ibid.* s.5(3), and *ibid.* s.102(5).
[5] *ibid.* s.5(5), and *ibid.* s.102(6) & (8).

are recoverable as a simple contract debt in any court of competent jurisdiction[6] although, as far as the 1984 Act is concerned, any sum of under £20 can be recovered summarily as a civil debt.[7] Moreover, under that Act, the authority may retain the vehicle until their charges in respect of it have been paid.[8]

---

[6] Refuse Disposal (Amenity) Act 1978, s.5(2).
[7] Road Traffic Regulation Act 1984, s.102(3).
[8] *ibid.* s.102(4).

8. WASTE COLLECTION

In England waste collection authorities are district councils and, in **8.01** Greater London, London borough councils or the Common Council of the City of London.[1] In Wales they are county or county borough councils.[2] In Scotland they are those councils constituted under section 2 of the Local Government, etc., (Scotland) Act 1994.[3] The functions and duties of these authorities are set out in paragraph 2.30. Local authorities now ensure that household and some other wastes in their area are collected but, in Great Britain, collection activities must be subject to the competition provisions of Part I of the Local Government Act 1988.[4]

When a person puts waste somewhere for collection by the authority he **8.02** does not abandon it in the sense that it is left for anyone to take away. It is put out for the authority to take and property in it only passes to them once they have collected it. It is in their constructive possession because it is taken by those employed by them.[5] Under section 45(8) of the Environmental Protection Act 1990 anything collected under arrangements made by the authority will belong to them although they will have to deal with it in accordance with section 48 of the Act. It will be an offence under section 60 of the Act to sort over or disturb anything deposited in a receptacle for waste provided by a collection authority unless the authority consents to the interference. A person committing such an offence will be liable on summary conviction to a fine not exceeding level 3 on the standard scale.[6]

THE COLLECTION OF WASTE

A collection authority must arrange for the collection of household waste, **8.03** as defined in regulation 3 of and Schedule 1 to the Controlled Waste Regulations 1992,[7] in their area unless it falls within the descriptions of such

---

[1] And Temple authorities.

[2] E.P.A. 1990, s.30(3) as amended by the Local Government (Wales) Act 1994, Sched. 9, para. 17(3).

[3] E.P.A. 1990, s.30(3) as amended by Local Government, etc., (Scotland) Act 1994, Sched. 13, para. 167(3).

[4] Local Government Act 1988, s.2(2)(a).

[5] *Williams v. Phillips* [1957] Cr.App.R. 5.

[6] E.P.A., s.60(1)(b), (2)(b) & (3).

[7] S.I. 1992 No. 588 and see para. 3.22 for "household wastes".

waste in Schedule 2 to the Regulations that is set out in paragraph 8.08.[8] For those listed wastes their duty only arises after they have been requested to collect them and on payment of any relevant charges. Further they do not have to collect waste which they consider is situated at a place that is so isolated or inaccessible that the costs of collection would be unreasonably high and which they are satisfied can be adequately disposed of by the person who controls it.[9]

8.04   The authority is only obliged to arrange the collection of commercial waste if the occupier of the relevant premises has requested them to do so.[10] They have no duty to arrange for the collection of industrial waste, but may do so on request and if, in England and Wales, the relevant waste disposal authority consents to the arrangement.[11] The authority must make a reasonable charge for the collection and disposal of such waste unless, in respect of commercial waste, they consider one would be inappropriate.[12] However, they may also contribute towards the costs of providing or maintaining plant or equipment to deal with commercial or industrial waste before it is collected under their arrangements.[13]

8.05   An authority must make arrangements for the emptying of privies— latrines with removable receptacles[14]—serving one or more private dwellings in their area free of charge. Further, if someone who controls a cesspool— which includes a settlement or other tank for the reception or disposal of foul matter from buildings[15]—serving only one or more private dwellings requests the relevant authority to empty it they must remove its contents, although they may make a reasonable charge for doing so.[16] The duty of the authority here is limited to removal of matter as the authority considers "appropriate." In *Leck v. Epsom Rural District Council*,[17] a prosecution under section 43 of the Public Health Act 1875, it was held that an authority had a reasonable excuse for not emptying a cesspool of which the owner made an excessive use. Similarly here it is submitted that "appropriate" means appropriate to the needs of the community as a whole and not to the desire of an individual to make excessive use of his facilities. A collection authority may also remove matter from a privy or cesspool in their area that does not exclusively serve private dwellings if requested to do so by the person controlling it and on payment of any reasonable charge they may make.[18] In Scotland septic tanks are emptied by sewerage authorities under the provisions of section 10 of the Sewerage (Scotland) Act 1968[19] so the provisions relating to them in the 1990 Act do not apply there.[20]

[8] E.P.A., s.45(3)(a) and see para. 8.07 below.
[9] *ibid.* s.45(1)(a).
[10] *ibid.* s.45(1)(b).
[11] *ibid.* s.45(2).
[12] *ibid.* s.45(4).
[13] *ibid.* s.45(8).
[14] *ibid.* s.45(12).
[15] *ibid.*
[16] *ibid.* s.45(5).
[17] [1922] 1 K.B. 383.
[18] E.P.A. 1990, s.45(6).
[19] As substituted by the Local Government, etc., (Scotland) Act 1994, s.102.
[20] E.P.A. 1990, s.45(11).

An authority may construct, lay and maintain, in or outside of their area, **8.06** pipes and associated works for the collection of waste and contribute towards the expenses of others whose pipes or works connect with those of the authority.[21] The authority may need planning permission for the construction of the pipes but will not require it for their subsequent repair or maintenance.[22] In addition the provisions of the Pipelines Act 1962 will apply to the works.[23] Pipes and works constructed under this provision will probably be "drains" for the purposes of section 18 the Building Act 1984 and so protected against new construction that would interfere with access to them.[24] Under section 28 of the Control of Pollution Act 1974 authorities had to maintain maps of their pipes but this requirement has been repealed and not re-enacted. In Scotland these pipes will be treated as if provided and maintained under sections 2, 3, 4 and 41 of the Sewerage (Scotland) Act 1968—with slight amendments—and the Pipelines Act 1962 will not apply to them.[25]

CHARGES FOR THE COLLECTION OF HOUSEHOLD WASTE

While normally no charge can be made for the collection of household **8.07** waste, for a type or source of waste described in regulation 4 and Schedule 2 to the Controlled Waste Regulations 1992[26] the authority's duty to arrange collection of it only arises when the person who controls it requests them to do so. In such a case the authority may make a reasonable charge for that service from the person who requested the collection.[27] The government considers that charges should be realistic and have a direct relationship to the costs of collection. Authorities' procedures for determining and reviewing charges should also have regard to the recommendations of the Audit Commission on the best practices in charging.[28]

Schedule 2 sets out 18 categories of waste for which a charge may be **8.08** made. These are;

1. Any article weighing more than 25 kg.
2. An article that cannot be fitted into a receptacle provided in accordance with a requirement under section 46 or otherwise a cylindrical container 750mm in diameter and 1m long.
3. Garden waste.
4. Clinical waste from a private home—which includes a caravan or houseboat.
5. Waste from residential homes, hotels, boarding schools, hospitals or nursing homes.

[21] E.P.A. 1990, s.45(7).
[22] T.&C.P. Act 1990, s.55(2)(c).
[23] Pipelines Act 1962, s.65.
[24] Building Act 1984, s.18(5), but not amended by E.P.A. 1990.
[25] *ibid.* s.45(10) as amended by Local Government, etc., (Scotland) Act 1994, Sched. 13, para. 167(6).
[26] S.I. 1988 No. 588.
[27] E.P.A. 1990, s.45(3).
[28] D.o.E. Circular 13/88, para. 8.

6. Waste from self-catering holiday homes.
7. Dead domestic pets.
8. Where a section 46 notice has been served, any substance that is prohibited from being put into the receptacle required under it.
9. Litter and refuse collected under section 89(1)(f) of the E.P.A. 1990.
10. Waste from:

    (a) In England and Wales, domestic property forming part of a composite hereditament;

    (b) in Scotland, the residential part of residential subjects.[29]

11. Mineral or synthetic oil or grease.
12. Asbestos.
13. Waste from a caravan site on which the caravans can only be occupied for part of the year.
14. Waste from a camp site, other than waste from a private dwelling on the site.
15. Waste from premises occupied by a charitable organisation.
16. Waste from a prison or other penal institution.
17. Waste from a hall or other premises used wholly or mainly for public meetings.
18. Waste from a royal place.

RECEPTACLES FOR WASTE

8.09   Where a collection authority has a duty to collect household waste they may serve a notice, under section 46(1) of the Environmental Protection Act 1990, on the occupier of premises to which it applies requiring him to place his waste for collection in receptacles of a type and number specified. A receptacle for these purposes includes one that holds others.[30] The requirements of a notice must be reasonable, but it can specify that a receptacle should have separate compartments for waste that is to be recycled and that which is not.[31] A notice concerning household waste can provide that the authority will make the receptacles available free of charge or for an agreed sum.[32] If, within a specified time, no agreement can be reached the authority may require the occupier to provide them or the notice may simply require him to do so anyway.[33] It will be an offence for an occupier to fail to comply with a notice and any requirements it may contain[34]; which is punishable on summary conviction by a fine not exceeding level 3 on the standard scale.[35]

8.10 An authority may supply someone with receptacles for commercial or industrial waste if he so requests, and should make a reasonable charge for

---

[29] See D.o.E. Circular 14/92, Annex 1 paras 1.36–1.38.
[30] E.P.A. 1990, s.46(10).
[31] *ibid.* s.46(2).
[32] s.46(3)(a) & (b).
[33] *ibid.* s.46(3)(c) & (d).
[34] Except those concerning payments of an agreed sum.
[35] E.P.A. 1990, s.46(6).

doing so unless, in respect of a commercial waste receptacle, they consider it inappropriate.[36] They may require those holding commercial or industrial waste to store it in specified receptacles if they consider that if no such requirement is made the waste is likely to cause a nuisance or to be detrimental to the amenities of the locality. This requirement is made by a notice served on the occupier of the premises under section 47(2) of the Environmental Protection Act 1990. The notice may specify the kinds and numbers of receptacles he should provide to store the waste at those premises, but the requirement must be reasonable.[37] It will be an offence for the occupier to fail to comply with a notice or any requirements imposed by it; punishable on summary conviction by a fine not exceeding level 3 on the standard scale.[38]

Notices dealing with receptacles for any type of waste may make a number of general provisions. They may specify the size and construction of them and require their maintenance.[39] The placement of the receptacle to facilitate its emptying may be stipulated and the occupier required to give access to it for that purpose.[40] While a notice may require that receptacles should be placed on a highway—or on a road in Scotland—the relevant highway or roads authority must have consented to their being so placed and an arrangement made between them and the collection authority as to liability for any resulting damage.[41] A notice may specify what articles may or may not be put into particular receptacles or compartments of them and the precautions to be taken in respect of particular substances or articles.[42] Local Acts may also place a general restriction on the placing of explosive or corrosive substances in receptacles and anything that gives rise to a substantial risk of injury to someone dealing with the waste in it.[43] Finally, a notice may require the occupier to take steps, such as the provision of a dropped kerb to allow wheeled bins to be taken onto the road, to facilitate the collection of waste from the receptacles.[44] 8.11

An occupier served with a section 46(1) or 47(2) notice may appeal against any requirement of it to a magistrates court or, in Scotland, to the sheriff by way of summary application. The grounds of such an appeal are either that the requirement is unreasonable or, for household waste that the receptacles the occupier already uses are adequate, and, for commercial or industrial notices, that the waste is not likely to cause a nuisance or be detrimental to the locality.[45] An appeal must be made within 21 days from service of the notice or, in respect of a requirement under section 46(3)(c) within 21 days of the end of period within which the authority required 8.12

[36] E.P.A. 1990, s.47(1).
[37] *ibid.* s.47(2) & (3).
[38] *ibid.* s.47(6).
[39] *ibid.* ss.46(4)(a) and 47(4)(a).
[40] *ibid.* ss.46(4)(b) and 47(4)(b).
[41] *ibid.* ss.46(4)(c), and (5) and 47(4)(c) and (5) and see s.60(1)(c).
[42] *ibid.* ss.46(4)(d) and 47(4)(d).
[43] *e.g.,* Leicestershire Act 1985, s.39.
[44] E.P.A. 1990, ss.46(4)(e) and 47(4)(e).
[45] *ibid.* ss.46(7) and 47(7).

agreement to be reached on payment for receptacles.[46] Once an appeal is lodged the requirement has no effect until it is determined.[47] A court on hearing the appeal may quash or modify the requirement or dismiss the appeal.[48] Once an appeal has been determined no question may be raised in subsequent proceedings for an offence under section 46(6) or 47(6) as to the reasonableness of that requirement.[49]

DISPOSAL OF COLLECTED WASTE

8.13   By virtue of section 48 of the Environmental Protection Act 1990 waste collection authorities in England and Wales[50] will usually be required to deliver the waste they have collected to such places as the waste disposal authority for their area directs. However, if they intend to recycle household or commercial waste they may keep it unless the waste disposal authority objects to their action. The provisions concerning this are dealt with at paragraph 10.11 while those relating to plant and equipment for sorting and baling retained waste are discussed at paragraph 10.12.

PLACING SKIPS ON HIGHWAYS

8.14   The placing of skips on highways is controlled under sections 139 and 140 of the Highways Act 1980. The Act refers to "builder's skips" but this is defined in section 139(11) to mean a container designed to be carried on a road vehicle and to be placed on a highway or other land for the storage of builder's materials, or for the removal and disposal of builder's rubble, waste, household and other rubbish or earth. A person will not need a waste management licence for the deposit of less than 50 cubic metres of non-hazardous waste in a skip that is designed so that, as far as practicable, waste cannot escape from it and members of the public cannot gain access to it, if the deposit is for less than three months, is made with the consent of the skip's owner and is not at a site designed or adapted as a transfer station.[51]

8.15   A skip must not be deposited on the highway without the permission of the relevant highway authority.[52] A permission to deposit a skip, or to have it deposited, on a specified highway may be granted conditionally or unconditionally. Conditions may include requirements as to the size and siting of the skip, its marking for road safety and its lighting, the care and disposal of its contents and for its removal on the expiry of the permission.[53]

---

[46] ss.46(8) and 47(8).
[47] *ibid.* ss.46(9)(a) and 47(9)(a).
[48] *ibid.* ss.46(9)(b) & 47(9)(c).
[49] *ibid.* ss.46(9)(c) and 47(9)(c).
[50] *ibid.* s.48(9).
[51] Waste Management Licensing Regulations 1994, S.I. 1994 No. 1056, reg. 17 & Sched. 3, para. 40.
[52] Highways Act 1980, s.139(1).
[53] *ibid.* s.139(2).

It is the skip owner's responsibility to ensure that it is properly lighted during the hours of darkness and marked according to the Builder's Skips (Markings) Regulations 1984,[54] that it has his name, address and telephone number, that it is removed as soon as practicable after it has been filled and that any conditions of the permission are complied with.[55] Even though permission may have been granted to place the skip on a road the person depositing it may still be liable for any injury or damage caused by its presence[56]; although failure to comply with section 139 will not give rise to a cause of action in itself.[57] However, he may have a defence to a charge of obstructing the highway.[58]

If a skip is deposited[59] on a road without permission, or if the provisions of section 139(4) are not complied with, its owner will be guilty of an offence and liable on summary conviction to a fine not exceeding level 3 on the standard scale.[60] For these purposes the skip's owner is its actual owner or a person who has hired it for more than a month or has it under a hire purchase agreement.[61] It will be a defence for the defendant to show that the commission of the offence was due to the act or default of someone else and that he took all reasonable precautions and exercised all due diligence to avoid the offence being committed by himself or his employees.[62] A skip owner who stipulates that the hirer should be responsible for meeting the lighting requirements of section 139(4) may have a defence here,[63] while where a local authority officer gave the owner wrong information about the requirement for permission the offence was held to have been committed by the authority's default.[64] To be able to rely on this defence the prosecutor should be informed in writing of the identity of the other person, if known,[65] at least seven clear days before the hearing.[66] However, the bench may waive this requirement or allow the case to proceed on different information.[67] If another person is named he can be prosecuted for, and convicted of, the offence as well as or instead of the original defendant.[68] If the defendant is charged under another enactment for failing to ensure the skip was properly lighted he will have the same defence as that provided by section 139(6), although it would seem that he does not have to name the responsible person.[69]

8.16

---

[54] S.I. 1984 No. 1933, reg. 4(1) requires the markings to be clean and efficient and clearly visible for a reasonable distance to those using the highway it is on.

[55] Highways Act 1980, s.139(4) as amended by Transport Act 1982, s.65.

[56] *ibid.* s.139(10), and *Saper v. Hungate Builders* [1972] R.T.R. 380.

[57] *Drury v. Camden L.B.C.* [1972] R.T.R. 391.

[58] Highways Act 1980, s.139(9).

[59] Or left—*Craddock v. Green* (1983) 81 L.G.R. 235.

[60] Highways Act 1980, s.139(3) & (4), as amended by Criminal Justice Act 1982, ss.38 & 46.

[61] *ibid.* s.139(11).

[62] *ibid.* s.139(6).

[63] *Lambeth L.B.C. v. Saunders Transport* [1974] R.T.R. 319.

[64] *York District Council v. Poller* (1975) 73 L.G.R. 522.

[65] *P.G.M. Building Co. Ltd v. Kensington L.B.C.* [1982] R.T.R. 107.

[66] Highways Act 1980, s.139(7).

[67] *ibid.* and *Barnet L.B.C. v. S. & W. Transport* [1975] R.T.R. 211.

[68] *ibid.* s.139(5).

[69] *ibid.* s.139(8).

8.17   Even though a skip may have permission to be on the highway, the relevant authority or a constable in uniform, in person,[70] may require its removal or repositioning.[71] A person who fails to comply as soon as is practicable with such a requirement that is directed to him will be guilty of an offence and liable on summary conviction to a fine not exceeding level 3 on the standard scale.[72] The authority or a constable in uniform may also move the skip or cause it to be moved.[73] If they remove it completely the owner must be notified but if he cannot be traced or if he does not recover it within a reasonable time they can dispose of it and its contents[74]; the proceeds being dealt with in accordance with section 140(7). Any expenses reasonably incurred by the authority or police in moving, storing or disposing of the skip may be recovered from its owner in a civil court or in the magistrates court as a civil debt.[75] If, as a result of the movement of the skip, a condition of the original permission is not complied with no offence will be committed under section 139(4).[76]

8.18   In Scotland, under section 85(1) of the Roads (Scotland) Act 1984, a "builder's skip" may not be placed on a road without the permission of the relevant roads authority. The skip must be clearly and indelibly marked with its owner's name, address and telephone number. Conditions may be imposed on the authority's permission. For these purposes a "builder's skip" means a container designed to be carried on a road vehicle and to be placed on a road for the removal and disposal of builder's materials, rubble, waste, household and other rubbish or earth.[77] It will be an offence for the owner to use it, or to cause or permit it to be used, on a road in contravention of section 85(1). Such an offence is punishable on summary conviction by a fine not exceeding level 3 on the standard scale.[78] A skip that is causing a danger or obstruction to road users or that is placed on a road in contravention of section 85 can be removed or repositioned by the police or the roads authority under section 86 of the Act.

8.19   Similar provisions to the Roads (Scotland) Act 1984 are contained in articles 3 to 5 of the Roads and Road Traffic Order (Northern Ireland) 1978.[79]

[70] *R. v. Worthing Justices, ex p. Waste Management* [1989] R.T.R. 131.
[71] Highways Act 1980, s.140(1) & (2).
[72] *ibid.* s.140(3), as amended by Criminal Justice Act 1982, ss.38 & 46.
[73] *ibid.* s.140(4).
[74] *ibid.* s.140(5).
[75] *ibid.* s.140(6) & (8).
[76] *ibid.* s.140(9).
[77] Roads (Scotland) Act 1984, s.85(5).
[78] *ibid.* ss.85(3), 131 and Sched. 8, para. 17.
[79] S.I. 1978 No. 1051 (N.I. 21).

9. CARRIERS AND BROKERS OF WASTE

Article 12 of the Framework Waste Directive[1] requires establishments or **9.01** undertakings which collect or transport waste on a professional basis or which arrange for the disposal or recovery of waste on behalf of others (dealers or brokers) to be registered with a competent authority unless they are authorised by it. In Great Britain this requirement is implemented, as far as carriers are concerned, by the Control of Pollution (Amendment) Act 1989[2] and the Controlled Waste (Registration of Carriers and Seizure of Vehicles) Regulations 1991[3] (the "Carriers") Regulations) that were made under the Act. These provisions are explained in Department of the Environment Circular 11/91.[4]

1. REGISTRATION OF CARRIERS

UNREGISTERED CARRIERS

Under section 1(1) of the 1989 Act it is an offence for any person who is **9.02** not a registered carrier of controlled waste to transport such waste to or from any place in Great Britain in the course of his business or otherwise with a view to profit. For these purposes transport includes transportation by road, rail, air, sea or inland waterway but not its movement between two places through a pipeline or other apparatus.[5] The offence is committed by someone carrying waste in the course of his business or otherwise with a view to profit. The phrase "with a view to profit" would be interpreted by the courts as meaning with the motive or intention of making a profit[6]; and for these purposes it will be irrelevant that no profit was actually made. Nor does the business have to be a waste disposal or transport business. Controlled waste here means such waste as defined by the 1974 or 1990 Act

---

[1] Dir. 91/156 [1991] O.J. L78/32.
[2] As amended by E.P.A. 1990, Sched. 15, para. 31.
[3] S.I. 1991 No. 1624 as amended, S.I. 1992 No. 588 & S.I. 1994, No. 1056, Reg. 23.
[4] W.O. Circular 34/91; S.O. Circular 18/91.
[5] C.P.(A.) Act 1989, s.9(1).
[6] *Re Bird, ex p. Hill*, 23 Ch. D. 695, and *Re Holmes (Eric) (Property) Ltd* [1965] Ch. 1052.

and by regulations made under those Acts.[7] Where the identity of the transporter is in doubt the correct test is to ask in whose name and under whose control was the carriage being undertaken.[8]

9.03   However, both the Act and the Carriers Regulations provide exemptions from the requirement for registration. Under section 1(2) of the Act particular types of transport are exempted. Registration is not necessary for the internal transport of waste from place to place in the same premises. Neither need the importer or an exporter of controlled waste register for the initial transport into, or the final export from, Great Britain.

9.04   Section 1(3) of the Act allows the Secretary of State to make regulations granting further exemptions to particular transporters of waste. These are enacted in regulation 2 of the Carriers Regulations. However even if exempt here, there may be a need to register under paragraph 12 of Schedule 4 to the Waste Management Licensing Regulations 1994.[9] The first group of exemptions are granted to waste collection, disposal or regulation authorities constituted under Part II of the Environmental Protection Act 1990; but they must be registered under paragraph 12 of the 1994 Regulations. This exemption does not encompass waste disposal contractors.[10] Any wholly owned subsidiary[11] of the British Railways Board need not be registered to carry waste as long as it has applied to be so registered, is also registered under paragraph 12 of Schedule 4 to the 1994 Regulations and while its application for registration under the 1991 Regulations is pending.[12]

9.05   A ferry operator need not register to carry a vehicle containing controlled waste. Further, the operator of a vessel, aircraft, hovercraft or other form of transport that is disposing of waste at sea under the authority of the Food and Environment Protection Act 1986 does not have to be registered to carry out such operations. Charities and defined voluntary organisations[13] are also exempt from registration under the 1991 Regulations; although they must register under paragraph 4 of Schedule 12 to the 1994 Regulations. Someone who has applied for registration before April 1, 1992 will be exempted while his application is pending. An application will be pending for these purposes while it is being considered by the relevant Agency and, if registration is refused, during the time allowed for making an appeal against that decision or until any appeal is withdrawn, dismissed by the Secretary of State or the Agency complies with his decision that the applicant should be registered.[14]

9.06   An exemption is also granted to the producer of the controlled waste in question except where it is building or demolition waste.[15] For these

[7] C.P.(A.) Act 1989, s.9(2); E.P.A. 1990, Sched. 15, para. 31(5)(a) & (6) D.o.E. Circular 11/91, paras 1.7–1.9.
[8] *Cosmick Transport Services and Anor v. Bedfordshire County Council* [1996] Env. L.R. 78.
[9] S.I. 1994 No. 1056.
[10] S.I. 1991 No. 1624, reg. 2(1)(a) and D.o.E. Circular 11/91, para. 1.10.
[11] As defined in s.736 of the Companies Act 1985.
[12] S.I. 1991 No. 1624, reg. 2(1)(c) as substituted by S.I. 1994 No. 1056, reg. 23(2) & (4).
[13] Within the meaning of s.48(11) fo the Local Government Act 1985 or section 83D of the Local Government (Scotland) Act 1973.
[14] S.I. 1991 No. 1624, reg. 1(3), and C.P.(A.) Act 1989, s.4(8).
[15] *ibid.* reg. 2(1)(b).

purposes building or demolition waste is defined as waste arising from works of construction or demolition, including that arising from preparatory works.[16] However, what is building or demolition waste here is open to question. If a painter takes empty paint pots back to his depot in a van it is unlikely that a court would say they are building waste. "Construction" for these purposes includes improvement, repair or alteration.[17] A "producer" of waste is not defined but it is submitted that this means the person in whose hands the substance becomes waste. Thus a printer who needs to dispose of waste ink is the producer of the waste rather than the manufacturer of the ink. If the printer takes it to a disposal site he need not be registered to do so. If he hires someone else to take it away then that person must be registered. A householder who takes his own waste to an amenity site is exempt as the producer. If he takes a neighbour's waste for free he commits no offence because he is not carrying it with a view to profit, but if he charges the neighbour to take it then he should be registered.

The 1994 Regulations[18] added an exemption for a person who is the **9.07** holder of a knacker's yard licence—in England and Wales, granted under section 34 of the Slaughterhouses Act 1974, in Scotland under section 6 of the Slaughter of Animals (Scotland) Act 1980—or a licence under article 5(2)(c) or 6(2)(d) of the Animal By-Products Order 1992.[19] This exemption extends to those by-product rendering premises approved under article 8 of the Order and premises registered under articles 9 or 10. In each case the movement of animal by-products will be subject to the conditions set out in Schedule 2 to the Order.[20]

A person who commits an offence under section 1(1) will be liable on **9.08** summary conviction to a fine not exceeding level 5 on the standard scale.[21] The offence is committed by "any person." In relation to a body corporate this is considered to mean each company in a group of companies. One member of the group could not, therefore, rely on the registration of another.[22] Where a body corporate is guilty of an offence under the Act in respect of any act or omission which is shown to have been committed with the consent or connivance of any director, manager, secretary or other similar officer[23] of that body or is attributable to his neglect he will be guilty of the offence in addition to the company and liable to punishment accordingly.[24] In addition where the offence was committed through the act or default of someone else that person will also be guilty of it and may be convicted even if no proceedings are taken against the actual offender.[25]

Three defences are available in response to a charge of being an **9.09** unregistered carrier of waste. The first is that the waste was carried in an

---

[16] S.I. 1991 No. 1624, reg. 2(2).
[17] *ibid.* as substituted by S.I. 1992 No. 588.
[18] S.I. 1994 No. 1056, reg. 23(3).
[19] S.I. 1992 No. 3303.
[20] *ibid.* reg. 7.
[21] C.P.(A.) Act 1989, s.1(5).
[22] D.o.E. Circular 11/91, para. 1.3.
[23] See *Tesco Supermarkets Ltd v. Natrass* [1971] A.C. 153.
[24] C.P.(A.) Act 1989, s.7(6) & (7).
[25] C.P.(A.) Act 1989, s.7(5).

emergency and notice of that carriage was given to the waste regulation authority as soon as practicable after the journey.[26] For these purposes an emergency means any circumstances in which, in order to avoid, remove or reduce any serious danger to the public or serious risk of damage to the environment it was necessary to transport without using a registered carrier.[27] It will also be a defence for the accused to show that he did not know, nor had reasonable grounds for suspecting, that what he was transporting was controlled waste and that he took all reasonable steps to find out whether what he was carrying was waste and if so whether it was controlled waste.[28] The steps required may depend on the situation. A parcel carrier who takes a package from A that does not have anything on it to show it is waste may not need to take any steps here. A should be liable under section 7(5) of the Act. If however, A is involved in waste disposal there may be a duty on the carrier to inquire as to the contents of the parcel. Finally, it will be a defence for the driver of the vehicle or other person to show that he was acting under instructions from his employer.[29]

9.10   In addition to the specific requirements of the 1989 Act there is a duty under section 34(1)(c)(i) of the Environmental Protection Act 1990 on anyone holding waste to ensure that it is transferred only to an authorised person; amongst others a registered carrier. Anyone transferring waste to a carrier will have to comply with that duty and the other duties imposed by section 34(1) and the documentation requirements of regulations made under that section. In addition the carrier will be holding the waste during its transport and so be subject to the duty itself.

REGISTERS

9.11   Both Agencies must establish and maintain a register of carriers of controlled waste. The register must be open for public inspection at the principal offices of the Agency free of charge at all reasonable hours. Further, the public should be allowed to take copies of entries in the register for a reasonable charge. A register can be kept in any form, including on a computer, but it must be indexed and arranged so that the information it contains can be easily traced.[30]

9.12   Registers are also established under paragraph 12 of Schedule 4 to the Waste Management Licensing Regulations 1994. This is to ensure compliance with article 12 of the Framework Waste Directive requiring professional collectors, transporters and brokers of waste to be registered. Registration under paragraph 12 of Schedule 4 is a simplified system to register some of the bodies exempted under the 1991 Regulations.[31] Importantly no application for registration under this provision can be refused, so enabling the Agencies to register without difficulty.[32]

---

[26] C.P.(A.) Act 1989, s.1(4)(a).
[27] *ibid.* s.1(6), and see *Perka et al v. the Queen* (1985) 13 D.L.R. (4th) 1.
[28] *ibid.* s.1(4)(b).
[29] s.1(4)(c).
[30] S.I. 1991 No. 1624, reg. 3.
[31] See D.o.E. Circular 11/94, Annex 1, para. 1.70.
[32] *ibid.* para. 1. 77.

Those who should register under these provisions are waste collection, 9.13
disposal and regulation authorities, charities and voluntary organisations.
Wholly owned subsidiaries of the British Railways Board should also be
registered under this provision until they are registered under the 1991
Regulations.[33] These bodies need not register under paragraph 12 if they are
not an "establishment or undertaking" for the purposes of the Directive, or
are not otherwise authorised—for example, under a waste management
licence—or do not carry waste on a professional basis.[34] Failure to register is
an offence, punishable on summary conviction with a fine of up to level 2 on
the standard scale.[35]

Registration under this provision is with the Agency in whose area the 9.14
person registering has its principal place of business in Great Britain or, if it
has no such place, with either Agency.[36] Both Agencies must establish and
maintain a register of establishments and undertakings that are registered
with it under paragraph 12.[37] The register should show the name of the
establishment or undertaking, the address of its principal place of business
and the address of any place from which it carries on its business.[38] The
Agency should enter these particulars in the register if either it becomes
aware of them or is notified of them in writing. Registration, once effected,
remains valid indefinitely. The register can be kept in any form and must be
available for public inspection and copying.[39]

UNDESIRABLE CARRIERS

The purpose of the 1989 Act is to ensure that controlled waste is only 9.15
carried professionally by suitable carriers. Thus section 3(1)(b) and regulation
5(1) of the Carriers Regulations allow an Agency to prevent undesirable
applicants from being registered and section 3(2) and regulation 10(1)
permit them to revoke the registration of carriers who show themselves to be
undesirables. A carrier will be undesirable if he, or another relevant person,
has been convicted of an offence under an enactment listed in Schedule 1 to
the Regulations[40] and if the Agency considers it undesirable for him to be
authorised to transport controlled waste or to continue to do so.

"Another relevant person" is defined for these purposes in section 3(5) of 9.16
the Act. It is a person closely connected with the carrier's business. Thus an
employee of the carrier who is convicted of a listed offence in the course of
his employment with the carrier is "another relevant person." If the carrier
was in partnership with somebody who committed an offence in the course
of the partnership business, that conviction will be relevant. An offence

---

[33] S.I. 1994 No. 1056, Sched. 4, para. 12(1) & D.o.E. Circular 11/94, para. 1.73.
[34] *ibid.* para. 12(3) and D.o.E. Circular 11/94, para. 1.74.
[35] *ibid.* para. 12(8).
[36] *ibid.* para. 12(4).
[37] *ibid.* para. 12(5).
[38] *ibid.* para. 12(6).
[39] *ibid.* para. 12(7), (9) & (10).
[40] *ibid.* reg. 1(2).

committed by a company at a time when the carrier, a partner in the business or an officer of the carrier company was an officer of the convicted company must also be taken into account as must an offence by someone who is an officer of a company with whom the Agency is concerned.[41] For an individual operating as a carrier an authority should also consider that where a relevant person in relation to him has been convicted of a listed offence, whether he has been a party to the carrying on of a business in a manner involving the commission of prescribed offences.[42] No guidance is given in the Circular about the application of this provision, although it is one presenting considerable practical difficulties. Certainly an Agency acting under it must do so fairly and reasonably and should probably afford the carrier an opportunity to comment on any allegations made about the conduct of his business.[43] Whether or not this should involve a hearing will depend on the circumstances of the case.[44]

9.17   Schedule 1 to the Regulations lists a number of environmental offences that have been prescribed for this purpose. They are: failure to comply with a notice requiring the abatement of a statutory nuisance,[45] professional carriage of goods without an operator's licence,[46] waste disposal offences under either the Control of Pollution Act 1974 or the Environmental Protection Act 1990, dumping cars or rubbish,[47] cable burning,[48] water pollution offences,[49] dumping waste at sea,[50] breach of the Control of Pollution (Special Waste) Regulations 1980 or the Special Waste Regulations 1996,[51] the Transfrontier Shipments of Hazardous Waste Regulations 1988 or the Merchant Shipping (Prevention of Pollution by Garbage) Regulations 1988, offences relating to integrated pollution control or air pollution under section 23 of the Environmental Protection Act 1990 and any offence under the Control of Pollution (Amendment) Act 1989. These listed offences include those in the relevant Acts concerned with giving false information, failing to provide information, obstructing inspectors or failing to comply with notices. Offences in relation to the discharge of trade effluent to sewers are not included here.

9.18   Under the Rehabilitation of Offenders Act 1974 a conviction will become "spent" after a certain period, depending on the sentence imposed for it.[52] The most common penalty for an environmental offence is a fine. Under the 1974 Act a conviction for which only a fine was imposed will become spent

---

[41] And see D.o.E. Circular 11/91, para. 1.39.
[42] C.P.(A.) Act 1989, s.3(6).
[43] *Cinnamond v. British Airports Authority* [1980] 1 W.L.R. 582.
[44] *Lloyd v. McMahon* [1987] A.C. 625.
[45] Public Health (Scotland) Act 1897, s.22; Public Health Act 1936, s.95(1), and E.P.A. 1990, s.80(4).
[46] Transport Act 1968, s.60.
[47] Refuse Disposal (Amenity) Act 1978, s.2.
[48] Clean Air Act 1993, s.33.
[49] C.P.A. 1974, ss.31 & 32 (SCOT); Water Act 1989, ss.107, 118(4) & 175(1).
[50] Food and Environment Protection Act 1986, s.9(1).
[51] As added by S.I. 1996 No. 972, reg. 22.
[52] See D.o.E. Circular 11/91, Table 1.3 at p. 39.

five years after that date on which sentence was passed.[53] However, the fine must have been paid.[54] Once a conviction is spent the defendant should be treated as not having been convicted of that offence.[55] The Act is slightly confusing in that it uses the term "individual" and "person" interchangeably. If it was only concerned with individuals it is possible that it would not apply to bodies corporate but as section 4(1), for example, states "a person who has become a rehabilitated person . . ." it is submitted that the Act's provisions apply to any defendant.[56]

If the carrier or a relevant person has an unspent conviction the Agency 9.19 must still consider whether it is undesirable for him to be registered. Guidance as to this consideration is given in paragraph 1.38 of the Circular. This sets out four criteria; the type of carrier (individual, partnership or company); whether it was the carrier or "another relevant person" that was convicted, the nature and gravity of the offences committed and the number involved. In looking at these last two factors an Agency should regard an offence as grave if it concerned the unlawful treatment, keeping or disposal of special waste, if it caused serious pollution, harm to human health or serious detriment to amenities or was sufficiently serious to result in a term of imprisonment. An Agency might also pay particular regard to a conviction concerning controlled waste or a breach of the duty of care under the Environmental Protection Act 1990.[57]

APPLICATION FOR REGISTRATION

An application for registration as a carrier of controlled waste must be 9.20 made in accordance with regulation 4 of the "Carriers" Regulations. The application should be made to the Agency for the area in which the applicant has, or proposes to have, his headquarters in Great Britain. If the applicant does not have, or propose to have, any offices in the country then he may apply to either Agency.[58] An applicant cannot make multiple applications or be registered with both Agencies. If he has one application pending[59] he cannot make another nor may he apply if he is already registered.[60] However, this does not stop a registered applicant from applying for the renewal of his registration.[61] While there is nothing in the application form requiring an applicant to show concurrent applications, this fact would doubtless emerge. If an applicant has made simultaneous applications then it is submitted he should be required to elect to proceed with one of them. Otherwise all the applications could be refused because there has been a breach of regulation 4.[62]

---

[53] Rehabilitation of Offenders Act 1974, s.5(2) & Table A.
[54] *ibid.* s.1(2)(a).
[55] *ibid.* s.4(1).
[56] See also D.o.E. Circular 11/91, paras 1.32–1.35.
[57] D.o.E. Circular 11/91, para. 1.43.
[58] S.I. 1991 No. 1624, reg. 4(1).
[59] See reg. 1(3), and para. 9.04 above.
[60] *ibid.* reg. 4(2).
[61] *ibid.* reg. 4(3).
[62] S.I. 1991 No. 1624, reg. 5(1)(a).

9.21   Where the carriage business is run, or to be run, by a partnership all the partners or prospective partners must be registered[63]; although they will all be entered under one registration number in the register.[64] When they apply only one fee will be payable.[65] A prospective partner of a business that is already registered should apply for registration to the Agency with whom the other partners are registered.[66] Such an application will not contravene the rules preventing multiple applications or the duration of the existing registration. If it accepts the new partner the Agency should also amend the partnership's registration to reflect the change of circumstances.[67] If one of the partners ceases to be registered, or if a non-registered person becomes a partner, the registration of the business will cease to have effect.[68] However if a partner leaves the partnership this will not affect its registration.[69]

9.22   An initial application should be made on a form like that set out in Part I of Schedule 2 to the "Carriers" Regulations and must contain the information required by that form.[70] An Agency must supply a copy of the form free of charge to anyone who asks for it.[71] On returning the form the applicant must pay a fee of £95 and a copy of the application will be placed on the register.[72] That copy can be removed from the register six years after the application was made.[73] Similar provision is made for applications for renewal of registration. In such a case the application must be made on a form like that in Part II of Schedule 2 and contain the information required by that form.[74] The fee payable on such an application is £65 rather than £95.[75] Where an application is made to be registered as both a broker and a carrier of waste the fee is governed by paragraph 3(11) of Schedule 5 to the Waste Management Licensing Regulations 1994.[76] This requires a £95 fee for a combined application and £65 for the renewal of a combined registration. A broker registering to be a carrier will be charged a fee of £25.[77]

9.23   Under section 7(3)(b) of the 1989 Act it is an offence for a person to provide information in compliance with a requirement made by the regulations that he knows to be false in a material particular, or to recklessly provide information which is false in a material particular. On summary conviction for such an offence the defendant will be liable to a fine not exceeding level 5 on the standard scale.[78] In addition any registration

[63] S.I. 1991 No. 1624, reg. 4(4).
[64] *ibid.* reg. 6(2).
[65] D.o.E. Circular 11/91, para. 1.24.
[66] S.I. 1991 No. 1624, reg. 4(5).
[67] *ibid.* reg. 8.
[68] *ibid.* reg. 11(6).
[69] *ibid.* reg. 11(7).
[70] *ibid.* reg. 4(6), and D.o.E. Circular 11/91, para. 1.19.
[71] *ibid.* reg. 4(8).
[72] *ibid.* reg. 4(9)(a) & (10).
[73] *ibid.* reg. 4(11).
[74] *ibid.* reg. 4(7).
[75] *ibid.* reg. 4(9)(b).
[76] S.I. 1994 No. 1056, reg. 23(5).
[77] *ibid.* reg. 23(6).
[78] C.P.(A.) Act 1989 s.7(4).

obtained through this means can be revoked as the offence is one listed in Schedule 1 to the "Carriers" Regulations.[79]

Once an initial application has been made the Agency must determine it **9.24** within two months of the date it was lodged with it or such longer period as may be agreed between the applicant and the Agency. However, the two month period may not be extended in the case of an application to renew. If the Agency fails to determine the application within the time allowed the applicant may appeal to the relevant Secretary of State[80] within 28 days of the date by which it should have been dealt with; although the time for making an appeal may be extended by the Secretary.[81]

An applicant can only be refused registration if he has contravened any of **9.25** the requirements of regulation 4 of the "Carriers" Regulations or if the Agency consider that it is undesirable for him to carry controlled waste.[82] What is meant by undesirable is dealt with in paragraphs 9.15 to 9.19 above. As far as breach of regulation 4 is concerned the Agency may allow any omission to be corrected, for example by adding information to the form or making a late payment.[83] Where the Agency decides to refuse an application on either of these grounds it must send, ideally in accordance with section 160 of the Environmental Protection Act 1990,[84] the applicant written notice that it has been refused and giving reasons for its decision.[85] It should also inform him of his rights of appeal and the rules under which appeals are dealt with and any other relevant regulations concerning the matter.[86]

On appeal the Secretary of State has said that he does not consider that **9.26** operators should be forced out of business unnecessarily as a result of the registration system. He is of the opinion that Regulation Authorities (now the Agency) should normally register operators who have taken steps to ensure that there is no repetition of the offences for which they have been convicted. Powers are available to regulation authorities to revoke a person's registration if that person is convicted of a prescribed offence and the authority considers that it is undesirable for the registered carrier to continue to be authorised to transport controlled waste.[87] Thus, if there has been a conviction for minor offences only or if it is considered that the appellant has taken steps to ensure that operations are conducted lawfully then the appeal may well be allowed.

An appeal against a refusal[88] should be made within 28 days of the date **9.27** on which the applicant is given notice of the refusal or such longer time as

---

[79] S.I. 1991 No. 1624, reg. 10(1).
[80] C.P.(A.) Act 1989 s.4(1)(b).
[81] S.I. 1991 No. 1624, reg. 16(b).
[82] *ibid.* reg. 5(1).
[83] D.o.E. Circular 11/91, para. 1.31.
[84] S.I. 1991 No. 1624, reg. 26.
[85] *ibid.* reg. 5(2).
[86] D.o.E. Circular 11/91, para. 1.44.
[87] *Baxketh Ltd* December 22, 1992, D.o.E. ref: LEQ28/1/04.
[88] C.P.(A.) Act 1989, s.4(1)(a).

the Secretary of State may allow.[89] While the appeal is to the Secretary of State, he may delegate the matter under section 114 of the Environment Act 1995.[90] If an appeal is made, either against refusal or non-determination, this fact, and the result of the appeal, should be entered on the register. If no appeal is made the register should indicate that the application has not been accepted and no appeal has been made. Such entries on the register may be removed from the register six years after the application they relate to was made.[91]

REGISTRATION

9.28   Once the Agency decides to register a carrier, or it is required to do so following an appeal, it should enter the relevant details on its register.[92] The carrier will be allocated a registration number and it and the dates the registration starts and ceases to have effect will be entered. The entry will show the name and business name of the carrier, the address, telephone, telex and fax numbers of his headquarters and, in the case of an individual his date of birth.[93] For a body corporate the register must show the names and dates of birth of each of its officers, while for a registered company the registered number must be entered and, if it is not registered in Great Britain, the country of incorporation. If the carrier or "another relevant person" has been convicted of a prescribed offence the details of the offence, the date of conviction, the court and the sentence must be shown together with the date of birth of the defendant. Finally, where the carrier, or a company in the same group,[94] holds a waste management or waste disposal licence the name of the licence holder should be entered together with that of the Agency which granted the licence. A partnership will only be entered once in the register, with all the partners being registered under that one entry.[95]

9.29   When the carrier has been entered on the register the Agency must issue him with a certificate of registration free of charge. That certificate should be in a form similar to that in Schedule 3 to the "Carriers" Regulations and containing the information required by it. It should also provide him with a free copy of the entry in the register relating to him.[96] Paragraph 1.54 of the Circular recommends that at this stage the Agency should also advise the carrier of certain other matters relating to registration. Additional copies of the certificate can be obtained from the Agency for a reasonable charge. These will be marked and numbered to show that they have been properly supplied by the Agency.[97] The certificates and copies will need to be renewed if there is a change in the information contained on them.

---

[89] S.I. 1991 No. 1624, reg. 16(a).

[90] C.P.(A.) Act 1989, s.4(9) as added by Environment Act 1995, Sched. 22, para. 37(3).

[91] *ibid.* reg. 5(3)–(5).

[92] *ibid.* reg. 6(1), and D.o.E. Circular 11/91, para. 1.48.

[93] See D.o.E. Circular 11/91, para. 1.49.

[94] As defined in section 53(1) of the Companies Act 1989 (S.I. 1991 No. 1624, reg. 6(3)).

[95] S.I. 1991 No. 1624, reg. 6(2).

[96] *ibid.* reg. 6(3).

[97] *ibid.* reg. 9, and D.o.E. Circular 11/91, para. 1.65.

An entry will be amended on the renewal of the registration. It must be 9.30
changed to show the date on which the renewal takes effect and the new
expiry date. It should also record any change disclosed as a result of the
renewal and note the date on which the change was entered in the register.[98]
After the register has been altered a new certificate and a copy of the
amended entry in the register should be sent free of charge to the carrier.[99]
The carrier should return the old certificate and copies of it to the Agency.[1]

An entry will also have to be amended on a change of circumstances 9.31
affecting information in it relating to the carrier or the addition of another
partner to a partnership entry. It is the duty of every registered person to
notify the Agency of such a change of circumstances.[2] Failure to do so may
be an offence under section 7(3)(a) of the 1989 Act. Thus, someone who
does not notify the Agency of a relevant conviction risks further prosecution
under section 7(3). If a person notifies the Agency of a relevant conviction he
should give the details required by regulation 6(1)(f). If one Agency is
notified of the conviction by the other it should check those details with the
carrier.[3] Where the Agency is notified of a change of circumstances or
accept, or are directed to accept, the registration of a new partner on a
partnership's entry it must amend the register accordingly and note the date
of the amendment in the register. If the change affects the information on
the carrier's certificate a new one should be issued free of charge and in any
event a free copy of the new entry should be sent to him.[4] The carrier should
return the old certificates and copies of it to the Agency.[5]

ENFORCEMENT

The key enforcement provisions are found in section 5 of the Control of 9.32
Pollution (Amendment) Act 1989. However, some of the enforcement
provisions of the Environmental Protection Act 1990 are also available to
waste regulation authorities by virtue of section 7(1) of the 1989 Act.[6] These
provisions are those of section 71; power to obtain information. Those
provisions have effect for the purposes of the 1989 Act as if they were
incorporated in it and exercisable by an Agency.

Under section 5(1)(a) of the 1989 Act a constable or an authorised officer 9.33
of an Agency have power to stop anyone they consider to be or to have been
transporting controlled waste and to require him to produce his, or his
employers, authority to do so. However, a vehicle on a road may only be
stopped by a uniformed police officer.[7] Once any vehicle that it is considered

---

[98] S.I. 1991 No. 1624, reg. 7(1).
[99] ibid. reg. 7(2).
[1] ibid. reg. 13.
[2] ibid. reg. 8(1).
[3] D.o.E. Circular 11/91, para. 1.63.
[4] S.I. 1991 No. 1624, reg. 8(2).
[5] ibid. reg. 13.
[6] As amended by E.P.A. 1990, Sched. 15, para. 31(4)(a) & Environment Act 1995, Sched.
22, para. 37(5).
[7] C.P.(A.) Act 1989, s.5(2).

is or has been used for transporting controlled waste has been stopped or found it may be searched, tests conducted on anything found in it and samples taken for analysis.[8]

9.34   The powers of stop and search provided by section 5(1) can only be exercised if it reasonably appears to the officer that any controlled waste is being or has been transported by an unregistered carrier. Thus there is no power to stop or search at random. This may give rise to some difficulty as the officer will have to show some reasonable grounds for his view that the vehicle was being used by an unregistered carrier. This could arise from finding that the owner of the vehicle is not registered before stopping it.[9] A reasonable ground may arise from information received by the officer, but not from mere gossip.[10] The conduct of a vehicle or operator may also gives rise to reasonable grounds for stopping it,[11] or the vehicle can be stopped for other reasons by a police officer who can then discover reasonable grounds for requiring the production of the authority.[12] Even if the officer has no reasonable grounds to stop and search the vehicle this will not necessarily render any evidence obtained inadmissible.[13] In general courts should admit any relevant evidence unless it has been obtained by bad faith or oppression.[14] The question, both in common law and under section 78 of the Police and Criminal Evidence Act 1984 is whether admitting the evidence will prevent the defendant from having a fair trial.

9.35   Where someone is required to produce his authority to carry controlled waste he must produce either his certificate of registration or a copy of that certificate provided under regulation 9 of the "Carriers" Regulations.[15] When he is required to produce it he must either do so at that time to the person making the request or by sending it to, or producing it at, an office of the Agency no later than seven days after the day on which the requirement was made.[16]

9.36   Anyone who intentionally obstructs any authorised officer of an Agency or a constable exercising these stop and search powers will be guilty of an offence and liable on summary conviction to a fine not exceeding level 5 on the standard scale.[17] It would also be an offence, punishable in the same way, to fail without reasonable excuse to comply with a requirement imposed in the exercise of those powers.[18] It will be for the defendant to establish that he had a reasonable excuse for his failure to comply. However, the prosecution will have to show that the waste transported was controlled waste and that the defendant transported it to or from a place in Great Britain.[19]

---

[8] C.P.(A.) Act 1989, s.5(1)(b).

[9] *Baxter v. Oxford* [1980] R.T.R. 315.

[10] *Monaghan v. Corbett, The Times,* June 23, 1983.

[11] *Steel v. Goacher* [1982] Crim. L.R. 689.

[12] *Adams v. Valentine* [1975] R.T.R. 563.

[13] *Fox v. Chief Constable of Gwent* [1985] 3 All E.R. 392.

[14] *Thomas v. D.P.P., The Times,* October 7, 1989.

[15] C.P.(A.) Act 1989, s.5(6), and S.I. 1991 No. 1624, reg. 14(2).

[16] *ibid.* s.5(3), and *ibid.* reg. 14(1).

[17] *ibid.* s.5(4)(a) & (7).

[18] *ibid.* s.5(4)(b) & (7).

[19] *ibid.* s.5(4) & (5).

CESSATION OF REGISTRATIONS

Normally a person's registration as a carrier will cease to have effect at the **9.37** end of three years after the date of registry or the date on which it was last renewed.[20] However, a registered carrier may at any time write to the Agency requiring it to remove his name from the register.[21]

At least six months before a registration is due to expire the Agency must **9.38** send the carrier a reminder that his registration will expire on a specified date, an application form for its renewal and a copy of his current entry in the register. The reminder notice should also state the effect that applying for a renewal will have on the currency of the present registration. [22] Where an application is received in the final six months of the existing registration then that registration will remain in force until the application is withdrawn or accepted or, if it is refused or not processed within two months of receipt, until the time for making an appeal has expired. If an appeal is made the registration will continue until the appeal is disposed of or discontinued.[23] If a person fails to apply to renew his registration until after it has expired he will have to apply as if he was being registered for the first time and will not be registered until that application is determined in his favour.[24] Where an application is accepted the renewal will take effect from the time the current registration expires.[25]

A registration may be revoked if, and only if, an Agency considers that it **9.39** is undesirable for the carrier to carry controlled waste.[26] What is meant by "undesirable" is set out in paragraphs 9.15 to 9.19 above. Where the Agency decide to revoke a registration it must send the carrier a notice informing him of it and of the reasons for its decision.[27] Despite that the registration will continue to have effect until the time for making an appeal has elapsed, unless the carrier indicates he will not make one, or until an appeal is disposed of or is discontinued.[28] Any appeal must be brought within 28 days of the date the carrier was given notice of the revocation.[29]

Once a registration has ceased to have effect the Agency should record **9.40** this in the register with the date it expired. The entry relating to that carrier can be removed from the register six years after its expiry.[30] The carrier should immediately return his certificate of registration and any copies of it that were issued by the Agency to it.[31]

---

[20] S.I. 1991 No. 1624, reg. 11(2).
[21] C.P.(A.) Act 1989, s.3(2).
[22] S.I. 1991 No. 1624, reg. 11(3).
[23] *ibid.* reg. 11(4), and C.P.(A.) Act 1989, s.4(7) & (8).
[24] D.o.E. Circular 11/91, para. 1.73.
[25] S.I. 1991 No. 1624, reg. 11(8).
[26] C.P.(A.) Act 1989, s.3(2), and S.I. 1991 No. 1624, reg. 10(1).
[27] S.I. 1991 No. 1624, reg. 10(2).
[28] *ibid.* reg. 11(5), and C.P.(A.) Act 1989, s.4(7) & (8).
[29] *ibid.* reg. 16(c).
[30] *ibid.* reg. 12.
[31] *ibid.* reg. 13.

APPEALS

9.41   Rights of appeal are conferred by section 4(1) and (2) of the 1989 Act. An applicant for registration may appeal to the Secretary of State if his application is refused or is not determined within two months after it was made or, for an initial application, such longer period as may be agreed between him and the authority. An appeal must be made against a refusal within 28 days of the date the appellant is notified of the refusal or against non-determination within 28 days of the date on which the determination should have been made.[32] A carrier whose registration is revoked may appeal against that revocation within 28 days of which he received the notice of revocation.[33] By virtue of section 4(7) & (8) an existing registration will continue to be in force until the appeal is disposed of by being withdrawn, dismissed or until the Agency complies with a direction to register the appellant.

9.42   The hearing of an appeal may be delegated under section 114 of the Environment Act 1995.[34] Notice of appeal must be sent in writing to the Secretary of State, or whoever is delegated to deal with it, and accompanied by a statement of the grounds of appeal and any relevant correspondence between the appellant and the Agency, together with a statement of whether the appellant wishes the appeal to be dealt with by written representations or a hearing. For an appeal against a refusal or non-determination of an application the notice should also be accompanied by a copy of the application and any notice of the Agency's refusal. For an appeal against revocation a copy of the relevant entry in the register and of the notice of revocation should be sent. These appeal documents should be served on the Agency at the same time that notice of the appeal is given to the Secretary of State.[35]

9.43   If either party to an appeal requests a hearing, or if the Secretary of State decides one should be held, then the appeal must take the form of, or continue as, a hearing before someone appointed by the Secretary of State.[36] The government anticipates that most appeals will be dealt with by written representations but this may be wishful thinking. In these cases where the appeal is in effect about the character of the carrier it may be advisable in many cases for the appellant to appear in person before an inspector; although this may involve him being cross examined about his character. After the hearing the inspector must make a written report to the Secretary of State on his conclusions and recommendations or on his reasons for not making recommendations.[37]

9.44   On an appeal the Secretary of State may, as he thinks fit, either dismiss it, or give an Agency a direction either to register the appellant or to cancel its

[32] S.I. 1991 No. 1624, reg. 16(a) & (b).
[33] ibid. reg. 16(c).
[34] C.P.(A.) Act 1989, s.4(9) as added by Environment Act 1995, Sched. 22, para. 37(3).
[35] S.I. 1991 No. 1624, reg. 15.
[36] ibid. reg. 17(1).
[37] ibid. reg. 17(2).

revocation of his registration. In an appeal against non-determination he can also direct the Agency not to register the applicant. It will be the duty of the Agency to comply with any such direction given it by the Secretary.[38] Once the decision is made the Secretary of State must notify the appellant of it, the reasons for it and, if a hearing was held, send him a copy of the inspector's report. These documents must also be sent to the Agency at the same time.[39]

2. SEIZURE AND DISPOSAL OF VEHICLES

One of the problems of controlling fly tipping—unauthorised disposal of 9.45 lorry loads of waste—has been to trace the operator of the vehicle concerned. Section 6 of the Control of Pollution (Amendment) Act 1989 and regulation 19 to 24 of the "Carriers" Regulations are intended to enable an Agency to obtain the name and address of the user of such a vehicle by, if other methods have failed, obtaining a warrant to seize the vehicle and to dispose of it if no one comes forward to claim it.

Initially where an Agency has reason to believe that a vehicle has been 9.46 involved in unlawful waste disposal it should try to obtain the name and address of its registered keeper and user from the registration number allocated to it by the country with whom it is registered.[40] If necessary it should also serve a notice under section 71(2) of the Environmental Protection Act 1990[41] on any person it considers may be able to provide it with the name and address of the user of the vehicle at the time when the offence was committed, requiring him, if he is able to do so, to provide it with that name and address.[42] Such a notice can be served in accordance with section 160 of the 1990 Act.[43] Failure to comply with such a notice will be an offence under section 7(3) of the 1989 Act or the provision under which the notice was served.

Where, after taking these steps, an Agency has been unable to obtain the 9.47 required information it may apply, on sworn information in writing, to a justice of the peace or, in Scotland, to a sheriff or a justice, for a warrant to seize the vehicle under section 6(1) of the 1989 Act. The justice or sheriff may only grant such a warrant if he is satisfied that there are reasonable grounds for believing that an offence involving the unlicensed deposit, treatment or disposal of waste has been committed, that the vehicle was used in the commission of that offence, that proceedings for that offence have not yet been brought against anyone and that the Agency has failed, after taking the steps set out in paragraph 9.45 above, to ascertain the name and address of anyone who can give it the name and address of the person who was using the vehicle at the time when the offence was committed.[44]

---

[38] C.P.(A.) Act 1989, s.4(3)–(5).
[39] S.I. 1991 No. 1624, reg. 18.
[40] S.I. 1991 No. 1624, reg. 20(2) & (4).
[41] Or s.93(1) of C.P. Act 1974 if s.71 not in force.
[42] S.I. 1991 No. 1624, reg. 20(3).
[43] *ibid.* reg. 26.
[44] C.P.(A.) Act 1989, s.6(1) (as amended by E.P.A. 1990, Sched. 15, para. 31(3)) and S.I. 1991 No. 1624, reg. 19.

9.48   Once a warrant has been issued any authorised officer and any constable may stop the vehicle concerned and, on behalf of the Agency, seize it and its contents. However, only a constable in uniform may stop a vehicle on a road while an officer of the Agency may only seize any property if he is accompanied by a constable.[45] The warrant issued remains in force until its purpose has been fulfilled. Where it is executed the person seizing the property shall, if required to do so, make both the warrant and the authority under which he is acting, available for inspection.[46] Any person who intentionally obstructs someone exercising these powers will be guilty of an offence and liable on summary conviction to a fine not exceeding level 5 on the standard scale.[47]

9.49   Where property has been seized under section 6 of the Act it may be removed by the Agency to somewhere it considers it appropriate to keep it. A vehicle that is seized may be driven, towed or removed by other means that are reasonable in the circumstances and any necessary steps may be taken in relation to it to facilitate its removal. All or some of the contents of the vehicle may be removed separately where it is reasonable to do so to facilitate its removal, there is a good reason for storing them at a different place to it or their condition requires them to be disposed of without delay.[48] It will be an offence under section 6(9) to intentionally obstruct the removal. The vehicle and contents must be kept in the Agency's custody until it is returned to someone who establishes his entitlement to it or the Agency lawfully disposes of them.[49] In holding the property the Agency must take such steps as are reasonably necessary to ensure that it is not stolen or vandalised.[50]

9.50   Unless property seized has been disposed of under regulation 23 of the "Carriers" Regulations it must be returned by the Agency to the person entitled to it. This person is someone who produces satisfactory evidence of his entitlement and of his identity or address or, if he is recovering it as an agent for someone else, who satisfactorily identifies both himself and his principal and produces the principal's entitlement to the property and his authorisation to act on his principal's behalf. Where the property is a vehicle the owner or keeper of it must also produce the vehicle registration book.[51] Normally the person showing he is entitled to a vehicle will also be entitled to its contents unless they, or part of them, have been claimed by someone else. If there is more than one claim the Agency must determine the entitlement to it on the basis of the evidence provided to it.[52] This means that an Agency may take time to consider the evidence before handing the property over to the person it decides is entitled to it.[53]

---

[45] C.P.(A.) Act 1989, s.6(2) & (3).
[46] ibid. s.6(4) and R. v. Longman [1988] 1 W.L.R. 619 at 627B.
[47] ibid. s.6(9).
[48] S.I. 1991 No. 1624, reg. 21 and D.o.E. Circular 9/11, para. 2.19.
[49] C.P.(A.) Act 1989, s.6(5).
[50] ibid. s.6(8).
[51] S.I. 1991 No. 1624, reg. 22(1).
[52] ibid. reg. 22(2) & (3).
[53] See D.o.E. Circular 9/11. para. 2.24.

Before an Agency disposes of property seized it must comply with 9.51
regulations made under section 6(6) of the 1989 Act.[54] These require it to
publish a notice in a newspaper circulating in the area where the property
was received stating the Agency's name and describing the property
concerned and the time and place at which and powers under which it was
seized together with the vehicle's registration mark, if any. The notice should
state that the property may be claimed at the place and times specified and
that, if no-one establishes their right to it, it will be disposed of 28 days after
the notice was published unless its condition necessitates earlier disposal.[55]
The Agency should also serve a copy of that notice on anyone served with a
notice under section 71(2) of the Environmental Protection Act 1990 in
relation to the vehicle, the chief officer of the police force in whose area it
was seized, the Secretary of State for Transport and H.P. Information plc.[56]
Once this is done the Agency may then dispose of the property 28 days after
the notice was published, or, if later, after notices were served on the
relevant people, if no-one has established a valid claim to it. It may be
disposed of earlier than this if its condition so requires.[57]

After a vehicle has been disposed of, notice of disposal should be served on 9.52
the chief officer of the police force in whose area it was seized, the Secretary
of State for Transport and H.P. Information plc.[58] If property has been sold
the proceeds of sale should be applied to meeting the Agency's expenses in
seizing vehicles under the 1989 Act. Any surplus should be applied to
meeting the costs of investigating claims to the proceeds of sale. Such a
claim will be established if the claimant proves to the Agency that he would
have been entitled to the return of the property if it had not been sold.[59]
Thus the claimant must provide evidence of his name and address.

3. REGISTRATION OF BROKERS OF WASTE

Article 12 of the Framework Waste Directive requires that establishments 9.53
or undertakings[60] that arrange for the disposal or recovery of waste on behalf
of others (brokers or dealers) should be registered with the competent
authorities. This is given effect in Great Britain by regulation 20 of and
Schedule 5 to the Waste Management Licensing Regulations 1994.[61] These
provisions are explained in Annex 8 to D.o.E. Circular 11/94.

There is no definition of "broker" in the Regulations. However, that 9.54
provided in article 12 of the Directive is probably sufficient. Although the
courts have discussed the term generally,[62] it is likely that its context in

---

[54] As substituted by Environment Act 1995, Sched. 22, para. 37(4).
[55] S.I. 1991 No. 1624, reg. 23(1)(a) & (2).
[56] *ibid.* reg. 23(1)(b) and see D.o.E. Circular 11/91, para. 2.26.
[57] *ibid.* reg. 23(1)(c).
[58] *ibid.* reg. 24.
[59] *ibid.* reg. 25.
[60] See D.o.E. Circular 11/94, Annex 8, para. 8.11.
[61] S.I. 1994 No. 1056.
[62] *Milford v. Hughes* (1846) M. & W. 174.

article 12—"arrange for the disposal or recovery of waste on behalf of others"—will be crucial. It is suggested in paragraph 8.5 of Annex 8 to Circular 11/94 that holders and producers of waste who arrange for its disposal cannot also be brokers. This may raise a question as to who is the "producer." For example, the managing agents of a block of flats do not produce the waste generated by the occupants. Nevertheless they may be holders of the waste in a communal receptacle and thus need not register as brokers when they arrange for disposal of that waste.

9.55   Generally it is an offence for an establishment or undertaking to arrange (as a dealer or broker) for the disposal or recovery of controlled waste on behalf of someone else unless it is a registered broker of such waste.[63] Anyone found guilty of such an offence will be liable on summary conviction to a fine not exceeding level 5 on the standard scale.[64] Those responsible for the management of a company can also be convicted of this offence if it was committed with their consent or connivance or through their neglect.[65]

9.56   There are some exceptions to this general rule. Establishments or undertakings that are authorised to carry out the disposal or recovery of waste under the relevant legislation and those that recover waste under either an exemption conferred by regulation 17(1) of and Schedule 3 to the 1994 Regulations, or by article 3 of the Deposits in the Sea (Exemptions) Order 1985[66] do not need to register as brokers or dealers.[67] Nor does a registered carrier have to register as a broker if he makes arrangements for the disposal or recovery of controlled waste as part of the transport arrangements for the waste to or from any place in Great Britain.[68]

9.57   A further exemption is granted to establishments or undertakings that are either charities, voluntary organisations or waste collection, disposal or regulation authorities.[69] However, these bodies must be registered under paragraph 12 of Schedule 4 to the 1994 Regulations.[70] Failure to be so registered will be an offence rendering the establishment or undertaking concerned liable on summary conviction to a fine of up to level 2 on the standard scale.[71]

9.58   Registrations are dealt with under the provisions of Schedule 5 to the 1994 Regulations.[72] It is the duty of both Agencies to establish and maintain a register of waste brokers and to ensure that it is open to public inspection. The public should also be able to take copies of entries on the register for a reasonable charge.[73] The register may be kept in any form.[74]

[63] S.I. 1994 No. 1056.
[64] ibid. reg. 20(5).
[65] ibid. reg. 20(6).
[66] S.I. 1985 No. 1699 — see paras 21.96 et seq.
[67] S.I. 1994 No. 1056, reg. 20(2).
[68] ibid. reg. 20(3).
[69] ibid. reg. 20(4).
[70] See paras 9.12–9.14.
[71] S.I. 1994 No. 1056, Sched. 4, para. 12(2) & (8).
[72] ibid. reg. 20(7).
[73] ibid. Sched. 5, para. 2(1).
[74] ibid. para. 2(2).

Applications for registration, or renewal of registration, are made to the 9.59
Agency in whose area the broker has or proposes to have his principal place
of business in Great Britain. If he does not have or propose to have one here
then the application is made to either Agency.[75] Generally an application
should not be made whilst a previous application is pending or while the
applicant is registered[76]; although renewal applications can be made within
six months of a current registration expiring.[77] An application for registration by a partnership should be made by all partners or prospective partners;
new partners to a business already registered apply for registration to the
Agency with whom the partnership is registered as a partner in that
business.[78]

An application for initial registration is made on a form corresponding to 9.60
or similar to that set out in Part II of Schedule 5.[79] The Agency should
provide a copy of this form free of charge to anyone requesting one.[80] If the
applicant wishes to be registered as a carrier and broker of waste than a
combined application can be made on a form that sets out the information
required by the form prescribed in regulation 4(6) of the Carriers Regulation
and the Part II form in Schedule 5 of the 1994 Regulations.[81] On making
the application the applicant should pay the fee of £95 for registration as a
broker or a combined registration; although for an already registered carrier,
the broker's registration costs only £25.[82] There is no provision for refunds of
charges paid.[83] A copy of the application should be placed on the register by
the Agency.[84]

The Agency may only refuse an application on the grounds that either the 9.61
procedural requirements of paragraph 3 of the Schedule have not been
complied with, or if it is of the opinion that the applicant or another relevant
person is not a fit and proper person to be registered.[85] This latter provision
is similar to that in the Carriers Regulations. It has two elements. First, that
the applicant or another relevant person must have been convicted of a
"relevant offence", or, if the applicant has not been convicted of such an
offence, he was a party to the carrying on of a business in a manner
involving the commission of relevant offences.[86] "Relevant offences" here are
those listed in regulation 3 of the 1994 Regulations, a list similar to that set
out in paragraph 9.17. For these purposes "another relevant person" has the
same meaning as that set out at paragraph 9.16.[87] Secondly, the Agency

---

[75] S.I. 1994 No. 1056, Sched. 5, para. 3(1).
[76] *ibid.* Sched. 5, para. 3(2).
[77] *ibid.* Sched. 5, para. 3(3) & 7(5).
[78] *ibid.* Sched. 5, para. 3(4) & (5).
[79] *ibid.* Sched. 5, para. 3(6).
[80] *ibid.* Sched. 5, para. 2(10).
[81] *ibid.* Sched. 5, para. 3(8).
[82] *ibid.* Sched. 5, para. 3(11).
[83] See D.o.E. Circular 11/94, Annex 8, para. 8.35.
[84] S.I. 1994 No. 1056, Sched. 5, para. 3(12).
[85] *ibid.* Sched. 5, para. 3(13).
[86] *ibid.* Sched. 5, paras 3(13) and 1(2).
[87] *ibid.* Sched. 5, para. 1(3).

must consider that as a result of this it is undesirable for him to be registered.[88] Guidance has to how the Agency should reach its decision is contained in paragraphs 8.37–8.52 of Annex 8 to Circular 11/94.

9.62   The Agency has two months to reach its decision unless a longer period is agreed between the parties. If it fails to make its decision in this time then the applicant may appeal to the Secretary of State.[89] If the Agency decides not to register the applicant it must notify him that his application is refused, giving reasons for the decision.[90] Failure to give reasons will invalidate the decision.[91] There is also a right of appeal against the refusal to register.[92]

9.63   Where an application is accepted the appropriate Agency should make an entry in its register showing the applicant as a registered broker and allocating him a registration number. The entry should also show the date on which the registration takes effect and the date it expires, the business name, the address of the principal place of business, with any telephone, fax or telex number, and, in the case of an individual, his date of birth. The names of a body corporate's managers, secretary or similar officers should be listed, together with their dates of birth; while a company registered under the Companies Act should have its registration number listed. If the company is incorporated outside of Great Britain the country in which it is incorporated should be shown. If any registered person or another relevant person has been convicted of a relevant offence details of the conviction should be entered. In addition if the registered person is the holder of a waste management licence, or any company in the group of companies is, the name of the holder must be entered together with the name of the Agency that granted the licence.[93] If the business is, or is to be, carried on by a partnership all the partners must be registered under one entry and only one number will be allocated to the partnership.[94] On making the entry in the register the Agency must provide the registered person or partnership with a free copy of their entry in the register.[95]

9.64   A registered person should notify the Agency maintaining the relevant register of any change of circumstances that affects the information in the register relating to him.[96] This includes a conviction for a relevant offence.[97] On being notified of such a change or on accepting the entry of a prospective partner to a business that is already registered or on being directed to do so on appeal, the Agency should amend the relevant entry to reflect the new situation, note the date on which the amendment was made

---

[88] See appeal decision extract at para. 9.26.
[89] S.I. 1994 No. 1056, Sched. 5, paras 6(1)(b) & 1(1).
[90] ibid. Sched. 5, para. 3(14).
[91] Appeal Decision: *Sprigcourt Ltd*; June 29, 1993. D.o.E. Ref: LEQ28/1/18.
[92] S.I. 1994 No. 1056, Sched. 5, para. 6(1)(a).
[93] ibid. Sched. 5, para. 4(1).
[94] ibid. Sched. 5, para. 4(2).
[95] ibid. Sched. 5, para. 4(3).
[96] ibid. Sched. 5, para. 4(6).
[97] See D.o.E. Circular 11/94, Annex 8, para. 8.67.

and provide the person or partnership concerned with a free copy of the amended entry.[98] If the Agency is informed from another source about an unspent conviction it should seek full details from the broker concerned.[99]

The appropriate Agency may revoke a broker's registration. It may only 9.65 do so if the broker or another person has been convicted of a relevant offence and if as a consequence it is the opinion of the Agency that it is undesirable for him to continue as a broker of waste.[1] If the registration is revoked the Agency must notify the broker concerned, informing him of the revocation and the reasons for it.[2] The broker may appeal against that decision to the Secretary of State.[3] A revoked registration will continue in force until either the time for appeal has expired or until the broker tells the Agency that he does not intend to appeal. In the latter case his registration expires on the date he informed the Agency.[4]

Appeals against Agency decisions are dealt with in much the same way as 9.66 appeals under the Carrier's Regulations as set out in paragraphs 9.41–9.44 above. The time limit is 28 days after notice of refusal or revocation or after the time in which a decision should have been made has run out.[5] The procedure and decision making powers are identical. The making of an appeal and the results of it should be entered in the register.[6] The Agency has a duty to comply with any decision made by the Secretary of State.[7] If no appeal is made against a decision to refuse any application that fact should also be entered in the register.[8]

Registration usually lasts for three years from the date of the initial 9.67 registration or its renewal.[9] Combined broker/carrier registrations that were made separately can be brought to an end three years from the date of registration as either a broker or carrier if this is requested.[10] Partnership registrations end if any of the partners cease to be registered without leaving the firm or if a new, unregistered, partner joins the firm.[11] A registration can end at any time if the broker asks to be de-registered.[12]

At least six months before a registration is due to expire the relevant 9.68 Agency should notify the broker of the date of expiry. It should also send him an application form for the renewal of his registration and a copy of his current entry in the register.[13] The form for renewals should be the same as,

---

[98] S.I. 1994 No. 1056, Sched. 5, para. 4(7).
[99] D.o.E. Circular 11/94, Annex 8, para. 8.69.
[1] S.I. 1994 No. 1056, Sched. 5, para. 5(1) and see para. 9.61.
[2] *ibid.* Sched. 5, para. 5(2).
[3] *ibid.* Sched. 5, para. 6(2).
[4] *ibid.* Sched. 5, para. 7(7) and see para. 9.70 ,below.
[5] *ibid.* Sched. 5, para. 6(6).
[6] *ibid.* Sched. 5, para. 3(15).
[7] *ibid.* Sched. 5, para. 6(12).
[8] *ibid.* Sched. 5, para. 3(16).
[9] *ibid.* Sched. 5, para. 7(1).
[10] *ibid.* Sched. 5, para. 7(2) & (3).
[11] *ibid.* Sched. 5, para. 7(9) & (10).
[12] *ibid.* Sched. 5, para. 7(4).
[13] *ibid.* Sched. 5, para. 7(5).

or similar to, that set out in Part III of Schedule 5 and should contain the information required by that form.[14] If the applicant is renewing a combined broker's and carrier's registration his form may contain the information required under both regimes.[15] The fee for renewal applications, whether individual or combined, is £65, which must accompany the application.[16]

9.69   A copy of the renewal application should be placed in the register.[17] If it is made within six months from the expiry of the registration then that registration will remain in force until either the application is withdrawn or accepted or, if it is refused or left undetermined, until either the time for making an appeal has expired without an appeal being made or until the applicant tells the Agency that he will not be appealing. In the latter case the registration expires on the date that he informs the Agency.[18] If the application is accepted the renewal takes effect from the expiry date.[19]

9.70   Where an appeal is properly made by an applicant for the renewal of his registration, or by a broker whose registration has been revoked, the registration will continue in force until the appeal has been finally disposed of.[20] This means until either the appeal is withdrawn, or the appellant is notified that his appeal has been dismissed or the Agency complies with a direction to renew or restore the broker's registration.[21] If the broker applies to judicially review the Secretary of State's decision he will have to seek a stay to prevent the Agency from de-registering him.

9.71   Where a registration ends, this fact should be entered in the register together with the date on which it ceased to have effect. After six years the entry relating to that registration may be removed from the register altogether.[22] While in addition, to keep registers pruned, applications for registration or renewal and notes about appeals may also be removed from the register after six years from the time they were placed there.[23] Further, the department advises that registers should be regularly checked to remove details of "spent" convictions.[24]

---

[14] S.I. 1994 No. 1056, Sched. 5, para. 3(7).
[15] ibid. Sched. 5, para. 3(9).
[16] ibid. Sched. 5, para. 3(11)(b).
[17] ibid. Sched. 5, para. 3(12).
[18] ibid. Sched. 5, para. 7(6).
[19] ibid. Sched. 5, para. 7(11).
[20] ibid. Sched. 5, para. 7(8).
[21] ibid. Sched. 5, para. 1(5).
[22] ibid. Sched. 5, para. 8.
[23] ibid. Sched. 5, para. 3(17).
[24] D.o.E. Circular 11/94, Annex 8, para. 8.59.

10. WASTE RECYCLING

"Recycling" of waste is its treatment by subjection to any process, 10.01
including making it re-usable or reclaiming substances from it.[1] Under
section 29(7) of the Environmental Protection Act 1990 the Secretary of
State may make regulations that set out what constitutes the treatment of
waste for these purposes.

One of the aims of the Framework Waste directive[2] is to encourage the 10.02
recycling of waste and the re-use of waste as raw materials. To this end it
divides waste streams into disposal operations set out in Annex IIA and
operations which may lead to recovery that are listed in Annex IIB. This
latter annex is intended to list recovery operations as they are carried out in
practice. They include solvent reclamation, recycling or reclamation of
metals or metal compounds, oil re-fining or other re-uses of oil and other
such operations.

Under Article 3 of the directive Member States should take appropriate 10.03
measures to encourage the recovery of waste by means of recycling, re-use or
reclamation or any other process with a view to extracting secondary raw
materials or the use of waste as a source of energy. However, like waste
disposal, such recovery operations must be carried out in a way that ensures
protection of the environment.[3] Thus any establishment or undertaking that
carries out the recovery operations listed in Annex IIB must have a permit to
do so.[4] However they may be exempted from the full rigours of the licensing
system under Article 11 of the directive. Under that article they must be
registered with the relevant competent authority, comply with the general
rules laid down by that authority for the particular recovery operation and be
subject to periodic inspection to ensure compliance.[5]

Waste collection authorities take the lead in recycling operations by being 10.04
required to produce waste recycling plans and being able to separate waste
for recycling from the waste disposal stream. Bottle banks, etc., may be
provided by a waste collection authority but where, as in London and the
other metropolitan areas, the collection authority is also a disposal authority

[1] E.P.A. 1990, s.29(6).
[2] Dir. 75/442 as amended by [1991] O.J. L78/32.
[3] Amended Waste Directive, Art. 4.
[4] *ibid.* Art. 10.
[5] *ibid.* Art. 13.

they must do so through a contractor.[6] Disposal authorities, in entering into contracts, should have regard to the desirability of including terms to maximize the recycling of waste under the contract.[7] A system of recycling credits is established under section 52 of the 1990 Act.

WASTE RECYCLING PLANS

10.05   Under section 49 of the Environmental Protection Act 1990 it is the duty of each waste collection authority to draw up plans for the recycling of household and commercial waste arising in their area. This provision implements the sub-paragraph of Article 7 of the Framework Waste directive that is concerned with such plans. These plans will deal with the preparatory work, such as separation and packaging, that has to be undertaken before waste can be passed on to a recycler. The national waste strategies also have as one of their objectives the encouragement of recycling.[8] Guidance on the preparation of such plans is contained in Waste Management Paper No. 28.

10.06   To begin with the collection authority should make an investigation into ways in which household and commercial waste can be sorted and packaged for recycling.[9] As they must include in the plan information as to the kinds and quantities of all controlled waste that they expect to collect or purchase during the period the plan covers[10] they will need to ascertain this, initially by a survey of present recycling activity.

10.07   Once this information is available the authority must decide what arrangements they need to make to facilitate recycling in their area.[11] They should then, having considered the effect on amenity and the likely costs or savings of their proposals, prepare the plan, setting out the arrangements they intend to make for the separation, baling or packaging of waste for recycling purposes.[12] The plan should state the kinds and quantities of waste that it is expected will be dealt with in this way, any arrangements that will be entered into with other parties for this purpose, the plant or equipment that will be provided and an estimate of the costs or savings attributable to dealing with the waste as provided for in it.[13]

10.08   Before the plan is finally determined the authority must send a copy of the draft to the Secretary of State to enable him to determine whether its contents reflect the requirements of section 49(3). If he gives directions for that purpose the authority must comply with them.[14] Apart from this there

---

[6] E.P.A. 1990, s.48(6) & (7).
[7] *ibid.* Sched. 2, para. 19(1).
[8] *ibid.*, Sched. 2A, para. 5 as added by E.A. 1995, Sched. 12.
[9] *ibid.* s.49(1)(a).
[10] *ibid.* s.49(3)(a) & (b).
[11] *ibid.* s.49(1)(b).
[12] *ibid.* s.49(1)(c) & (2).
[13] *ibid.* s.49(3)(c)–(f).
[14] *ibid.* s.49(4).

is no statutory requirement to consult anyone about the plan but it is considered that an adequate one could not be prepared without proper consultation.[15] Waste Management Paper No. 28 gives guidance on the consultation process.

The Secretary of State may give an authority directions, with which they **10.09** must comply, to complete a plan within a certain time.[16] Once the plan has been finalised it must be publicised locally and copies sent to the relevant waste disposal and regulation authorities.[17] A copy of it must be kept available for free public inspection at all reasonable times at the authority's principal offices. Copies of it should be supplied on request to anyone who pays the price asked, although this must be reasonable.[18]

The authority should update the information on which the plan is based **10.10** from time to time and, where they consider it necessary, modify the plan in the light of that information. Again they must take the effect on amenities and any likely costs or savings into account in making the modification.[19] This duty can be enforced by the Secretary of State under section 49(7). The same procedures set out in paragraphs 10.08 and 9 above must be followed for a modification as for a plan.

WASTE RECYCLING IN ENGLAND AND WALES

Waste collection authorities have, for recycling purposes, first refusal on **10.11** waste they collect. If they decide, having had regard to their recycling plan, that they will make arrangements for recycling some household and commercial waste they are not required to deliver it to the disposal authority in accordance with their duty under section 48(1).[20] As soon as practicable after they have taken that decision they should write to inform the disposal authority of the arrangements they intend to make.[21] If the disposal authority have already entered into a contract with a waste disposal contractor[22] for the recycling of all or some the waste they may serve a notice on the collection authority objecting to their decision. The notice may object to all of the waste being recycled or only some of it or certain types of it.[23] The notice means that the collection authority will have to deliver the waste specified in it to the disposal authority rather than keeping it for their own recycling operation.[24]

A collection authority may provide plant or equipment for sorting and **10.12** baling the waste they retain, but, if they are also a disposal authority, they may only do so through a contractor.[25] They may allow others to use their

---

[15] See *Hansard*, H.L. col. 1477, (June 26, 1990).
[16] E.P.A. 1990, s.49(7).
[17] *ibid.* s.49(5).
[18] *ibid.* s.49(6).
[19] *ibid.* s.49(1)(d) & (e) and (2).
[20] *ibid.* s.48(2).
[21] *ibid.* s.48(3).
[22] As defined *ibid.* s.30(5).
[23] *ibid.* s.48(4).
[24] *ibid.* s.48(5).
[25] *ibid.* s.48(6) & (7).

sorting and baling facilities or provide them for others to use. They should make a reasonable charge for the use of those facilities for the recycling of commercial, but not household, waste; unless they consider charges inappropriate. Anything delivered to them by someone who uses the facilities will belong to the authority and may be dealt with accordingly.[26] A waste collection authority may buy or otherwise acquire waste so as to recycle it and use or otherwise dispose of any of their waste, or anything produced from it, to another.[27]

10.13   A waste disposal authority may make arrangements with contractors for the recycling of waste it would otherwise be their duty to dispose of and for them to use waste to produce heat or electricity or both from it.[28] They may also buy or otherwise acquire waste for recycling purposes and use, sell or otherwise dispose of, despite their duty under section 51(1), waste belonging to them or anything derived from it.[29]

WASTE RECYCLING IN SCOTLAND

10.14   In Scotland waste disposal authorities double as collection authorities. They may provide, within or outside their areas, depots at which waste can be recycled and the plant or equipment to recycle it.[30] They may allow others to use those facilities and provide such facilities for others to use. Except when dealing with household waste, a charge should normally be made to people using these facilities unless the authority does not consider one appropriate. Anything delivered to the authority in the course of use of the facilities will belong to them and may be dealt with accordingly.[31]

10.15   In addition to the above powers a disposal authority may also do anything they consider appropriate to enable their waste, or waste of anyone who asks the authority to deal with it, to be recycled or used to produce heat or electricity or both. They may buy or otherwise acquire waste in order to recycle it and use, sell or otherwise dispose of their waste and anything produced from it.[32]

WASTE RECYCLING IN NORTHERN IRELAND

10.16   District councils have power under article 23 of the Pollution Control and Local Government Order (Northern Ireland) 1978[33] to recycle waste they receive or to enable substances to be reclaimed from it. They may also buy in waste for recycling or reclamation and sell the products of such operations. In addition, they may use waste for the production of heat and electricity and operate generating and other installations for this purpose.[34]

[26] E.P.A. 1990, s.48(8).
[27] ibid. s.55(3).
[28] ibid. s.55(2)(a) & (b).
[29] ibid. s.55(2)(b) & (c).
[30] ibid. s.53(1)(b).
[31] ibid. s.53(3).
[32] ibid. s.56(1).
[33] S.I. 1978 No. 1049 (N.I. 19).
[34] ibid. Art. 24.

FACILITIES FOR RECYCLING WASTE

Authorities in Great Britain may provide recycling facilities under the 10.17 powers set out in the last two sections. In addition waste collection authorities may stipulate that either receptacles they provide for household waste should have separate compartments for certain wastes or that separate receptacles should be used for different types of waste.[35] This power is not extended to receptacles provided by the authority for commercial waste as much of that waste is collected privately and confusion might arise through different systems operated.[36]

Ordinarily a facility for recycling waste will require a waste management 10.18 licence under section 35 of the Act as the waste is being kept or treated there or plant or equipment is provided for that purpose. However, certain exemptions from licensing are granted under Schedule 3 to the Waste Management Licensing Regulations 1994.[37] Thus the baling, sorting, etc., of waste paper, textiles and other materials for their recovery or re-use does not require a licence if the occupier of the land consents to it and the amounts dealt with in any week do not exceed the amounts set out in Table 3 to Schedule 3.[38] Similarly such materials can be stored in the amounts set out in Table 4 to the Schedule without a licence if they are going to be recovered or re-used.[39] Bottle banks and similar secure storage facilities of a capacity of less than 400 cubic metres will escape licensing provisions by virtue of paragraph 18 of Schedule 3.

RECYCLING CREDITS

Increased recycling of waste is expected to lead to savings through an 10.19 authority having less waste to collect or dispose of. Under section 52 of the Environmental Protection Act 1990 the authority making the saving will pay a credit to the person who has allowed them to make it, so providing an incentive for recycling initiatives. These provisions will not apply where the same authority is responsible for both waste collection and disposal.

A waste disposal authority must give a recycling credit to a collection 10.20 authority in their area that retains waste collected for recycling purposes. The credit, in respect of the retained waste, should represent the net saving that the disposal authority consider they would make from not having to dispose of it.[40] A similar credit may be made to others collecting waste for recycling in the disposal authority's area.[41] While at present this latter power is discretionary, the Secretary of State may make regulations requiring the

---

[35] S.I. 1978 No. 1049, s.46(2).
[36] See *Hansard*, H.L., col. 1154, (June 21, 1990).
[37] S.I. 1994 No. 1056, reg. 17.
[38] *ibid.* Sched. 3, para. 11 and see para. 16.34.
[39] *ibid.* Sched. 3, para. 17 and see para. 16.31.
[40] E.P.A. 1990, s.52(1).
[41] *ibid.* s.52(3).

payment of credits in specified circumstances.[42] A net saving for these purposes is the amount that the disposal authority, but for the waste being retained, would have incurred in having it disposed of, less any sum it has to pay to anyone in consequence of the waste being recycled rather than disposed of.[43]

10.21   A waste collection authority must give a recycling credit to their waste disposal authority where that authority has provided refuse amenity sites under section 51 of the Act. The credit should represent the net saving the collection authority consider they have made in not having to collect the waste delivered to the sites; based on the costs they would have incurred in collecting it.[44] However, it is not intended to bring this provision into force for the time being through fears that it might have adverse effects on the use of amenity sites. A similar credit may be made to others who collect waste for recycling that would otherwise have to be collected by the authority.[45]

10.22   While the credit system covers all controlled waste, the Department of the Environment considers that credits should only be paid in relation to household waste because charges can be made for the collection and disposal of other forms of waste.[46] Credits should be paid under section 52(1) and (3) on recycling activities that were operating on April 1, 1992, whether or not they were paid in the past.[47]

10.23   Payments are made under section 52(3) to the person collecting the waste for recycling. This person is the one who establishes the service or facility to collect the waste, not a member of the public using it.[48] However, before a payment is made the collection authority will have to be satisfied that the waste is being recycled after collection. It is therefore advised that credits should only be paid on production of a receipt from an "approved recycler." Approval would be given to those on a collection authority's list and it is recommended that they operate a system of registration for this purpose.[49]

10.24   The way in which the value of credits should be calculated for the purposes of section 52(1) and (3) is set out in the Environmental Protection (Waste Recycling Payments) Regulations 1992.[50] They are based on the net saving a disposal authority have made as a result of the recycling operation. The net saving is defined as an amount equal to half the expense they would have incurred in disposing of the waste at a cost per tonne equal to their average cost per tonne of disposing of similar waste at the relevant time, using their most expensive method. However, if it can be shown that the net saving was more than obtained on this basis the higher figure will be used.[51]

[42] E.P.A. 1990, s.52(5).
[43] ibid. s.52(6).
[44] ibid. s.52(2) & (7).
[45] ibid. s.52(4).
[46] D.o.E. Circular 4/92, Environmental Protection Act 1990; Environmental Protection (Waste Recycling Payments) Regulations, para. 16 (W.O. Circular 10/92).
[47] ibid. para. 24.
[48] ibid. paras 19 & 20.
[49] ibid. paras 40–47.
[50] S.I. 1992 No. 462.
[51] ibid. reg. 2(2).

In determining the average cost per tonne, the disposal authority must 10.25
take into account the market value of any of their assets used in connection
with the disposal of that waste, any costs incurred in operating the site or
transfer station, the transport costs, potential site closure and aftercare
expenses in relation to those sites and any other expenses incurred in relation
to that waste. However, they cannot take account of their expenses in
determining the amount of the credit.[52]

Alternatively, where there is insufficient information to make this 10.26
calculation, or it would be too expensive to do so, the representative table in
the Schedule can be used.[53] This gives an estimated figure of savings based
on the type of disposal authority concerned. Thus for an outer London
borough council the saving is estimated at £38.87 per tonne. The Schedule
was replaced by the Waste Management (Miscellaneous Provisions) Regulations 1997.[54]

PRODUCER RESPONSIBILITY

Sections 93 to 95 of the Environment Act 1995 introduce the concept of 10.27
producer responsibility to United Kingdom law. The principle behind this is
to make industry responsible for producing products that minimise waste
throughout their life by meeting specified targets. The concept stems from
the E.C. "priority waste streams" programme and the Packaging and
Packaging Waste Directive.[55]

The concept will be implemented by regulations made under section 93. 10.28
This allows the Secretary of State, for the purpose of promoting or securing
an increase in the re-use, recovery or recycling of products or materials, to
make regulations to impose producer responsibility obligations on such
persons, and in respect of such products or materials, as may be prescribed.[56]
For these purposes "producer responsibility obligation" means the steps
which are required to be taken under the regulations to secure the
attainment of the targets specified or described in them.[57]

Two types of regulation will introduce producer responsibility. The first 10.29
are those that give effect to E.C. or International obligations; such as the
Packaging and Packaging Waste Directive. This type of regulation will not
be subject to the consultation required by section 93(2) of the Act.[58] "Home
grown" regulations will have to be discussed with representatives of parties
whose interests appear to the Secretary of State to be, or are likely to be,
substantially affected by the proposed regulations.[59]

[52] S.I. 1992 No. 462, reg. 2(3) & (4).
[53] *ibid.* reg. 2(5).
[54] S.I. 1997 No. 351.
[55] Dir. 94/62; [1994] O.J. L365/10.
[56] E.A. 1995, s.93(1).
[57] *ibid.* s.93(8).
[58] *ibid.* s.93(3).
[59] *ibid.* s.93(2).

10.30   Before making "home grown" regulations the Secretary of State must be satisfied that the criteria set out in section 93(6) will be met by them. These are that they will be likely to result in an increase in the re-use, recovery or recycling of the products or materials in question, that this increase will produce environmental or economic benefits, that these benefits will be significant as against the costs imposed by them and that any burdens imposed on businesses are the minimum necessary to secure those benefits. In addition any burdens must be imposed on those most able to make a contribution to the achievement of the relevant targets, having regard to the desirability of acting fairly between the businesses involved (although the regulations may impose a burden on, for example, manufacturers as opposed to retailers) and to the need to ensure that the regulations are effective.[60]

10.31   The requirements of section 93(6) are modified for regulations made to comply with E.C. or international obligations. First, the Secretary of State need only have regard to those criteria as opposed to having to be satisfied that they are met in respect of domestic initiatives.[61] Secondly, he can, in making such regulations merely require that they sustain a minimum level of recycling.[62] Therefore in applying the section 93(6) criteria to them he can adapt it to the sustainment of such minimum levels.[63]

10.32   Any regulations will be made by statutory instrument which will usually have to be approved by a resolution of each House of Parliament.[64] Details of the contents of such regulations are set out in section 94(1). In particular the section provides that regulations may exempt members of a "registered exemption scheme" from the need to comply with the producer responsibility requirements of the regulations.[65] Such schemes may be subject to competition scrutiny to ensure fair competition is being achieved and that there is no abuse of market power. Further, operators acting under such a scheme may be required to act, or desist from acting, in a way that ensures compliance with an E.C. or international obligation.[66] Achievement of the requirements of the regulations will be shown by the issuing of certificates of compliance by persons approved for that purpose by the appropriate Agency. The regulations may also make the contravention of a prescribed provision an offence in accordance with section 95 of the Act.

PACKAGING AND PACKAGING WASTE

10.33   The E.C. Directive on Packaging and Packaging Waste[67] came into force on June 30, 1996. Its aim is to harmonise national measures on the management of packaging and packaging waste for both environmental ends

---

[60] E.A. 1995, s.93(6).
[61] ibid. s.93(5).
[62] ibid. s.93(4).
[63] ibid. s.93(5).
[64] ibid. s.93(9)–93(12).
[65] ibid. s.94(1)(j)–94(1)(o).
[66] ibid. ss.94(2) & (8)(b).
[67] Dir. 94/62 [1994] O.J. L365/10.

and to insure the proper functioning of the common market.[68] The Directive covers all packaging marketed in the Community and all packaging waste produced there.[69] The principles behind its measures are to prevent the production of packaging waste, to re-use packaging, to recycle or recover packaging waste in other ways and to reduce the final disposal of such waste.[70] While the Directive came into force on June 30, 1996, packaging manufactured before that date can be marketed until June 30, 2001.[71]

Prevention of packaging waste is required under Article 4 of the Directive. This is linked to the provisions of Article 9 which sets out essential requirements in relation to packaging, particularly those of Annex II. Annex II is concerned with requirements for the manufacturing and composition of packaging and its reusable or recoverable nature. Thus, for example, packaging must be designed, produced and commercialised in such a way as to permit its reuse or recovery, including recycling, and to minimise its impact on the environment when packaging waste or residues from packaging waste management operations are disposed of. In addition to these requirements for packaging, the heavy metal content present in packaging is regulated by Article 11. **10.34**

Targets for recovery and recycling of packaging must be set by Member States under Article 6. Thus by June 30, 2001 between 50 per cent as a minimum and 65 per cent as a maximum by weight of packaging waste must be recovered; while, within that time, a minimum of 25 per cent by weight of the totality of packaging materials contained within packaging waste must be recycled, with at least 15 per cent recycling being achieved for each packaging material. To this end return, collection and recovery systems should be established in accordance with Article 7. **10.35**

The Commission will, by June 1997, have established, based on the provisions of Annex I to the Directive, the numbering and abbreviations for an identification system for packaging. This system must be approved by the Council by June 29, 1998. When this marking system is in place, all packages must show the relevant mark clearly and visibly.[72] The Commission will also prepare and promote European standards relating to packaging, for example for recycling methods.[73] Member States must also establish information systems on packaging and packaging waste under Article 12. **10.36**

By June 1998 Member States must also establish systems to enable consumers to obtain information on the return, collection and recovery systems available to them and their role in contributing to the reuse, recovery and recycling of packaging and packaging waste. They should also be able to find out the meanings of the markings on packages and how a Member State is implementing the Directive through its waste management plan.[74] **10.37**

---

[68] Dir. 94/62 [1994] O.J. L365/10, Art. 1.1
[69] *ibid.* Art. 2.1
[70] *ibid.* Art. 1.2.
[71] *ibid.* Art. 22.5
[72] *ibid.* Art. 8.
[73] *ibid.* Art. 10.
[74] *ibid.* Art 13.

10.38   The Directive is implemented in Great Britain by the Producer Responsibility Obligations (Packaging Waste) Regulations 1997.[75] Guidance on compliance with the Regulations has been issued by the Department of the Environment.

OTHER RECYCLING MEASURES

10.39   The E.C. Directive on batteries and accumulators containing certain dangerous substances[76] came into effect on September 18, 1992. Its aim is to recover and/or ensure the controlled disposal of spent batteries and accumulators containing lead, mercury and cadmium. The Directive is implemented in Great Britain by the Batteries and Accumulators (Containing Dangerous Substances) Regulations 1994.[77]

[75] S.I. 1997 No. 648.
[76] Dir. 91/157 [1991] O.J. L78/38.
[77] S.I. 1994 No. 232.

11. SPECIAL WASTE

Some wastes need particular attention because of their toxic or other 11.01
dangerous properties. In the United Kingdom these are "special wastes" but
for the E.C. and the Basel Convention they are "hazardous" wastes. The
purpose of the extra controls imposed on the handling of these wastes is to
ensure that they are disposed of properly and that, if necessary, they can be
retrieved from landfill sites.

THE HAZARDOUS WASTE DIRECTIVE

The E.C. Council adopted a new Directive on hazardous waste on 11.02
December 12, 1991.[1] This replaced the 1978 Directive on toxic and
dangerous waste[2] from June 27, 1995.[3] The object of the new Directive is to
approximate the laws of the Member States on the controlled management
of hazardous waste. It was drawn up pursuant to Article 2.2 of the
Framework Waste Directive[4] which enables the Council to adopt individual
Directives on the management of particular categories of waste. Thus much
of the new Directive is taken up with applying rules contained in the Waste
Directive to the Management of Hazardous Waste.

The classification of hazardous wastes is closely tied to controls on the 11.03
packaging and labelling of dangerous goods. Thus a note to Annex III of the
Hazardous Waste Directive states that attribution of the hazard properties
"toxic" (and "very toxic") "harmful", "corrosive" and "irritant" should be
made on the basis of the criteria contained in the E.C. Directive on the
approximation of laws, regulations and administrative provisions relating to
the classification, packaging and labelling of dangerous goods, as amended.[5]
This will also apply, with reference to the to the latest information, to the
properties "carcinogenic", "teratrogenic" and "mutagenic." While some of
these properties, such as "harmful", are not defined in the same way in both
Directives this may not be important.

The definition of "waste" will be the same as in the waste directive.[6] 11.04
However, under the new Directive a list should have been drawn up by June

[1] Dir. 91/689 [1991] O.J. L377/20.
[2] Dir. 78/319 [1978] O.J. L84/43.
[3] Dir. 91/689, Art. 11 as amended Dir. 94/31.
[4] Dir. 75/442 amended by Dir. 91/156.
[5] Dir. 67/548 and see in particular Dir. 92/32 [1992] O.J. L154/1.
[6] *ibid.* Art. 1.3.

12, 1993 at the latest to set out what wastes are hazardous. The criteria for drawing up the list are prescribed in the Annexes to the Directive. The key is Annex III which lists properties which make wastes hazardous. These range from "explosive" to "flammable" and include "ecotoxic"—substances and preparations which present or may present immediate or delayed risks for one or more sectors of the environment—and "harmful" substances and preparations which, if they are inhaled or ingested or if they penetrate the skin, may involve limited health risks.

11.05   Only wastes listed in Annex IA such as inks, paints, or photographic chemicals, or those listed in Annex IB, such as ashes or cinders that contain any of the constituents in Annex II such as cadmium or cadmium compounds will be hazardous wastes and then only if they have one or more of the properties in Annex III. The list specifies those wastes to which the definition applies. The list must be periodically reviewed[7] while the Annexes themselves may be revised under article 9 of the Directive. The list was established on December 31, 1994 through Council Decision 94/904/EEC.[8]

11.06   If a Member State wants to classify a waste that is not on the list as hazardous it must notify the Commission accordingly. The Commission will then review the matter through the committee set up under Article 18 of the Framework Waste Directive and if it considers it necessary will amend the list to add the proposed waste.

11.07   Domestic waste does not come within the terms of the new Directive. Instead the Council should have established special rules taking into consideration the particular nature of domestic waste by the end of 1992.[9] So far it has not yet done so.

11.08   Member States must take necessary measures to ensure that those who dispose of, recover, collect or transport hazardous waste do not mix different categories of it or mix hazardous with non-hazardous wastes.[10] However, mixing is allowed to improve safety during disposal or recovery where this can be done without risk to the environment as long as there is a permit under Articles 9, 10 or 11 of the Framework Waste Directive.[11] "Where waste is already mixed with other waste . . ."—which presumably means where *hazardous* waste is already— . . . "it should be separated where this is technically and economically feasible and where necessary to protect the environment."[12]

11.09   In general, disposal to landfill will require a permit under Article 9 of the Framework Waste Directive. Where hazardous waste is involved the waste must be recorded and identified on site.[13] It is not clear whether this means that the final resting place of the waste must be recorded and identified or simply the fact that certain waste is on the site.

---

[7] Dir. 91/689 Art. 1.4.
[8] [1994] O.J. L356/14.
[9] Dir. 91/689, Art. 1.5.
[10] *ibid.* Art. 2.2.
[11] *ibid.* Art. 2.3.
[12] *ibid.* Art. 2.4.
[13] *ibid.* Art. 2.1.

Article 11 of the Framework Waste Directive provides exemptions to the 11.10 strict permit regime. Under Article 11.1(a) undertakings or establishments carrying out their own waste disposal at the place of production can merely be registered with the relevant authority. This exemption will not apply if they dispose of hazardous wastes.[14] Article 11.1(b) allows registration of those carrying out waste recovery. This exemption continues for the recovery of hazardous wastes but subject to stricter rules for registration. In particular specific conditions will have to be imposed in general rules covering such registration as to the amount of hazardous substances in the waste, etc.[15] These rules must be sent to the Commission at least three months before they are due to take effect and agreed by it in accordance with Article 18 of the Waste Directive.[16]

The requirements in the Framework Waste Directive as to the inspection 11.11 of waste management facilities and the keeping of records also apply to producers of hazardous wastes and record keeping requirements are further extended to those transporting them.[17] Records must be kept for at least three years except in the case of transporters who need only retain them for twelve months. Documentary evidence that management operations such as the collection, transport, recovery or disposal of waste[18] have been carried out must be supplied at the request of the relevant authority or of a previous holder of the hazardous waste concerned.[19]

Member States must ensure that, in the course of collection, transport and 11.12 temporary storage, waste is properly packaged and labelled in accordance with the international standards and Community standards in force.[20] Inspections of collection and transport operations should concentrate particularly on the origin of any hazardous waste and on its destination.[21] Where hazardous waste is transferred it should be accompanied by a note identifying it as prescribed by Section A of Annex I to the transfrontier shipment of hazardous waste directive (84/631/EEC as amended).[22]

The relevant authorities in the Member States should either include 11.13 details of hazardous waste arisings and recovery or disposal in their general waste management plans or should draw up separate plans for this purpose. These plans should be available to the public. The Commission will make a comparative study of these plans and in particular the methods of disposal or recovery set out in them. Relevant authorities in the Member States can obtain the study from the Commission.[23]

In cases of emergency or grave danger, Member States may take all 11.14 necessary steps to ensure that hazardous waste is so dealt with as not to

---

[14] Dir. 91/689, Art. 3.1.
[15] ibid. Art. 3.2 & 3.3.
[16] ibid. Art. 3.4.
[17] ibid. Art. 4.1 & 4.2.
[18] Dir. 75/442, Art. 1(d) as amended by Dir. 91/156.
[19] Dir. 91/689, Art. 4.3.
[20] ibid. Art. 5.1.
[21] ibid. Art. 5.2.
[22] ibid. Art. 5.3.
[23] ibid. Art. 6.

become a threat to the population or the environment. This may necessitate temporary derogations from the terms of the Directive. The Member State must inform the Commission of any such derogation.[24]

11.15   Reports to the Commission on the implementation of the Directive are provided for in Article 8. This requires a report every three years based on a questionnaire drawn up by the Commission. Further, by December 12, 1994 Member States will have to send the Commission a list of every establishment or undertaking within their jurisdiction that disposes of or recovers hazardous waste either wholly or mainly on behalf of others. The list should include the name and address of each organisation, the method it uses to treat waste and the types and quantities of waste that can be treated. This information should be updated annually. Competent authorities in each Member State will be able to obtain a copy of this list from the Commission.

UNITED KINGDOM CONTROLS

SPECIAL WASTE

11.16   The Hazardous Waste Directive is implemented in Great Britain by the Special Waste Regulations 1996.[25] These entered into force on September 1, 1996. The Regulations are mainly concerned with controls on shipments of waste within Great Britain, or those imported or exported from and to Gibraltar and Northern Ireland. In addition they set out rules for deposits of special waste at landfills and restrict the mixing of special waste. The system set up under the Regulations is explained in D.o.E. Circular 6/96.[26]

11.17   Special wastes are dealt with by section 62[27] of the Environmental Protection Act 1990. This provides powers for the Secretary of State to make regulations concerning special waste, and, in particular, to enable records made under the regulations to be kept in the public register under section 64 of the Act and for waste that is consigned in breach of the regulations to be redelivered to the consignee. In addition the Agencies will be able to recover the costs of operating the consignment note system from those involved in it.

11.18   Section 62 also provides that regulations may be made to ensure that special waste is kept on a site in prescribed quantities and conditions while awaiting treatment or disposal[28]; this matter will usually be dealt with in the conditions of a waste management licence, but regulations may be made to ensure a uniform approach across the country. Finally, provision is made in section 62 for appeals against decisions of the Agencies, although the regulations do not contain powers for the Agency to make decisions that could be the subject of any appeal.

[24] Dir. 91/689, Art. 7.
[25] S.I. 1996 No. 972 as amended by S.I. 1996 No. 2019.
[26] W.O. Circular 21/96; S.O. Circular 13/96.
[27] As amended by E.A. 1995, Sched. 22, para. 80 & S.I. 1996 No. 972, Sched. 3.
[28] E.P.A. 1990, s.62(2)(b).

THE DEFINITION OF SPECIAL WASTE

To see if waste is "special", the consignor must know its chemical 11.19
components and the risks associated with it. Under the Chemicals (Hazard
Information and Packaging for Supply) Regulations 1994[29] (known as
CHIP2) an Approved Guide to Classification and Labelling (AG) is
established and, under regulation 4, an Approved Supply List (3rd edition)
(ASL) is issued by the Health and Safety Commission. These set out a
classification system for substances and preparations that are considered
"dangerous for supply". This classification system is also used under the
Special Waste Regulations 1996.[30]

The first route to a waste being "special" is if it is in a category that is 11.20
assigned a six digit code in Part I of Schedule 2 to the 1996 Regulations,[31]
which is reproduced at the end of this chapter. In the Schedule oil wastes
other than certain edible oils are assigned the code number 13, so they are
not special wastes. Bilge oils have the code number 1304 so they are not
special. But bilge oils from inland navigation are assigned the code 130401
so that they may be special wastes. Wastes on the list in Part I of Schedule 2
can usefully be compared with the European Waste Catalogue to see if the
descriptions match. If the EWC gives a more detailed description then the
material may not be special waste.[32]

Being on the list is not enough in itself to make the waste special. It must 11.21
also have one of the hazardous properties specified in Part II of Schedule 2.[33]
Part II, reproduced at the end of this chapter, sets out a list of fifteen
hazardous properties, numbers H1 to H14 (with an H3A and an H3B.)
Thus, for asbestos the hazardous property would be H7—"carcinogenic".
Usually being on the list and having a defined hazardous property will make
waste special. But if the property is one of H4 to H8 then, to be special, it
must be at or over the threshold set out in Part III of Schedule 2.[34] Thus,
while a waste may contain a substance that is "harmful" the total proportion
by weight of that substance in the waste must be equal to or greater than 25
per cent for it to be special waste.[35]

Alternatively wastes that used to be special under the Control of Pollution 11.22
(Special Wastes) Regulations 1980[36] but are not on the list will remain
special by virtue of regulation 2(2). These wastes are prescription only
medicines,[37] and any controlled wastes that are highly flammable liquid
substances and preparations with a flash point below 21 degrees Celsius or
such wastes that fall into hazard categories H4 to H8 and are at or above
the thresholds set out in Part III of Schedule 2.[38]

[29] S.I. 1994 No. 3247 as amended by S.I. 1996 No. 1092.
[30] See D.o.E. Circular 6/96, Annex B3.
[31] S.I. 1996 No. 972, reg. 2(1)(a) as amended.
[32] See D.o.E. Circular 6/96, Annex B, para. 6 and Annex B1.
[33] S.I. 1996 No. 972, reg. 2(1)(b) as amended.
[34] *ibid.* reg. 2(3).
[35] See *ibid.* Sched. 2, Part IV, para. 3, as added by S.I. 1996 No. 2019.
[36] S.I. 1980 No. 1709 as amended.
[37] S.I. 1996 No. 972, reg. 2(2)(b) as amended and see Circular 6/96, Annex B, para. 19.
[38] *ibid.* reg. 2(2)(a) & 2(3) as amended.

11.23   Wastes may be inherently hazardous but pose no risks to man or the environment. Risks therefore have also to be considered. The potential risks are set out in what are known as "risk phrases"—for example R1, explosive when dry; R22, harmful if swallowed and R57, toxic to bees. Not all these risk phrases are definites—for example R45, may cause cancer or R60, may impair fertility. A list of risk phrases forms Annex B6 of D.o.E. Circular 6/96. Waste will be deemed to have the hazardous properties if it is so classified in the ASL of CHIP2, if it is not in that list it should be classified using the test methods set out in Annex V of E.C. Directive 67/548/EEC as amended.[39] References in Part III of the Schedule to a substance being categorised as having hazardous properties, a risk phrase or being placed in a particular category of classification is a reference to its listing either in Part V of the ASL or in criteria laid down in the AG. Guidance on use of the ASL for these purposes is set out in Annex B5 of D.o.E. Circular 6/96.

11.24   The producer of special waste will need to know its source, either by process or by industry, its main components and their physio-chemical properties. Some wastes may develop special characteristics through the use to which they have been put. A main source of information will be the Safety Data Sheets that a supplier of dangerous substances or preparations must compile under CHIP2. Guides to assessing whether waste has any one or more of the properties H1 to H14 are set out in Annex B to Circular 6/96. Annex B2 indicates the sort of substances that may give rise to a waste being special if any one of them is present in the waste.

11.25   Household waste is not special waste.[40] However, some wastes emanating from domestic use, namely mineral or synthetic oil or grease, asbestos and clinical waste are excluded from the definition of "household waste" by Regulation 3 of the Controlled Waste Regulations 1992.[41] Such wastes therefore may be special. In addition the 1996 Regulations state that asbestos, waste from a laboratory and waste from a hospital other than that from a self contained part of one that is used wholly for the purposes of living accommodation will not be "household waste" for these purposes.[42]

11.26   Radioactive wastes are not controlled wastes. However, the Secretary of State can make provision for waste which, were it not radioactive, would be controlled waste so that its "special" characteristics can be dealt with. Any radioactive waste that has "special" characteristics within the meaning of regulations 2(1) or 2(2) of the 1996 Regulations will therefore be "special waste".[43]

CONSIGNMENTS OF SPECIAL WASTE

11.27   The Regulations impose duties on the consignor of special waste. This is the person who, in relation to a particular consignment, causes that waste to be removed from the premises at which it is being held.[44] The consignor will

[39] S.I. 1996 No. 972, Sched. 2, Part IV, para. 1.
[40] *ibid.* reg. 2(1), as amended by S.I. 1996 No. 2019.
[41] S.I. 1992 No. 588.
[42] S.I. 1996 No. 972, reg. 1(4) — definition added by S.I. 1996 No. 2019.
[43] S.I. 1996 No. 972, reg. 3.
[44] S.I. 1996 No. 972, reg. 1(4).

usually be the person in a factory or other premises who arranges for the waste to be removed and his duties can only be delegated to an agent such as a waste disposal contractor to a certain extent.[45] While no definition is given of a "consignment", it is considered that it should be regarded as those wastes carried on one vehicle that are transported to the same consignee.[46] The "consignee" is the person to whom the waste is transported.[47]

Consignments of special waste must usually be accompanied by a **11.28** consignment note once they have left the premises at which the waste is being held.[48] A "consignment note" means a note in the form set out in Schedule 1 to the 1996 Regulations, or substantially similar to it, giving the details required by the Regulations in respect of the consignment.[49] The Agency will produce these notes and must supply them if required.[50] Annex C to D.o.E. Circular 6/96 sets out guidance that should be printed with the notes; although this is not a requirement of the Regulations. The consignment note system is intended to notify the Agency of the movement of waste in advance so that it can object to it if necessary. It also provides a record of what has happened to a particular consignment, enabling remedial action to be taken if required. Finally, it ensures that such waste is dealt with properly rather than being dumped on an unlicensed site.

Where a consignor has a shipment of special waste to send he must first **11.29** ensure that there is a code assigned to it by the Agency. The code will be assigned by the office of the Agency that is concerned with the place where the shipment is to be taken. The code, which will be unique to the consignment, can consist of letters, numbers or symbols and may, in the future, be a bar code to allow electronic identification.[51] Codes may be put on consignment notes by the Agency before they supply them, but otherwise the consignor will have to obtain a code for each consignment.[52] Fees are payable when a code is assigned. If no fee is paid the Agency can wait for payment before assigning a code or require the person who requested the code to pay the fee within two months of the date he requested it.[53]

A consignment note (the note) has five parts, A, B, C, D and E. Before **11.30** the consignment is moved five copies of the note must be prepared. This will usually be done by way of a five-copy self-carbonating pad with different coloured sheets for each copy.[54] Where the Agency must be notified before the shipment is made—which is the normal procedure—Parts A and B should be completed on each copy and the relevant code entered.[55] Part A gives details of the consignment and Part B is a description of the waste.

[45] See Circular 6/96, para. 33.
[46] *ibid.* Annex A, para. 35.
[47] S.I. 1996 No. 972, reg. 1(4).
[48] *ibid.* reg. 5(1).
[49] *ibid.* reg. 1(4).
[50] D.o.E. Circular 6.96, Annex A, para. 25.
[51] S.I. 1996 No. 972, reg. 4.
[52] D.o.E. Circular 6/96, Annex A, para. 31.
[53] S.I. 1996 No. 972, regs 4(3) and 14(3) as added by s.I. 1996 No. 2019.
[54] D.o.E. Circular 6/96, Annex A, para. 28.
[55] S.I. 1996 No. 972, reg. 5(2)(a).

The details in Part B will also be required for the duty of care under section 34 of the Environmental Protection Act 1990. The carrier or some other person may complete these Parts, but it is the consignor who will have to sign Part A and for the accuracy of Part B. In addition the consignor must ensure that one completed copy is sent to the appropriate Agency office at least three clear working days, but not more than one month, before the waste is removed.[56]

11.31 Part C on the remaining four copies is to be completed by the carrier.[57] If the information provided on Part A and B has changed, or is disputed, agreed amendments should be indicated on Part C.[58] The consignor should then complete Part D to certify that the information in Parts B and C is correct, that the carrier is a registered carrier under the 1989 Act or is exempt from registration and has been advised of relevant precautionary measures. He then keeps one completed copy and hands the remaining three to the carrier.[59] The carrier must ensure that his three copies travel with the consignment and are given to the consignee on delivery.[60] Unless the consignee rejects the consignment[61] he should complete Part E to certify the date of receipt and the amount of waste received and that he is authorised to deal with it. He then keeps one copy of the note himself, gives one to the carrier—which the carrier must retain—and sends one copy to the local office of the Agency.[62]

11.32 The Agency does not have to be notified before every shipment. Exceptions to the pre-notification requirement are contained in regulations 6, 8 and 9. The exemption under regulation 6 is mainly for repetitive consignments.[63] The first consignment in the series must follow the rules in regulation 5. After that the second and subsequent loads can follow the shortened procedure in regulation 7, which dispenses with the copy of the note that otherwise would have to be sent to the Agency in advance.[64] However, to take advantage of this exception in each consignment the waste must be of the same description, sent by the same consignee from the same premises to the same consignor at his same premises and be within the same 12 month period.[65]

11.33 Regulation 7 procedure can also be followed where off-specification products or materials are returned to the supplier or manufacturer.[66] In such a case the consignor (the person supplied with the product) must be satisfied that the product or material does not meet a standard he expected it to meet[67] and, if so, no fee will be payable for a code assigned to the

---

[56] S.I. 1996 No. 972, regs 5(2)(b) and 12(1) and see D.o.E. Circular 6/96, Annex A, para. 40 *et seq.*

[57] *ibid.* reg. 5(2)(c).

[58] D.o.E. Circular 6/96, Annex A, para. 37.

[59] S.I. 1996 No. 972, reg. 5(2)(d).

[60] *ibid.* reg. 5(3).

[61] See *ibid.* reg. 10.

[62] *ibid.* reg. 5(4) & 5(5).

[63] See D.o.E. Circular 6/96, Annex A, para. 52.

[64] S.I. 1996 No. 972, reg. 6(1)(a) & 7.

[65] *ibid.* reg. 6(2)(a).

[66] *ibid.* regs 6(1)(b) & 7(c).

[67] *ibid.* regs 6(2)(b) & 7(b)(i).

consignment.[68] Consignments that consist entirely of lead acid batteries do not require pre-notification, and attract a reduced fee of £10 for their code.[69] In addition where waste is moved for storage purposes between companies in the same group[70] regulation 7 procedure also applies, if the consignee either has a waste management licence or carries on an activity that is exempt from licensing.[71]

CARRIER'S ROUNDS

A "carrier's round" means a journey made by a carrier during which he **11.34** collects more than one consignment of special waste and delivers them all to the same consignee as specified on the consignment note.[72] Under regulation 8 a consignment note is prepared for the round itself rather than for individual consignments that make it up. The form of the note is adapted by Part II of Schedule I to the Regulations to allow each consignor to certify for his waste on the round and the carrier to certify that he collected it.

The special waste to be transported must be of the description specified in **11.35** the note.[73] This is considered by the government to mean that on the round several types of waste from a common source can be collected. However, it would not allow collections from a large number of premises carrying out unrelated types of business and producing unrelated wastes.[74] This view seems difficult to justify in the context of the regulations, given that Part B of the form allows for the description of a number of types of waste. This is provided for in regulation 8(2A)[75] which requires either separate entries for each description of waste collected from each consignor describing that waste, or, in the carrier's schedule, the different descriptions should be shown and the quantity of each that is to be collected.

The procedure set out in Regulation 8 applies to a round, or a succession **11.36** of rounds by the same carrier within a 12 month period. Each consignor and the premises from which the load originates must be specified in the note; although people and places may be added to the round as long as at least 72 hours notice is given to the Agency before the round starts.[76] All premises on the round must be subject to the same Agency, so that a round could not take waste from premises in both England and Scotland. Generally a round should not take more than 24 hours to complete.[77]

The carrier must ensure that the first load—unless it consists entirely of **11.37** lead acid motor vehicle batteries—is pre-notified in the usual way.[78] For

[68] S.I. 1996 No. 972, reg. 14(2)(b).
[69] *ibid.* regs 6(1)(e) & 14(1)(a).
[70] See *ibid.* reg. 6(3).
[71] *ibid.* reg. 6(1)(c) & 6(2)(c) and see D.o.E. Circular 6/96, Annex A, para. 66.
[72] *ibid.* reg. 1(4).
[73] *ibid.* reg. 8(1)(c).
[74] D.o.E. Circular 6/96, Annex A, para. 60.
[75] As added by S.I. 1996 No. 2019.
[76] S.I. 1996 No. 972, reg. 8(1)(a) & (b).
[77] *ibid.* reg. 8(1)(d).
[78] *ibid.* regs 8(2)(a) — as amended by S.I. 1996 No. 2019 — & 12(2).

subsequent loads he need only complete three main copies, and the schedule in Part II of Schedule 1, plus one copy for each of the consignors on the round.[79] A consignor must complete his box on the schedule before waste is removed from his site.[80] The carrier then completes his part of the schedule for that load, enters the time at which he completed it and ensures he has the schedule as completed by both parties.[81] The consignor keeps a copy of the note and a copy of his part of the completed schedule. Before the last collection he must complete Part C on his three copies of the note.[82] After the last pick up, the carrier must ensure that the completed note travels with the waste and is given to the consignee.[83] Unless he rejects the consignment, the consignee should complete his Part of the note on all three copies, keep one copy and its schedule, give one copy plus a schedule to the carrier—which the carrier must retain—and forthwith send the third copy and schedule to the office of the Agency for his area.[84]

11.38   Only one code is assigned to all rounds undertaken in the 12 month period. Thus only one £10 or £15 fee is payable.[85] Usually for subsequent rounds another fee would be required. However, no fee may be required for a second or subsequent round in which a single vehicle is used if the carrier is also the consignee for each load, no more than one consignment is collected from any consignor during the rounds, the total weight of special waste collected from each round is no more than 400kg. All such rounds must be completed within a week of the first one.[86]

REMOVAL OF WASTE FROM SHIPS

11.39   For the purposes of the Special Waste Regulations 1996, a "ship" means a vessel of any type whatsoever operating in the marine environment. It includes oil rigs and other structures which are fixed or floating platforms.[87] Where special waste is removed from a ship in a harbour area to either reception facilities in the harbour or via a pipeline to facilities outside the harbour, the procedure in regulation 9 applies. It is intended to apply to special waste originating in the ship, rather than imported consignments of waste; although this is not clear from the regulations.[88]

11.40   Under regulation 9 only three copies of a consignment note are required as there is no carrier involved and pre-notification is not required. The operator of the reception facility will take the lead here, the master being required to complete Part D on the copies and keeping one copy for himself.

[79] S.I. 1996 No. 972, reg. 8(2)(b) as amended by S.I. 1996 No. 2019.
[80] *ibid.* reg. 8(3).
[81] *ibid.* reg. 8(4) as amended by S.I. 1996 No. 2019.
[82] *ibid.* reg. 8(5A) as added by S.I. 1996 No. 2019.
[83] *ibid.* reg. 8(6) as amended by S.I. 1996 No. 2019.
[84] *ibid.* reg. 8(7) & (8).
[85] See *ibid.* reg. 14(1).
[86] *ibid.* reg. 14(2) as amended by S.I. 1996 No. 2019.
[87] *ibid.* reg. 1(4).
[88] D.o.E. Circular 6/96, Annex A, para. 67.

Otherwise the operator will probably fill in the rest of the note and obtain the code. He must complete Part E, keep one copy of the note and send the other copy forthwith to the local office of the Agency.[89] No fee is payable for the code assigned to such transactions.[90]

Where an imported consignment of special waste arrives by ship its 11.41 removal from the ship to a conveyance for transportation beyond the harbour need not be pre-notified to the Agency.[91] However, the procedure under regulation 6 will otherwise apply. No fee is payable for a code assigned for this type of consignment.[92]

SHIPMENTS FROM OR TO GIBRALTAR OR NORTHERN IRELAND

Special waste transferred between Great Britain and Gibraltar or 11.42 Northern Ireland must comply with the rules in regulation 13 where they are shipments of waste within a Member State for the purposes of the Transfrontier Shipment of Waste Directive.[93] This requires the usual procedure to be followed except that, for imports, the "consignor" is the person importing the load, the "premises" concerned are the place where the waste first enters Great Britain and it is deemed to be removed from that place at the time when it first enters the country. However, these procedures do not apply to shipments covered by regulations 6(1)(d) or 9.[94] For exports any reference to the consignee is to the person exporting the waste who is deemed to receive it at the place where and the time when it leaves the country.[95]

DUTIES IN RESPECT OF CONSIGNMENT NOTES

Where the Agency is to be pre-notified of a consignment, notification 11.43 should be given not more than a month and at least 72 hours before the start of the journey; although for carrier's rounds earlier notification is possible.[96] The Agency can be notified within 72 hours where a consignment is to be rejected by the consignee who intends to send it to an authorised waste management facility or it cannot lawfully remain where it is for 72 hours.[97] Copies can be sent by fax or E-mail if done within the time limits. If it is posted it should be sent by first class post five days before the load is removed.[98]

---

[89] S.I. 1996 No. 972, reg. 9(2) & (3) & see reg. 11(2).
[90] *ibid.* reg. 14(1)(c).
[91] *ibid.* reg. 6(1)(d) and see reg. 11(2).
[92] *ibid.* reg. 14(1)(c).
[93] *ibid.* reg. 13(4).
[94] *ibid.* reg. 13(1) & (3).
[95] *ibid.* reg. 13(2).
[96] *ibid.* reg. 12(1) & (2) and see 12(6) for calcualtion of time.
[97] *ibid.* reg. 12(3).
[98] *ibid.* reg. 12(4) & D.o.E. Circular 6/96, Annex A, para. 45.

11.44   Where the regulations require copies of notes or other documents to be furnished "forthwith" this requirement will be met if the relevant material is delivered to the Agency or posted first-class to it, within one day of the transaction concerned.[99]

11.45   The office of the Agency in the consignee's area is responsible for most matters concerning the shipment. However, where it receives documents that are required to be sent to it under the regulations, it should, within two weeks of receipt, send the office of the Agency for the consignor's area a copy.[1]

REJECTED CONSIGNMENTS

11.46   A consignee who rejects a load must follow the procedure in regulation 10 instead of that in regulations 5(4) or 8(7). This requires him to enter, on Part E of the consignment note, that he does not accept the consignment and setting out his reasons for not doing so.[2] He then keeps one copy of the note, ensures that one copy, and (if the load came from a carrier's round) the schedule supplied with it, is sent forthwith to his local Agency office and returns one copy to the carrier. If no copy of the note was given to him he must send a written explanation for his actions, including what he knows about the consignment and the carrier, to his local Agency office forthwith.[3]

11.47   On being informed that the consignee will not accept delivery, the carrier should inform the Agency and seek instructions from the consignor.[4] The consignor may propose either that the load be returned to the place where it was collected from or where the waste was produced or he can arrange for its delivery to specified premises that are authorised under waste management legislation to accept it.[5] Once he has come to a decision he should forthwith inform the Agency and the carrier of his intentions. It is the duty of the carrier to take all reasonable steps to ensure that his intentions are fulfilled.[6]

REGISTERS

11.48   Every consignor of special waste must keep a register at each site from which waste has been removed containing a copy of the consignment note, and, if applicable, the schedule that accompanies a carrier's round, in respect of each consignment. These documents must be kept for three years from the date on which the waste was removed.[7] Carriers are required to keep the same documents for the same period in respect of each consignment which they have transported.[8] Registers may be in any form as long as they comply with the requirements of the Regulations.[9]

---

[99] S.I. 1996 No. 972, reg. 12(5).
[1] *ibid.* reg. 11(1).
[2] *ibid.* reg. 10(3)(a).
[3] *ibid.* reg. 10(4).
[4] *ibid.* reg. 10(5)(a).
[5] *ibid.* reg. 10(6) and see D.o.E. Circular 6/96, Annex A, paras 77 & 78.
[6] *ibid.* reg. 10(5)(b) & (c).
[7] *ibid.* reg. 15(1) & (4).
[8] *ibid.* reg. 15(2) & (4).
[9] *ibid.* reg. 15(8).

Consignees must also keep a copy of notes and, where applicable, 11.49
schedules, in respect of each consignment they have received, unless they
have rejected a consignment under the provisions of regulation 10.[10] Their
documents must be kept until any waste management licence for the site is
surrendered or revoked entirely—or if the site is authorised under Part I of
the Environmental Protection Act 1990 that authorisation is surrendered or
revoked entirely. At that time the register should be sent to the Agency who
must keep it for at least three years.[11] If the activities conducted by the
consignee are exempt from licensing or authorisation each note or schedule
must be kept for at least three years from the date the waste was received at
that site.[12] The register may be kept in any form.[13] These records are not
available to the general public while with the consignor or consignee, but
once they have been handed in to the Agency they will be placed on the
relevant Agency's public register.[14]

HANDLERS OF SPECIAL WASTE

The Agencies are under a duty to carry out periodic inspections of any 11.50
establishment or undertaking which produces special waste.[15] The government considers that this can be integrated with other inspection visits by the
Agency.[16] Producers of waste are entitled to access to the records of those
who have subsequently handled it to be able to ensure that it has been
properly treated or disposed of.[17]

All those who handle special waste are subject to the duty of care imposed 11.51
by section 34 of the Environmental Protection Act 1990. However, the
consignment note for special waste, and the schedule in the case of carriers'
rounds, replace the transfer note required by regulation 2 of the Environmental Protection (Duty of Care) Regulations 1991.[18]

An establishment or undertaking which carries out the disposal or 11.52
recovery of special waste, or collects it or transports it, must not mix
different categories of special waste or mix it with waste that is not special.[19]
There is no definition of what is a "category" of special waste. The
government considers that this should be interpreted in its broadest sense.[20]
A category of waste could well be interpreted by the courts as waste
allocated a six figure number in Part I of Schedule 2 to the 1996
Regulations.

Mixing is allowed if this is authorised by a waste management licence or 11.53
authorisation under Part I of the 1990 Act. It may also be possible under an

[10] S.I. 1996 No. 972, reg. 15(3).
[11] *ibid.* reg. 15(5) & (6).
[12] *ibid.* reg. 15(7).
[13] *ibid.* reg. 15(8).
[14] S.I. 1994 No. 1056, reg. 10(1)(k) as amended by S.I. 1996 No. 972, Sched. 3.
[15] S.I. 1994 No. 1056, Sched. 4, para. 1391) as amended by S.I. 1996 No. 972, Sched. 3.
[16] D.o.E. Circular 6/96, Annex A, para. 87.
[17] S.I. 1994 No. 1056, Sched. 4, para. 14(1)(b) as amended by S.I. 1996 No. 972, Sched. 3.
[18] S.I. 1996 No. 972, reg. 23.
[19] *ibid.* reg. 17(1).
[20] D.o.E. Circular 6/96, Annex A, para. 89.

exemption granted by Regulation 17 of the Waste Management Licensing Regulations 1994.[21] Producers may mix special wastes but may need to consider the Code of Practice on the Duty of Care before doing so.[22] The restriction on mixing does not prevent carriers from picking up different wastes on a round as long as they are kept in separate containers.[23]

11.54　Where special waste is recovered or disposed of the establishment or undertaking doing so must keep a record of the operation and, in the case of disposal, of the aftercare of the disposal site.[24] Deposits of special waste must be recorded in accordance with regulation 17 of the 1996 Regulations. For these purposes waste will be "deposited" when it is dumped on the site with no realistic prospect of further examination or inspection.[25]

11.55　Any person who makes a deposit of special waste in or on any land must record the location of each such deposit.[26] The record should consist of either a site plan marked with a grid or a site plan with overlays on which deposits are shown in relation to the contours of the site.[27] They should be described by reference to the register of consignment notes the consignee is required to keep. If there is no such note because there has been a disposal by pipeline or by the producer at the place of production the deposit should be described by reference to a record of the quantity and composition of the waste and the date of its disposal.[28] Liquid wastes which are discharged into underground strata or disused workings should be recorded by a written statement of the quantity and composition of the waste and the date of discharge.[29]

11.56　These records must be kept until the depositor's waste management licence is surrendered or revoked. They should then be sent to the Agency for the site.[30] Records made by the depositor under regulation 14 of the Control of Pollution (Special Waste) Regulations 1980 should be kept with records under regulation 16 of the 1996 Regulations and accompany them when they are sent to the Agency.[31] Both sets of records will be put on the relevant Agency's public register.[32]

OFFENCES

11.57　It will be an offence for a person, other than an authorised member, officer or employee of the Agency, to fail to comply with any provision of the 1996 Regulations—and those record keeping requirements of paragraph

---

[21] S.I. 1996 No. 972, reg. 17(2).
[22] D.o.E. Circular 6/96, Annex A, para. 90.
[23] *ibid.* para. 91.
[24] S.I. 1994 No. 1056, Sched. 4, para. 14 (1A) as added by S.I. 1996 No. 972, Sched. 3.
[25] *Leigh Land Reclamation v. Walsall Borough Council* [1991] Crim.L.R. 298.
[26] S.I. 1996 No. 972, reg. 16(1).
[27] *ibid.* reg. 16(2).
[28] *ibid.* reg. 16(3).
[29] *ibid.* reg. 16(4).
[30] *ibid.* reg. 16(1).
[31] *ibid.* reg. 16(5).
[32] S.I. 1994 No. 1056, reg. 10(1)(k), as amended by S.I. 1996 No. 972, Sched. 3.

14 of Schedule 4 to the Waste Management Licensing Regulations 1994[33]—that impose any obligation or requirement on him.[34] The only defence available to a person charged with such an offence is that he was unable to comply with the Regulations by reason of an emergency or grave danger and that he took all steps as were reasonably practicable in the circumstances to minimise the hazards posed and to ensure that the relevant provision was complied with as soon as reasonably practicable after the event.[35] There is no "due diligence" defence to such a charge.

It will also be an offence to furnish false information in purported **11.58** compliance with the Regulations—or paragraph 14 of Schedule 4 to the 1994 Regulations—or to make a statement that is knowingly or recklessly false or misleading in a material particular.[36] In addition a person who intentionally makes a false entry in any record or register that is required to be kept by the Regulations—or paragraph 14—also commits an offence.[37]

Where such an offence has been committed by a body corporate directors, **11.59** managers or similar officers of that body—or if the body is managed by members, those members—may also be liable for it if it was committed with their consent or connivance or through their neglect. In Scotland provision is made for the liability of partners and officers of unincorporated associations.[38]

Any person who commits an offence under these provisions will be liable, **11.60** on summary conviction, to a fine not exceeding level 5 on the standard scale or, on conviction on indictment, to a fine or to imprisonment for up to two years or to both.[39]

CARRIAGE OF SPECIAL WASTES

Special wastes will be classified as dangerous for supply under regulation 5 **11.61** of the Chemicals (Hazard Information and Packaging) Regulations 1994[40] or Chemicals (Hazard Information and Packaging for Supply) Regulations (Northern Ireland) 1993)[41] Substances so classified may only be supplied in packages in accordance with the provisions of the Regulations. Substances dangerous for supply are listed in an Approved Supply List under the regulations. Substances in the list are named and given an identification number and Safety Data Sheets must be prepared in respect of them.

A different regime exists for transporting packages of dangerous goods. **11.62** This is done under the supervision of the Health and Safety Executive who

---

[33] S.I. 1994 No. 1056 Sched. 4, para. 14(4) as added by S.I. 1996 No. 972, Sched. 3.
[34] S.I. 1996 No. 972, reg. 18(1).
[35] *ibid.* reg. 18(2) & S.I. 1994 No. 1056, Sched. 4, para. 14(5).
[36] *ibid.* reg. 18(3); *ibid.* 14(6).
[37] *ibid.* reg. 18(4); *ibid.* para. 14(7).
[38] *ibid.* reg. 18(6)–(8); *ibid.* para. 14(8).
[39] *ibid.* reg. 18(9); *ibid.* para. 14(8).
[40] S.I. 1994 No. 3247 as amended by S.I. 1996 No. 1092.
[41] S.R. 1995 No. 60.

are responsible for enforcing the Carriage of Dangerous Goods (Classification, Packaging and Labelling) and Use of Transportable Pressure Receptacles Regulations 1996 (the CDGCPL Regulations).[42] The Regulations apply to the carriage of any dangerous goods subject to the exceptions in regulation 3. These exceptions include the carriage of explosives which are dealt with in the Packaging of Explosives for Carriage Regulations 1991.[43]

11.63 "Dangerous goods" are defined in regulation 2(1) as goods which have one or more of the hazardous properties set out in column 2 of Part I of Schedule 1. The list of such properties is different from that in the Special Waste Regulations 1996 but is likely to encompass all special wastes that present a risk to the health and safety of people. Goods that are hazardous to the environment are classified under "Miscellaneous dangerous goods." It is the duty of the consignor of the waste to ascertain whether these 1996 Regulations apply to it.[44] Regulations to deal with environmentally harmful loads can be made under the Health and Safety at Work, etc., Act 1974 (Application to Environmentally Hazardous Substances) Regulations 1996.[45]

11.64 Wastes must be contained in a package suitable for their transport. In particular the receptacle in which they are contained, and any associated packaging, should be designed, constructed, maintained, filled and closed in a way that prevents any of its contents from escaping when subjected to the stresses and strains of normal handling and conditions encountered in carriage; although a suitable safety device may be fitted. In addition the receptacle and associated packagings should not be adversely affected by the waste or be liable to combine with it to cause a health and safety risk. If the receptacle has a replaceable closure the closure should be designed to ensure that it can be used without the contents escaping.[46] Special conditions for packaging of the goods set out in the approved carriage list must also be complied with. Packagings, unless exempted under regulation 6(3), should have been properly tested and given an ADR mark, an RID mark, a UN mark or a joint ADR and RID mark by a competent authority and should bear that mark.[47]

11.65 The labelling of packages for carriage is dealt with by regulations 7 and 8 of the Regulations. The consignor is responsible for ensuring that these labelling requirements, and any set out in the approved carriage list for the relevant waste, are complied with. The label should show the designation of the goods, the UN number, the danger sign for the goods as shown in column 6 of Part I of Schedule 1 to the regulations, and the subsidiary hazard sign, if any.[48] The consignor should also ensure compliance with the requirements as to the methods of marking or labelling packages set out in regulation 11.

11.66 The Regulations are enforced under the provisions of the Health and Safety at Work Act 1974 as if they were a health and safety regulation made

---

[42] S.I. 1996 No. 2092. For N.I. see S.R. 1995 No. 47.
[43] S.I. 1991 No. 2097.
[44] S.I. 1996 No. 2092, reg. 5(1).
[45] S.I. 1996 No. 2075.
[46] S.I. 1996 No. 2092, reg. 6(1)(a)–(c).
[47] *ibid.* reg. 6(1)(d)–(e).
[48] *ibid.* reg. 8(2).

under section 15 of that Act. Thus contravention of them, or of any requirement or prohibition imposed under them, will be an offence under section 33(1)(c) of the Act. Anyone charged with such an offence will be liable on summary conviction to a fine not exceeding level 5 on the standard scale or on conviction on indictment to imprisonment for up to two years or a fine or both.[49] If it is alleged that the offence was committed through the act or default of someone else the provisions of section 36 of the Act will apply. Further, under regulation 15 of the Regulations it will be a defence for the accused to show that he took all reasonable precautions and exercised all due diligence to avoid the commission of the offence. In Northern Ireland the 1995 Regulations will be enforced under the Health and Safety at Work Order (N.I.) 1978.[50]

The carriage of special wastes on a road will be governed by the Carriage **11.67** of Dangerous Goods by Road Regulations 1996[51] or the Road Traffic (Carriage of Dangerous Substances in Packages, etc.) Regulations (Northern Ireland) 1988.[52] The Regulations will apply to the carriage of any dangerous goods, subject to the exception in Schedule 2.[53]

Part III of the 1996 Regulations sets out provisions as to the construction **11.68** of vehicles, tanks and freight containers used for carrying dangerous substances. Before such a substance is carried the vehicle operator should have obtained sufficient information about it from the consignor or his agent to enable the operator to fulfil his duties under the Regulations. The operator in turn must provide the driver with written information about the nature of the substance and the hazards it creates and the action to be taken in an emergency. The driver must ensure that this information is kept on the vehicle, is readily available while the substance is being carried and that it does not become confused with information that is irrelevant to the load he is carrying.[54]

Drivers must be adequately trained so that they understand the nature of **11.69** the hazards created by the substance being carried, the action to be taken in an emergency and their duties under the Regulations. Drivers must also have a vocational training certificate, unless exempt, and produce it to a police officer on demand. Records of the training of drivers must be kept by the operator. These requirements are set out in the Carriage of Dangerous Goods by Roads (Driver Training) Regulations 1996.[55] In addition the operator and anyone involved in the carriage of a dangerous substance must take all such steps as it is reasonable for someone in their position to take to ensure that nothing in the loading, stowage or unloading of the substance is liable to create a hazard to the health and safety of anyone.[56] Precautions

[49] H. & S. Act 1974, ss.33(3) & (4)(a) and 56(1).
[50] S.I. 1978 No. 1039 (N.I. 9)
[51] S.I. 1996 No. 2095.
[52] S.R. 1988 No. 415 as amended by S.R. 1990 No. 32.
[53] S.I. 1996 No. 2095, reg. 3.
[54] *ibid.* regs 13–15.
[55] S.I. 1996 No. 2094 [N.I. S.R. 1992 No. 262 as amended S.R. 1993 No. 240].
[56] S.I. 1996 No. 2095, reg. 19(1).

must be taken against the risk of fire or explosion.[57] and vehicles carrying dangerous goods must be parked in a safe place and supervised at all times.[58]

11.70 The 1996 Regulations are concerned with the marking of vehicles carrying dangerous substances. A container, tank or vehicle that is being used for the carriage of dangerous goods to which the Regulations apply, must be marked in accordance with regulation 17. The marking must be displayed at all relevant times and be kept clean, visible and free from obstruction.

11.71 These Regulations were made under section 15 of the Health and Safety at Work Act 1974 or the 1978 (N.I.) Order and thus are enforceable in the same way as the Classification, Packaging and Labelling Regulations set out above.[59] By virtue of regulation 17 it will be a defence in any proceedings under the 1992 Regulations for a defendant to show that he took all reasonable precautions and exercised all due diligence to avoid the offence being committed.

11.72 As far as rail transport is concerned the Convention concerning Carriage by Rail 1980,[60] the Annexes to it and official amendments to it has the force of law in the United Kingdom by virtue of section 1 of the International Transport Conventions Act 1983. The Annex to the Convention sets out regulations concerning the transport of dangerous substances. Dangerous Goods, as defined in regulation 2(1) of the CDGCPL Regulations, must be carried in accordance with the Carriage of Dangerous Goods by Rail Regulations 1996.[61] Railway operators will have their own conditions and regulations dealing with the carriage of dangerous substances on the railway. It will be an offence under Byelaw 19 of the British Railways Board Byelaws to carry special waste on the railway unless those conditions and regulations are complied with. In addition it is an offence under section 105 of the Railways Clauses Consolidation Act 1845 to send dangerous goods on a train without the package being marked to show their nature.

11.73 The Air Navigation (Dangerous Goods) Regulations 1994[62] set out the provisions under which dangerous goods, as defined in the "Technical Instructions", may be carried by air. In essence the written permission of the Civil Aviation Authority is required before goods may be taken and any conditions the Authority impose must be complied with.[63] It will be an offence to contravene, or permit the contravention of or fail to comply with any of the provisions of the Order by virtue Article 52(2) of the Air Navigation Order 1995.[64] Anyone found guilty of such an offence will be liable on summary conviction to a fine not exceeding £2,000 or on indictment to imprisonment for up to two years or a fine or both.[65]

---

[57] S.I. 1996 No. 2095, reg. 23.
[58] *ibid.* reg. 24.
[59] See para. 11.66.
[60] Cm. 8535.
[61] S.I. 1996 No. 2089.
[62] S.I. 1994 No. 3187, as amended by S.I. 1996 No. 3100.
[63] *ibid.* reg. 4(1).
[64] S.I. 1955 No. 1038.
[65] *ibid.* Art. 111(6) & Sched. 12 Part B.

When dangerous substances are to be carried through the area of a 11.74
harbour the provisions of the Dangerous Substances in Harbour Areas
Regulations 1987[66] must be complied with. Under regulation 6 the harbour
master and the berth operator must be notified in advance of the dangerous
substance to be brought into the harbour area from inland or from the sea.
Enough information must be given to enable the recipient to evaluate the
risks to health and safety from the substance. The Regulations also include
provision for the marking and navigation of vessels, for the handling and
storage of the substance while in the harbour and for its packaging and
labelling. A Code of Practice for the Dangerous Substances in Harbour Areas
Regulations 1987 has been approved under section 16(1) of the Health and
Safety at Work Act 1974. In addition byelaws may also control the carriage
of dangerous substances through a particular harbour.

Ships carrying dangerous goods in bulk or packaged form must comply 11.75
with the Merchant Shipping (Dangerous Goods and Marine Pollutants)
Regulations 1990.[67] Dangerous Goods here are goods classified as such in
accordance with the 1990 edition of the International Maritime Dangerous
Goods Code. Under regulation 7 the shipper or forwarder must give the
shipowner or master a declaration stating that the goods are dangerous and
all relevant information about them. If the goods are in a freight container
or vehicle the person responsible for packing them must also supply a
certificate that they have been packed in accordance with the provisions of
the Code.[68] All such goods must be packaged in accordance with those
provisions and if necessary the packaging should be performance tested.[69]
The packages should be durably marked and labelled in accordance with the
Code. The marking and labelling should be sufficient to remain legible or
fixed to the package after three months' immersion in the sea.[70]

NORTHERN IRELAND

In Northern Ireland special wastes are dealt with in accordance with the 11.75A
Pollution Control (Special Waste) Regulations (Northern Ireland).[71]

---

[66] S.I. 1987 No. 37 as amended by S.I. 1990 No. 2605 and S.I. 1994 No. 669, Sched. 4 and
S.I. 1996 No. 2092, Sched. 5.
[67] S.I. 1990 No. 2605,
[68] *ibid.* reg. 8.
[69] *ibid.* reg. 10.
[70] *ibid.* reg. 11.
[71] S.R. 1981 No. 252.

11.76 SPECIAL WASTE REGULATIONS 1996

SCHEDULE 2

SPECIAL WASTE

PART 1

HAZARDOUS WASTE LIST

Waste code (6 digits) Chapter Heading (2 and 4 digits)	Description
02	WASTE FROM AGRICULTURE, HORTICULTURAL, HUNTING, FISHING AND AQUACULTURE PRIMARY PRODUCTION, FOOD PREPARATION AND PROCESSING
0201	PRIMARY PRODUCTION WASTE
020105	agrochemical wastes
03	WASTES FROM WOOD PROCESSING AND THE PRODUCTION OF PAPER, CARDBOARD, PULP, PANELS AND FURNITURE
0302	WOOD PRESERVATION WASTE
030201	non-halogenated organic wood preservatives
030202	organochlorinated wood preservatives
030203	organometallic wood preservatives
030204	inorganic wood preservatives
04	WASTES FROM THE LEATHER AND TEXTILE INDUSTRIES
0401	WASTES FROM THE LEATHER INDUSTRY
040103	degreasing wastes containing solvents without a liquid phase
0402	WASTES FROM TEXTILE INDUSTRY
040211	halogenated wastes from dressing and finishing
05	WASTES FROM PETROLEUM REFINING, NATURAL GAS PURIFICATION AND PYROLYTIC TREATMENT OF COAL
0501	OILY SLUDGES AND SOLID WASTES
050103	tank bottom sludges
050104	acid alkyl sludges
050105	oil spills
050107	acid tars
050108	other tars
0504	SPENT FILTER CLAYS
050401	spent filter clays
0506	WASTE FROM THE PYROLYTIC TREATMENT OF COAL
050601	acid tars
050603	other tars
0507	WASTE FROM NATURAL GAS PURIFICATION
050701	sludges containing mercury
0508	WASTES FROM OIL REGENERATION
050801	spent filter clays
050802	acid tars
050803	other tars

Waste code (6 digits) Chapter Heading (2 and 4 digits)	Description	
050804	aqueous liquid waste from oil regeneration	11.77
06	WASTES FROM INORGANIC CHEMICAL PROCESSES	
0601	WASTE ACIDIC SOLUTIONS	
060101	sulphuric acid and sulphurous acid	
060102	hydrochloric acid	
060103	hydrofluoric acid	
060104	phosphoric and phosphorous acid	
060105	nitric acid and nitrous acid	
060199	waste not otherwise specified	
0602	ALKALINE SOLUTIONS	
060201	calcium hydroxide	
060202	soda	
060203	ammonia	
060299	wastes not otherwise specified	
0603	WASTE SALTS AND THEIR SOLUTIONS	
060311	salts and solutions containing cyanides	
0604	METAL-CONTAINING WASTES	
060402	metallic salts (except 0603)	
060403	wastes containing arsenic	
060404	wastes containing mercury	
060405	wastes containing heavy metals	
0607	WASTES FROM HALOGEN CHEMICAL PROCESSES	
060701	wastes containing asbestos from electrolysis	
060702	activated carbon from chlorine production	
0613	WASTES FROM OTHER INORGANIC CHEMICAL PROCESSES	
061301	inorganic pesticides, biocides and wood preserving agents	
061302	spent activated carbon (except 060702)	
07	WASTES FROM ORGANIC CHEMICAL PROCESSES	
0701	WASTE FROM THE MANUFACTURE, FORMULATION, SUPPLY AND USE (MFSU) OF BASIC ORGANIC CHEMICALS	
070101	aqueous washing liquids and mother liquors	
070103	organic halogenated solvents, washing liquids and mother liquors	
070104	other organic solvents, washing liquids and mother liquors	
070107	halogenated still bottoms and reaction residues	
070108	other still bottoms and reaction residues	
070109	halogenated filter cakes, spent absorbents	
070110	other filter cakes, spent absorbents	
0702	WASTE FROM THE MFSU OF PLASTICS, SYNTHETIC RUBBER AND MAN-MADE FIBRES	
070201	aqueous washing liquids and mother liquors	
070203	organic halogenated solvents, washing liquids and mother liquors	
070204	other organic solvents, washing liquids and mother liquors	
070207	halogenated still bottoms and raction residues	
070208	other still bottoms and reaction residues	
070209	halogenated filter cakes, spent absorbents	
070210	other filter cakes, spent absorbents	
0703	WASTE FROM THE MFSU FOR ORGANIC DYES AND PIGMENTS (EXCLUDING 0611)	

Waste code (6 digits) Chapter Heading (2 and 4 digits)	Description
11.78 070301	aqueous washing liquids and mother liquors
070303	organic halogenated solvents, washing liquids and mother liquors
070304	other organic solvents, washing liquids and mother liquors
070307	halogenated still bottoms and reaction residues
070308	other still bottoms and reaction residues
070309	halogenated filter cakes, spent absorbents
070310	other filter cakes, spent absorbents
0704	WASTE FROM THE MFSU FOR ORGANIC PESTICIDES (EXCEPT 020105)
070401	aqueous washing liquids and mother liquors
070403	organic halogenated solvents, washing liquids and mother liquors
070404	other organic solvents, washing liquids and mother liquors
070407	halogenated still bottoms and reaction residues
070408	other still bottoms and reaction residues
070409	halogenated filter cakes, spent aborbents
070410	other filter cakes, spent absorbents
0705	WASTE FROM THE MFSU OF PHARMACEUTICALS
070501	aqueous washing liquids and mother liquors
070503	organic halogenated solvents, washing liquids and mother liquors
070504	other organic solvents, washing liquids and mother liquors
070507	halogenated still bottoms and reaction residues
070508	other still bottoms and reaction residues
070509	halogenated filter cakes, spent absorbents
070510	other filter cakes, spent absorbents
0706	WASTE FROM THE MFSU OF FATS, GREASE, SOAPS, DETERGENTS, DISINFECTANTS AND COSMETICS
070601	aqueous washing liquids and mother liquors
070603	organic halogenated solvents, washing liquids and mother liquors
070604	other organic solvents, washing liquids and mother liquors
070607	halogenated still bottoms and reaction residues
070608	other still bottoms and reaction residues
070609	halogenated filter cakes, spent absorbents
070610	other filter cakes, spent absorbents
0707	WASTE FROM THE MFSU OF FINE CHEMICALS AND CHEMICAL PRODUCTS NOT OTHERWISE SPECIFIED
070701	aqueous washing liquids and mother liquors
070703	organic halogenated solvents, washing liquids and mother liquors
070704	other organic solvents, washing liquids and mother liquors
070707	halogenated still bottoms and reaction residues
070708	other still bottoms and reaction residues
070709	halogenated filter cakes, spent absorbents
070710	other filter cakes, spent absorbents
08	WASTES FROM THE MANUFACTURE, FORMULATION, SUPPLY AND USE (MFSU) OF COATINGS (PAINTS, VARNISHES AND VITREOUS ENAMELS), ADHESIVE, SEALANTS AND PRINTING INKS
0801	WASTES FROM MFSU OF PAINT AND VARNISH
080101	waste paints and varnish containing halogenated solvents

Waste code (6 digits) Chapter Heading (2 and 4 digits)	Description
080102	waste paints and varnish free of halogenated solvents
080106	sludges from paint or varnish removal containing halogenated solvents
080107	sludges from paint or varnish removal free of halogenated solvents
0803	WASTES FROM MFSU OF PRINTING INKS
080301	waste ink containing halogenated solvents
080302	waste ink free of halogenated solvents
080305	ink sludges containing halogenated solvents
080306	ink sludges free of halogenated solvents
0804	WASTES FROM MFSU OF ADHESIVE AND SEALANTS (INCLUDING WATER-PROOFING PRODUCTS)
080401	waste adhesives and sealants containing halogenated solvents
080402	waste adhesives and sealants free of halogenated solvents
080405	adhesives and sealants sludes containing halogenated solvents
080406	adhesives and sealants sludges free of halogenated solvents
09	WASTES FROM THE PHOTOGRAPHIC INDUSTRY
0901	WASTES FROM PHOTOGRAPHIC INDUSTRY
090101	water based developer and activator solutions
090102	water based offset plate developer solutions
090103	solvent based developer solutions
090104	fixer solutions
090105	bleach solutions and bleach fixer solutions
090106	waste containing silver from on-site treatment of photographic waste
10	INORGANIC WASTES FROM THERMAL PROCESSES
1001	WASTES FROM POWER STATION AND OTHER COMBUSTION PLANTS (EXCEPT 1900)
100104	oil fly ash
100109	sulphuric acid
1003	WASTES FROM ALUMINIUM THERMAL METALLURGY
100301	tars and other carbon-containing wastes from anode manufacture
100303	skimmings
100304	primary smelting slags/white drosses
100307	spent pot lining
100308	salt slags from secondary smelting
100309	black drosses from secondary smelting
100310	waste from treatment of salt slags and black drosses treatment
1004	WASTES FROM LEAD THERMAL METALLURGY
100401	slags (1st and 2nd smelting)
100402	dross and skimmings (1st and 2nd smelting)
100403	calcium arsenate
100404	flue gas dust
100405	other particulates and dust
100406	solid waste from gas treatment
100407	sludges from gas treatment
1005	WASTES FROM ZINC THERMAL METALLURGY
100501	slags (1st and 2nd smelting)
100502	dross and skimmings (1st and 2nd smelting)
100503	flue gas dust

11.79

Waste code (6 digits) Chapter Heading (2 and 4 digits)	Description
11.80 100505	solid waste from gas treatment
100506	sludges from gas treatment
1006	WASTES FROM COPPER THERMAL METALLURGY
100603	flue gas dust
100605	waste from electrolytic refining
100606	solid waste from gas treatment
100607	sludges from gas treatment
11	INORGANIC WASTE WITH METALS FROM METAL TREAT-MENT AND THE COATING OF METALS; NON-FERROUS HYDRO-METALLURGY
1101	LIQUID WASTES AND SLUDGES FROM METAL TREATMENT AND COATING OF METALS (*e.g.* GALVANIC PROCESSES, ZINC COATING PROCESSES, PICKLING PROCESSES, ETCH-ING, PHOSPHATIZING, ALKALINE DE-GREASING)
110101	cyanidic (alkaline) wastes containing heavy metals other than chromium
110102	cyanidic (alkaline) wastes which do not contain heavy metals
110103	cyanide-free wastes containing chromium
110105	acidic pickling solutions
110106	acids not otherwise specified
110107	alkalis not otherwise specified
110108	phosphatizing sludges
1102	WASTES AND SLUDGES FROM NON-FERROUS HYDRO-METALLURGICAL PROCESSES
110202	sludges from zinc hydrometallurgy (including jarosite, goethite)
1103	SLUDGES AND SOLIDS FROM TEMPERING PROCESSES
110301	wastes containing cyanide
110302	other wastes
12	WASTES FROM SHAPING AND SURFACE TREATMENT OF METALS AND PLASTICS
1201	WASTES FROM SHAPING (INCLUDING FORGING, WELD-ING, PRESSING, DRAWING, TURNING, CUTTING AND FILING)
120106	waste machining oils containing halogens (not emulsioned)
120107	waste machining oils free of halogens (not emulsioned)
120108	waste machining emulsions containing halogens
120109	waste machining emulsions free of halogens
120110	synthetic maching oils
120111	machining sludges
120112	spent waxes and fats
1203	WASTES FROM WATER AND STEAM DEGREASING PRO-CESSES (EXCEPT 1100)
120301	aqueous washing liquids
120302	steam degreasing wastes
13	OIL WASTES (EXCEPT EDIBLE OILS, 0500 AND 1200)
1301	WASTE HYDRAULIC OILS AND BRAKE FLUIDS
130101	hydraulic oils, containing PCBs and PCTs
130102	other chlorinated hydraulic oils (not emulsions)

144

Waste code (6 digits) Chapter Heading (2 and 4 digits)	Description
130103	non-chlorianted hydraulic oils (not emulsions)
130104	chlorianted emulsions
130105	non-chlorinated emulsions
130106	hydraulic oils containing only mineral oil
130107	other hydraulic oils
130108	brake fluids
1302	WASTE ENGINE, GEAR AND LUBRICATING OILS
130201	chlorinated engine, gear and lubricating oils
130202	non-chlorinated engine, gear and lubricating oils
130203	other machine, gear and lubricating oils
1303	WASTE INSULATING AND HEAT TRANSMISSION OILS AND OTHER LIQUID
130301	insulating or heat transmission oils and other liquids containing PCBs or PCTs
130302	other chlorinated insulating and heat transmission oils and other liquids
130303	non-chlorinated insulating and heat transmission oils and other liquids
130304	synthetic insulating and heat transmission oils and other liquids
130305	mineral insulating and heat transmission oils
1304	BILGE OILS
130401	bilge oils from inland navigation
130402	bilge oils from jetty sewers
130403	bilge oils from other navigation
1305	OIL/WATER SEPARATOR CONTENTS
130501	oil/water separator solids
130502	oil/water separator sludges
130503	interceptor sludges
130504	desalter sludges or emulsions
130505	other emulsions
1306	OIL WASTE NOT OTHERWISE SPECIFIED
130601	oil waste not otherwise specified
14	WASTE FROM ORGANIC SUBSTANCES EMPLOYED AS SOLVENTS (EXCEPT 0700 AND 0800)
1401	WASTES FROM METAL DEGREASING AND MACHINERY MAINTENANCE
140101	chlorofluorocarbons
140102	other halogenated solvents and solvent mixes
140103	other solvents and solvent mixes
140104	aqueous solvent mixes containing halogens
140105	aqueous solvent mixes free of halogens
140106	sludges or solid wastes containing halogenated solvents
140107	sludges or solid wastes free of halogenated solvents
1402	WASTES FROM TEXTILE CLEANING AND DEGREASING OF NATURAL PRODUCTS
140201	halogenated solvents and solvent mixes
140202	solvent mixes or organic liquids free of halogenated solvents
140203	sludges or solid wastes containing halogenated solvents

11.81

Waste code (6 digits) Chapter Heading (2 and 4 digits)	Description
11.82 140204	sludges or solid wastes containing other solvents
1403	WASTES FROM THE ELECTRONIC INDUSTRY
140301	chlorofluorocarbons
140302	other halogenated solvents
140303	solvents and solvent mixes free of halogenated solvents
140304	sludges or solid wastes containing halogenated solvents
140305	sludges or solid wastes containing other solvents
1404	WASTES FROM COOLANTS, FOAM/AEROSOL PROPELLANTS
140401	chlorofluorocarbons
140402	other halogenated solvents and solvent mixes
140403	other solvents and solvent mixes
140404	sludges or solid wastes containing halogenated solvents
140405	sludges or solid wastes containing other solvents
1405	WASTES FROM SOLVENT AND COOLANT RECOVERY (STILL BOTTOMS)
140501	chlorofluorocarbons
140502	halogenated solvents and solvent mixes
140503	other solvents and solvent mixes
140504	sludges containing halogenated solvents
140505	sludges containing other solvents
16	WASTES NOT OTHERWISE SPECIFIED IN THE CATALOGUE
1602	DISCARDED EQUIPMENT AND SHREDDER RESIDUES
160201	transformers and capacitors containing PCBs and PCTs
1604	WASTE EXPLOSIVES
160401	waste ammunition
160402	fireworks waste
160403	other waste explosives
1606	BATTERIES AND ACCUMULATORS
160601	lead batteries
160602	Ni-Cd batteries
160603	mercury dry cells
160606	electrolyte from batteries and accumulators
1607	WASTE FROM TRANSPORT AND STORAGE TANK CLEANING (EXCEPT 0500 AND 1200)
160701	waste from marine transport tank cleaning, containing chemicals
160702	waste from marine transport tank cleaning, containing oil
160703	waste from railway and road transport tank cleaning, containing oil
160704	waste from railway and road transport tank cleaning, containing chemicals
160705	waste from storage tank cleaning, containing chemicals
160706	waste from storage tank cleaning, containing oil
17	CONSTRUCTION AND DEMOLITION WASTE (INCLUDING ROAD CONSTRUCTION)
1706	INSULATION MATERIALS
170601	insulation materials containing asbestos
18	WASTES FROM HUMAN OR ANIMAL HEALTH CARE AND/ OR RELATED RESEARCH (EXCLUDING KITCHEN AND RESTAURANT WASTES WHICH DO NOT ARISE FROM IMMEDIATE HEALTH CARE)

Waste code (6 digits) Chapter Heading (2 and 4 digits)	Description
1801	WASTE FROM NATAL CARE, DIAGNOSIS, TREATMENT OR PREVENTION OF DISEASE IN HUMANS **11.83**
180103	other wastes whose collection and disposal is subject to special requirements in view of the prevention of infection
1802	WASTE FROM RESEARCH, DIAGNOSIS, TREATMENT OR PREVENTION OF DISEASE INVOLVING ANIMALS
180202	other wastes whose collection and disposal is subject to special requirements in view of the prevention of infection
180204	discarded chemicals
19	WASTES FROM WASTE TREATMENT FACILITIES, OFF-SITE WASTE WATER TREATMENT PLANTS AND THE WATER INDUSTRY
1901	WASTES FROM INCINERATION OR PYROLYSIS OF MUNICIPAL AND SIMILAR COMMERCIAL, INDUSTRIAL AND INSTITUTIONAL WASTES
190103	fly ash
190104	boiler dust
190105	filter cake from gas treatment
190106	aqueous liquid waste from gas treatment and other aqueous liquid wastes
190107	solid waste from gas treatment
190110	spent activated carbon from flue gas treatment
1902	WASTES FROM SPECIFIC PHYSICO/CHEMICAL TREATMENTS OF INDUSTRIAL WASTES (e.g. DECHROMATATION, DECYANIDATION, NEUTRALIZATION)
190201	metal hydroxide sludges and other sludges from metal insolubilization treatment
1904	VITRIFIED WASTES AND WASTES FROM VITRIFICATION
190402	fly ash and other flue gas treatment wastes
190403	non-vitrified solid phase
1908	WASTES FROM WASTE WATER TREATMENT PLANT NOT OTHERWISE SPECIFIED
190803	grease and oil mixture from oil/waste water separation
190806	saturated or spent ion exchange resins
190807	solutions and sludges from regeneration of ion exchangers
20	MUNICIPAL WASTES AND SIMILAR COMMERCIAL INDUSTRIAL AND INSTITUTIONAL WASTES INCLUDING SEPARATELY COLLECTED FRACTIONS
2001	SEPARATELY COLLECTED FRACTIONS
200112	paint, ink, adhesives and resins
200113	solvents
200117	photo chemicals
200119	pesticides
200121	fluorescent tubes and other mercury containing waste

PART II

HAZARDOUS PROPERTIES

11.84 H1   "Explosive": substances and preparations which may explode under the effect of flame or which are more sensitive to shocks or friction than dinitrobenzene.

H2   "Oxidizing": substances and preparations which exhibit highly exothermic reactions when in contact with other substances, particularly flammable substances.

H3–A   "Highly flammable":
—liquid substances and preparations having a flash point below 21°C (including extremely flammable liquids), or
—substances and preparations which may become hot and finally catch fire in contact with air at ambient temperature without any application of energy, or
—solid substances and preparations which may readily catch fire after brief contact with a source of ignition and which continue to burn or to be consumed after removal of the source of ignition, or
—gaseous substances and preparations which are flammable in air at normal pressure, or
—substances and preparations which, in contact with water or damp air, evolve highly flammable gases in dangerous quantities.

H3–B   "Flammable": liquid substances and preparations having a flash point equal to or greater than 21°C and less than or equal to 55°C.

H4   "Irritant": non-corrosive substances and preparations which, through immediate, prolonged or repeated contact with the skin or mucous membrane, can cause inflammation.

H5   "Harmful": substances and preparations which, if they are inhaled or ingested or if they penetrate the skin, may involve limited health risks.

H6   "Toxic": substances and preparations (including very toxic substances and preparations) which, if they are inhaled or ingested or if they penetrate the skin, may involve serious, acute or chronic health risks and even death.

H7   "Carcinogenic": substances and preparations which, if they are inhaled or ingested or if they penetrate the skin, may induce cancer or increase its incidence.

H8   "Corrosive": substances and preparations which may destroy living tissue on contact.

H9   "Infectious": substances containing viable micro-organisms or their toxins which are known or reliably believed to cause disease in man or other living organisms.

H10   "Teratogenic": substances and preparations which, if they are inhaled or ingested or if they penetrate the skin, may induce non-hereditary congenital malformations or increase their incidence.

H11   "Mutagenic": substances and preparations which, if they are inhaled or ingested or if they penetrate the skin, may induce hereditary genetic defects or increase their incidence.

H12   Substances and preparations which release toxic or very toxic gases in contact with water, air or an acid.

H13   Substances and preparations capable by any means, after disposal, of yielding another substance, *e.g.* a leachate, which possesses any of the characteristics listed above.

H14   "Ecotoxic": substances and preparations which present or may present immediate or delayed risks for one or more sectors of the environment.

PART III

THRESHOLDS FOR CERTAIN HAZARDOUS PROPERTIES

In the waste:

—the total concentration of substances classified as irritant and having assigned to them **11.85** any of the risk phrases R36 ("irritating to the eyes"), R37 ("irritating to the respiratory system") or R38 ("irritating to the skin") is equal to or greater than 20%;

—the total concentration of substances classified as irritant and having assigned to them the risk phrase R41 ("risk of serious damage to eyes") is equal to or greater than 10%;

—the total concentration of substances classified as harmful is equal to or greater than 25%;

—the total concentration of substances classified as very toxic is equal to or greater than 0.1%;

—the total concentration of substances classified as toxic is equal to or greater than 3%;

—the total concentration of substances classified as carcinogenic and placed by the approved classification and labelling guide in category 1 or 2 of that classification is equal to or greater than 0.1%;

—the total concentration of substances classified as corrosive and having assigned to them the risk phrase R34 ("causes burns") is equal to or greater than 5%; and

—the total concentration of substances classified as corrosive and having assigned to them the risk phrase R35 ("causes severe burns") is equal to or greater than 1%.

"PART IV

RULES FOR THE INTERPRETATION OF THIS SCHEDULE

1. Except in the case of a substance listed in the approved supply list, the test methods to **11.86** be used for the purposes of deciding which (if any) of the properties mentioned in Part II of this Schedule are to be assigned to a substance are those described in Annex V to Council Directive 67/548/EEC, as amended by Commission Directive 92/69/EEC.

2. Any reference in Part III of this Schedule to a substance being classified as having a hazardous property, having assigned to it a particular risk phrase, or being placed within a particular category of a classification is a reference to that substance being so classified, having that risk phrase assigned to it or being placed in that category—

(i) in the case of a substance listed in the approved supply list, on the basis of Part V of that list;

(ii) in the case of any other substance, on the basis of the criteria laid down in the approved classification and labelling guide.

3. Any reference in Part III of this Schedule to the total concentration of any substances being equal to or greater than a given percentage is a reference to the proportion by weight of those substances in any waste being equal to or, as the case may be, greater than that by percentages."

12. TRANSBOUNDARY SHIPMENT OF WASTE

The provisions concerning the import and export of waste initially concerned only shipments of hazardous wastes. This changed on May 6, 1994 with the entry into force of the E.C. Regulation on the shipment of wastes[1] which covers most wastes, whether hazardous or not. However, the international conventions in this area are still primarily targeted at the trade in hazardous wastes.

12.01 The control of movement of wastes internationally has become a controversial issue. It is a particularly difficult problem at the interface between the principle of free movement of goods on the one hand and that of sustainable development on the other. It is one of the issues in the discussions on trade and the environment in the World Trade Organisation. However, such discussions, while they may have an eventual effect on transboundary shipment rules are outside the scope of this book; nevertheless they do form the background from which the law emerges.

12.02 The legality of transfrontier shipments of wastes and national rules concerning them was discussed in *Commission of the European Communities v. Kingdom of Belgium.*[2] The court looked at the issue of whether waste was a "good" covered by Community rules on the free movement of goods and concluded that even non-recyclable wastes must be considered as products the movements of which cannot be prevented by virtue of Article 30 of the E.C. Treaty. However, a ban of the import of hazardous wastes was upheld. This case is no longer important in the light of the Transfrontier Shipment Regulation. It does, however, highlight the problems that can arise in the conflict between the issues of free trade and environmental protection.

---

[1] Reg. 259/93, [1993] O.J. L30/1.
[2] [1995] 1 J.Env.L. 133.

150

1. THE BASEL CONVENTION

The United Nations Convention on the Control of Transboundary **12.03**
Movements of Hazardous Wastes and their Disposal[3] (the Basel Convention)
was signed in 1989 and entered into force in 1992. The aim of the
Convention is to encourage disposal of wastes in the State where they were
generated and to ensure that where transboundary movements take place
they do not cause any environmental harm or pose a threat to human health.
To this latter end it provides controls on shipments so as to ensure they are
made in an environmentally sound manner and to prevent illegal traffic in
hazardous and other wastes.

The scope of the Convention is set out in Article 1 which categorises those **12.04**
wastes that are subject to transboundary movement as "hazardous wastes"
for its purposes. These are either wastes of the descriptions in Annex I,
unless they are harmless, or wastes considered to be hazardous wastes under
the domestic legislation of the relevant party. Each party must provide the
Secretariat to the Convention with lists of wastes its legislation defines as
hazardous.[4] Wastes in Annex II are "other wastes" under the Convention,
while radioactive wastes and wastes derived from the normal operations of a
ship do not come within its scope. In 1993 the OECD drew up lists of
wastes commonly moved for recovery purposes—the Green, Amber and Red
lists. The purpose of these lists is to ensure that "recovery" wastes are not
impeded unnecessarily. The lists, as revised in 1994, are incorporated in the
E.C. Waste Shipments Regulation.

A State has a sovereign right to ban the entry or disposal of foreign wastes **12.05**
in its territory. If a party to the Convention decides to exercise this right in
respect of any or all wastes it must inform the other parties of this decision
in accordance with Article 13.[5] On receipt of such notification the other
parties must prevent the export of any banned wastes to the state imposing
the ban[6] and must take appropriate measures to ensure that the ban is not
broken by exports from their territories.[7] Further, a party may not allow
hazardous or other wastes to be imported from or exported to a non-party.[8]
In 1994 a decision of the Convention parties banned the export of hazardous
wastes to non-OECD countries.[9]

Otherwise exports of waste to another state must have the written **12.06**
consent of the state to receive the specific shipment.[10] Notifications of
consignments will be made under Article 6 by the generator or exporter or
the exporting State, through its competent authority to the competent

---

[3] For text see [1990] 2 J.Env.L. 255.
[4] Basel Convention, Art. 3.
[5] *ibid*. Art. 4.1(a).
[6] *ibid*. Art. 4.1(b).
[7] *ibid*. Art. 4.2(e).
[8] *ibid*. Art. 4.5.
[9] Decision II/12, March 1994.
[10] Basel Convention, Art. 4.1.(c).

authority of the importing state. Such a notification should include all the information set out in Annex VA to the Convention.[11] The importing state may consent to the consignment, conditionally or unconditionally, refuse it, or request further information.[12] The consignment may only go ahead if the exporting state has had written confirmation that the notifier has received the written confirmation of the importing state and also confirmation that there is a contract between the exporter and disposer specifying environmentally sound management of the wastes in question.[13] Notifications must also be sent to states through which the waste will pass, who must also consent to the waste being shipped through their territory.[14] General notifications are allowed in the case of regular shipments of wastes of the same character by the same route to the same disposer; although such a notification will only be valid for 12 months.[15]

12.07   Where a properly authorised consignment of waste cannot be made in accordance with the terms of the contract, the exporting state must ensure that it is brought back to its territory by the exporter unless alternative arrangements can be made to dispose of it in an environmentally sound manner. The re-import or safe disposal must take place within either ninety days of the date the importing state informed the exporting state and the Secretariat or such longer period as the states agree. In order that this provision can be given effect no party may oppose, hinder or prevent the return of the consignment to the state of export.[16]

12.08   For the purposes of the Convention any shipment of wastes without notification to all states concerned or their consent or where that consent is obtained through falsification, misrepresentation or fraud is illegal traffic. In addition shipments that do not conform in a material way with the documents, or that results in the deliberate disposal of wastes in contravention of the Convention and of the general principles of international law, is also illegal traffic.[17] Illegal traffic in waste is criminal and each party must take the appropriate legal, administrative and other measures to implement and enforce the Convention and to prevent and punish contravention[18] and illegal traffic.

12.09   Where a consignment is deemed to be illegal traffic as a result of conduct by the exporter or generator the state of export must ensure that the wastes are returned to its territory by the exporter or generator or, if necessary, by it itself or, if that is impracticable, that the waste is otherwise disposed of in accordance with the provisions of the Convention. These measures must be taken within 30 days from the date when the state of import became aware of the illegal traffic or such longer time as the states may agree. To this end

[11] Basel Convention, Art. 6.1.
[12] *ibid*. Art. 6.2. and see Art. 7.
[13] *ibid*. Art. 6.3.
[14] *ibid*. Art. 6.4.
[15] *ibid*. Art. 6.6–6.8.
[16] *ibid*. Art. 8.
[17] *ibid*. Art. 9.
[18] *ibid*. Art. 4.3 & 4.4.

the parties concerned must not oppose, hinder or prevent the return of the consignment to the state of export.[19]

If the consignment is illegal due to conduct on the part of the importer or disposer, the state of import must ensure that the wastes in question are disposed of in an environmentally sound manner by the importer or disposer or, if necessary, by itself within 30 days of the time it became aware of the illegality or within such other time as the states concerned may agree. To this end the parties concerned must co-operate in ensuring that the consignment is disposed of in an environmentally sound manner.[20] Where responsibility for the illegality cannot be assigned the parties concerned or, if appropriate other parties, must co-operate to ensure that the wastes are disposed of in an environmentally sound manner either in the exporting or importing state or elsewhere, as appropriate.[21] **12.10**

Any consignment of waste must be covered by insurance, bond or other guarantee that may be required by the state of import or a state through which the consignment will pass.[22] Later on a Protocol to the Convention will be agreed on liability and compensation for damage resulting from the transboundary movement and disposal of waste.[23] In addition a fund may be established for dealing with emergencies.[24] **12.11**

2. THE LOME CONVENTION

Relations between the European Community and a number of African, Caribbean and Pacific (A.C.P.) states are guided by the Lome Conventions. The Lome IV Convention was signed on December 15, 1989. Article 39 of Lome IV requires the Community to prohibit all direct or indirect export of hazardous or radioactive waste to the A.C.P. states while at the same time the A.C.P. states must prohibit the direct and indirect import into their territory of such waste from the Community or any other country. This provision took effect in 1992. Hazardous waste is not defined in detail but would probably have the same meaning as in the Basel Convention. The controls under Lome IV may be substituted for those of the Basel Convention provided that they are compatible with the requirements of that Convention that wastes are handled and disposed of in an environmentally sound manner.[25] **12.12**

3. THE E.C. "WASTE SHIPMENTS" REGULATION AND UNITED KINGDOM PRACTICE

The E.C. Regulation on the supervision and control of shipments of waste within, into and out of the European Community (259/93)[26] replaces the earlier Transfrontier Shipments of Waste Directive. The Regulation gives **12.13**

---

[19] Basel Convention, Art. 9.2.
[20] *ibid.* Art. 9.3.
[21] *ibid.* Art. 9.4.
[22] *ibid.* Art. 6.11.
[23] *ibid.* Art. 12.
[24] *ibid.* Art. 14.2.
[25] Basel Convention, Art. 11.2.
[26] [1993] O.J. L30/1.

effect to the Basel Convention and enables the E.C. to comply with Article 39 of Lome IV. E.C. Regulations have direct effect in Community law so that they apply without the necessity of implementing legislation in member states. However, in the United Kingdom Transfrontier Shipment of Waste Regulations 1994[27] expand the Waste Shipment Regulation to ensure it operates in practice. The United Kingdom Regulations are explained in D.o.E. Circular 13/94.[28] They are supplemented by the *United Kingdom Management Plan for Exports and Imports of Waste*[29] which sets out the policies that will be followed in administering the Regulation.

12.14   The Regulation is mainly concerned with "wastes for disposal". This gives effect to the OECD decision on the movement of "wastes for recovery" dealt with above. For the purposes of the Regulation "waste" is defined in the same way as in the Framework Directive on Waste (75/442 as amended) and "disposal" in the same way as in Article 1(e) of that directive.[30] Annexes II, III and IV of the Regulation have been amended to set out the OECD Green, Amber and Red lists of wastes as revised in July 1994.[31]

12.15   The Regulation applies to all shipments of waste within, into and out of the Community except those excluded under Article 1.2.[32] These wastes are shipping and offshore installation wastes falling under the MARPOL regime, civil aviation wastes, radioactive wastes and wastes imported into the Community in accordance with the Protocol on Environmental Protection to the Antarctic Treaty. In addition shipments of wastes exempted from the Framework Directive by Article 2.1(b) of that directive are also excluded from the provisions of the Waste Shipments Regulation where they are already covered by other relevant legislation.[33] It is not clear whether this means that the movement of such wastes has to be covered by other legislation or if any control regime will suffice.

COMPETENT AUTHORITIES

12.16   Under Article 36 of the Regulation Member States must designate a competent authority or authorities for its implementation. There are three types of such authority—the authority of dispatch, the authority of destination and the authority of transit. Where wastes are loaded on board a ship for disposal at sea the relevant authority will that of destination.[34] A state may only designate one authority of transit.

12.17   In the United Kingdom authorities have been designated under regulation 3 of the Transfrontier Shipment of Waste Regulations 1994. In Great Britain they are Waste Regulation Authorities (the Agencies) and in

---

[27] S.I. 1994 No. 1137.
[28] W.O. Circular 44/94; S.O. Circular 21/94; D.o.E. (N.I.) W.M. Circular 1/94.
[29] May 1996, D.o.E., etc.
[30] E.C. Reg. 259/93, Art. 2(a), 2(i).
[31] Comm. Dec. 94/721 [1994]; O.J. L288/36.
[32] E.C. "Shipments" Regulation, Art. 1(1).
[33] *ibid*. Art. 1(2)(d).
[34] *ibid*. Art. 2(d).

Northern Ireland, district councils. The Secretary of State will be the competent authority of transit. Usually the relevant authority will be the local Agency office in whose area a load for shipment originates or is to be disposed of or recovered. However, in cases where waste is shipped without proper notification or otherwise contrary to the provisions of the Regulation the relevant authority will be the one for the area in which the waste arrives.[35]

The principle behind the Regulation is that transfers of waste should be 12.18 subject to the "prior informed consent" of the relevant authorities in the countries involved. To this end shipments must be notified to those authorities through a consignment note and only when they consent to a shipment can it proceed. For waste shipments destined for recovery a modified procedure applies to encourage such operations.

When deciding which regime applies to a particular shipment there are 12.19 three main criteria to consider. The first is whether the shipment is of "waste". The second is whether the shipment is for disposal or recovery. The third is whether the shipment will be made between Member States or be exported from the European Community or imported into it.

Notification is effected by means of a standard consignment note that has 12.20 been drawn up under Article 42 of the Regulation which was adopted by Commission Decision 94/774/EEC.[36] The standard consignment note is both a movement/tracking form in respect of waste shipments and also serves as a certificate of disposal and recovery. Information in consignment notes should be treated as confidential although there may be a conflict here with the Environmental Information Directive.[37] Where requested by a competent authority, the notifier should provide a translation of a note in a language acceptable to that authority.[38]

The general notification procedure under Article 28 of the Regulation 12.21 may apply where waste having the same physical and chemical characteristics is shipped regularly to the same consignee by the same route. A general notification can only last for one year at the most.[39]

Wastes which are the subject of different consignment notes may not be 12.22 mixed during shipment.[40] Anyone in the United Kingdom who does mix such wastes will commit an offence.[41]

SHIPMENTS OF WASTE FOR DISPOSAL

(a) Made between Member States

Under Article 3 of the Regulation, where the notifier intends to ship 12.23 waste for disposal from one E.C. state to another and/or send it through other states he must notify the competent authority of the state that will

---

[35] D.o.E. Circular 13/94, para. 33.
[36] [1994] O.J. L310/70.
[37] E.C. "Shipments" Regulation, Arts 3.7, 6.7, 11.2 & 15.12 and see D.o.E. Circular 13/94, paras 162–164.
[38] *ibid.* Art. 31.1.
[39] See D.o.E. Circular 13/94, paras 105 & 106.
[40] E.C. "Shipments" Regulation, Art. 29.
[41] S.I. 1994 No. 1137, reg. 12(6).

receive it and send copies of the notification to the authorities of the states from where it is sent and through which it will pass and to the consignee.

12.24　For these purposes the "notifier" of the waste is the natural person or body corporate who proposes to ship the waste or to have it shipped. This can either be the producer of the waste, the person who collected it or brokers the arrangements, the person with control over the waste or, where the waste comes into the E.C. from outside it, the person designated by the state of dispatch or the person with control of the waste. The D.o.E. Circular considers that there is a strict hierarchy to be followed here so that the notifier should normally be the producer of the waste.[42] but this may not be correct. Article 34 of the Regulations, which puts a "duty of care" on a producer, does not necessarily mean that the producer has to move into the waste disposal business.

12.25　It is the notifier's duty to complete the note and furnish any further information that may be required by any relevant authority. However, the competent authority of dispatch can decide to make notifications itself in respect of any shipments.[43] The procedure by which an authority takes or revokes such a decision and publicises its decision is set out in regulation 6 of the Transfrontier Shipment of Waste Regulations 1994.[44] Shipments subject to this procedure can be effectively banned by the authority deciding not to proceed with notification but the authority can only decide not to proceed on the grounds set out in Article 4.3.

12.26　Before the shipment can proceed the notifier must have a contract with the consignee for the disposal of the waste.[45] This must include terms to ensure that the notifier will take the waste back if the shipment is not completed as planned or if there is a breach of the Regulation's requirements. Further it must require the consignee to certify, within 180 days of receipt of the waste, that it has been disposed of in an environmentally sound manner. A copy of the contract must be supplied to the competent authority on request. Failure to have such a contract in place will be an offence under the Transfrontier Shipment of Waste Regulations 1994.[46]

12.27　The principal authority here is the authority in the state of destination. When it receives a notification it must acknowledge receipt within three days. It then has 30 days to make a decision on whether to authorise the shipment with or without conditions or to refuse it. It may also request further information.[47] It should authorise it unless it has proper objections to it or it has received objections from other authorities. For this reason it must wait for at least 21 days from acknowledgment before authorising unless the other authorities give their written consent to the shipment.

12.28　Proper objections to a shipment are those set out in Article 4.3. The first reason may be that the Member State concerned has taken a general view

[42] D.o.E. Circular 13/94, para. 26.
[43] E.C. "Shipments" Regulation. Art. 6.8.
[44] As amended by S.I. 1996, Sched. 2, para. 11.
[45] ibid. Art. 3.6.
[46] S.I. 1994 No. 1137, reg. 12(7).
[47] E.C. "Shipments" Regulation, Art. 4.2(a).

about the need for states to be self-sufficient in waste disposal and therefore refuses all shipments or all shipments of a particular type. A transit authority could object on this ground to all shipments passing through its territory. However, this ground cannot be applied in the case of a hazardous waste for which it would be uneconomic for the state of dispatch to provide a new specialised disposal installation. Any disputes on this ground will be resolved by the Commission. In the United Kingdom this policy will be contained in the waste management plan drawn up by the Secretary of State under regulation 11 of the Transfrontier Shipment of Waste Regulations 1994. The plan bans exports for disposal including the re-export of wastes sent to the United Kingdom for disposal, and limits imports of waste for disposal.[48]

The second ground for objection is that to ensure the proper operation of 12.29 disposal facilities in the state of dispatch or destination, for example that capacity is not exhausted too quickly, the shipment should be refused.[49] The third reason is that the shipment would be unlawful as against relevant domestic or international legislation or that the notifier or consignee has previously been convicted of illegal trafficking.[50]

Approval is signified by the authority of destination stamping the 12.30 consignment note.[51] Once this has been sent to the notifier he can complete his part of the note and make the shipment. When the consignee receives it he must complete his part and, within three working days send copies to the notifier and all relevant authorities. Finally he must send a certificate of disposal, which forms part of the standard consignment note, to all parties. This should be done as soon as possible and not later than 180 days after receipt of the waste.[52] Failure to send such a certificate within the right time, or to falsify one, is an offence under the 1994 Regulations.[53]

If the shipment will go via a non-EEC state then the provisions of Article 12.31 12 will also apply. This requires the notifier to send a copy of the consignment note to the competent authority of the non-EEC state (the third state) and the state of destination must ask that authority if it wants to send it its written consent for the shipment. If it does there is an extended timescale of 60 days for parties to the Basel Convention or an agreed time for non-Basel Convention parties.

There are two types of conditions that can be attached to an approval. 12.32 The first are conditions of transport. These may not be more stringent than conditions normally applying to similar shipments in the Member State in question. They should also take account of relevant international conventions.[54] Where any of the international transport conventions listed in Annex 1 to the Regulation apply to a shipment the terms of that convention must

---

[48] *U.K. Management Plan for Exports and Imports of Waste*, May 1996.
[49] Transfrontier Shipments Regulation, Art. 4.3(b).
[50] *ibid*. Art. 4.3(c).
[51] *ibid*. Art. 4.5.
[52] *ibid*. Art. 5.
[53] S.I. 1994 No. 1137, reg. 12(3).
[54] E.C. "Shipments" Regulation, Art. 4.2(d).

be complied with as far as it covers the waste involved.[55] The second type of condition are those falling under Article 4.3 which sets out the criteria on which objections can be based. The most likely ground for conditions here is Article 4.3(c) to ensure compliance with national laws and regulations concerning environmental protection, public order, public safety or health and safety requirements. It will be an offence for anyone in the United Kingdom to handle waste in contravention of a condition imposed under the Regulation[56] unless he can show that he was not reasonably able to comply with it by reason of an emergency.[57]

(b) Exports of waste for disposal

12.33   All exports of waste for disposal are prohibited unless made to EFTA countries that are also parties to the Basel Convention.[58] However, even these exports will be banned if the EFTA country of destination prohibits imports or has not given its specific written consent to the shipment. In addition no shipment can be made if the competent authority of dispatch in the EEC has reason to believe that the waste will not be dealt with in an environmentally sound manner during shipment and in the state of destination.[59] If a shipment is allowed it must be a requirement of authorisation that it is dealt with in an environmentally sound manner at all times.[60] The procedure for authorising shipments is similar to that for those made between EEC states but with longer timescales. The relevant provisions are set out in Article 15 of the Regulation. The United Kingdom Management Plan for Exports and Imports of Waste bans all exports for disposal.

12.34   Exports of waste to ACP states are prohibited.[61] This does not, however, prevent Member States from returning waste imported into the EEC by an ACP state. In such a case a specimen of the consignment note together with the stamp of authorisation, must accompany each shipment.[62]

(c) Imports into the Community

12.35   Waste can only be imported into the European Community from countries that are either parties to the Basel Convention or with which the Community or individual Member States have concluded bilateral agreements.[63] In any case the authority of the importing country can ban a shipment if it has reason to believe that the waste will not be managed in an environmentally sound manner in its area.[64] The United Kingdom Manage-

[55] E.C. "Shipments" Regulation, Art. 32.
[56] S.I. 1994 No. 1137, reg. 12(2).
[57] ibid. Reg. 14(2).
[58] E.C. "Shipments" Regulation, Art. 14.1.
[59] ibid. Art. 14.2.
[60] ibid. Art. 14.3.
[61] ibid. Art. 18.1.
[62] ibid. Art. 18.2 & 18.3.
[63] ibid. Art. 19.
[64] ibid. Art. 19.4.

ment Plan for Exports and Imports of Waste severely limits the wastes that can be imported to the United Kingdom for disposal and restricts the disposal options available for imported wastes.

Article 20 of the Regulation sets out the procedure to be followed by the 12.36 sender[65] of the waste. This involves the same cycle of notification, acknowledgment, objection or setting conditions, consent or refusal as for other shipments. However, the timescale is longer than for traffic between Member States with the competent authority in the importing state having 70 days from acknowledging receipt of notification to determine the matter.[66]

SHIPMENTS OF WASTE FOR RECOVERY

(a) Between Member States

The OECD "green" list forms Annex II to the Regulation and shipments 12.37 of such wastes need only be accompanied by their usual commercial description, the quantity involved, the name and address of the consignee and the specific recovery operation involved.[67] If this information is not available with the shipment in the United Kingdom an offence will be committed under regulation 12(8) of the 1994 Regulations. In addition the waste must go to a facility that is licensed or registered and be dealt with in accordance with the "Waste" Directive.[68] If a waste in the green list has hazardous properties to bring it within the hazardous wastes directive it may be subject to the full rigours of the Waste Shipment Regulation. Further, in exceptional cases a green list waste may be treated as if it were in the amber or red lists for public health or environmental reasons.[69] If a green list waste is shipped in contravention of the Regulation it may be ordered to be returned to the sender or treated as illegal traffic.[70]

The OECD "amber" list forms Annex III to the Regulation. Shipments 12.38 between Member States of such waste for recovery must comply with the provisions of Articles 6 to 9. These follow the same pattern of notification, the requirement for a contract, the provision of information, approval or rejection, and, if approved, conditions of shipment as for shipments for disposal. However, the information to be provided must include the recovery operation concerned, the way in which the residual waste from the operation will be disposed of, the amount of recycled material in relation to the residual waste and the estimated value of the recycled material.[71] This information will be used in considering an additional objection that can be raised, namely that the ratio of recoverable and non-recoverable waste, the

[65] See E.C. "Shipments" Regulation, Art. 2(g)(iv).
[66] *ibid.* Art. 19.4.
[67] *ibid.* Arts 1.3(a) & 11.
[68] *ibid.* Art. 1.3(b).
[69] *ibid.* Art. 1.3(c) & (d).
[70] *ibid.* Art. 1.3(e).
[71] *ibid.* Art. 6.5.

estimated value of the recovered materials or the costs of the operation, including the disposal of residues, do not justify the recovery on economic and environmental considerations.[72] Confidentiality of such information is required by Article 6.7 and the relationship between this requirement and the Environmental Information directive is discussed in D.o.E. Circular 13/94.[73]

12.39   Article 9 allows states under whose jurisdiction recovery facilities operate to give a general approval for certain types of shipment going to a particular facility. This does not preclude authorities in the states of dispatch or transit from raising objections or imposing conditions on any shipment.[74] The requirements as to notification and contract must still be complied with and the shipment itself must be conducted in accordance with Article 8.2 to 8.6.

12.40   "Red" list waste, as set out in Annex IV to the Regulation, and waste not allocated to either Annex II, III, or IV, will be treated in the same way as waste for disposal. However, the consent of the competent authorities concerned must be obtained in writing before the shipment is commenced.[75]

(b) Exports from the Community

12.41   All exports of waste for recovery from the Community are prohibited unless they are to countries to which the March 30, 1992 OECD decision applies or which are parties to the Basel Convention and with which bilateral or regional agreements have been concluded.[76] However, exports cannot be made to states that have banned them or to states in which it is considered by the competent authority of dispatch that the waste will not be dealt with in an environmentally sound manner.[77] No waste may be exported to an ACP state, although this does not prevent the return of processed waste.[78]

12.42   Any agreement about the export of waste for recovery must guarantee that the recovery operation is carried out at an authorised centre and in an environmentally sound manner. It must also make arrangements for the treatment of the non-recoverable element in any shipment, if necessary requiring the notifier to take that residue back. Provision should be made for enforcement of the agreement and all agreements should be subject to periodic review by the Commission.[79] Exports of hazardous wastes for recovery to non-OECD countries are to be phased out by December 31, 1997 unless covered by an "exceptional cases" exemption. Exports of hazardous wastes to OECD countries must comply with the provisions of the Regulation. "Hazardous wastes" are those that are not on the OECD "Green List".[80]

---

[72] See E.C. "Shipments" Regulation, Art. 7.4(a).
[73] D.o.E. Circular 13/94, paras 162–164.
[74] E.C. "Shipments" Regulation, Art. 9.3.
[75] ibid. Art. 10.
[76] ibid. Art. 16.1.
[77] ibid. Art. 16.3.
[78] ibid. Art. 18.
[79] ibid. Art. 16.2.
[80] U.K. Management Plan for Exports and Imports of Waste, May 1996.

The procedure for exports is set out in Article 17 of the Regulation. Green 12.43
list waste can be export ed without any notification procedure as long as the
state of destination is listed by the Commission as not requiring such control
or is a country to which the OECD decision applies. If the state either has its
own controls or requests the E.C. to control exports to it or if controls apply
under Article 3 of the Basel Convention then the control procedures
applying to exports will be determined by the Commission.[81] Any "green"
list waste must be treated in an authorised recovery centre. The system
adopted for supervision of the centre must require automatic export licensing
and copies of licences must be forwarded to the exporting state.[82]

For "amber" or "red" list waste or wastes that have not been assigned to 12.44
a list most of the procedures in Articles 6 to 8 of the Regulation will apply.[83]
"Amber" list wastes can however, be sent under a general approval in
accordance with Article 9.[84] Extra controls are provided by Article 17.7 to
both types of export ensuring increased monitoring of the shipment. Further,
the contract under which the shipment is made must stipulate that if a
consignee issues an incorrect certificate of recovery and the financial
guarantee is then released any costs incurred in re-shipment will be borne by
him. If these categories of waste are sent to states to which the OECD
decision does not apply then the procedures in Article 15—except paragraph
3—are applicable and reasoned objections can only be made as provided for
in Article 7.4.[85]

(c) Imports into the Community

Imports can only be made from countries to which the OECD decision 12.45
applies or from those that are parties to the Basel Convention and/or with
which either the Community or a Member State has concluded bilateral
agreements in accordance with Article 21 of the Regulation.[86] The procedures to be applied for "amber" or "red" list wastes or unassigned wastes
are similar to those for exports.[87] Guidance in the United Kingdom
Management Plan for the Exports and Imports of Waste is in favour of
importing wastes for beneficial and environmentally sound recovery.

SHIPMENTS IN TRANSIT THROUGH THE COMMUNITY

Where shipments of waste for disposal are sent through the Community 12.46
the provisions of Article 23 will apply. This requires notification to
competent authorities of each state through which the load will pass and to
the customs offices through which the shipment will enter and leave the
Community. The key authority is the competent authority of the state from

---

[81] Transfrontier Shipment Regulation, Art. 17.1 & 17.3, and see Com. Dec. 94/575.
[82] *ibid.* Art. 17.2.
[83] *ibid.* Art. 17.4 & 17.6.
[84] But see Art. 17.5.
[85] *ibid.* Art. 17.8.
[86] *ibid.* Art. 21.
[87] *ibid.* Art. 22.

which the shipment will leave the Community. It will authorise the shipment; although the other authorities can comment to that authority about the shipment or impose conditions on its transport. A copy of the consignment note must be given to the customs office through which the waste leaves the Community who will send it to the key authority. The notifier must, within 42 days of the load leaving the Community, inform that authority that the shipment has arrived at its intended destination.

12.47   Transit controls do not apply to shipments of wastes on the green list for recovery.[88] They do apply to amber or red list wastes for recovery in transit from and to countries to which the OECD decision applies. Under Article 24 shipments of these wastes must be notified to the states through which it will pass; who have up to 30 days after acknowledging the notification to object on any of the grounds set out in Article 7.4. If there is no objection then written consent to the shipment should be given. For "red" list and unassigned wastes this consent must be given before the shipment is commenced. Any other shipments of waste for recovery will be dealt with under Article 23.

FINANCIAL GUARANTEES

12.48   All shipments of waste made under the Regulation—except most "green" list waste shipments for recovery—must be covered by either a financial guarantee or have appropriate insurance.[89] The purpose of this requirement is to ensure that funds are available for completion of the disposal or recovery of the waste concerned or for the competent authority to recover costs of dealing with illegal traffic. Circular 13/94 sets out the type of guarantee that will be acceptable and a method of calculating the size of it in relation to a particular shipment.[90]

12.49   The guarantee must be returned when the competent authority has been provided with proof that the object of the shipment has been met. This is either by a certificate of disposal or recovery in an environmentally sound manner or, in the case of base metal waste or scrap in transit through the Community, a copy of the T5 control document confirming that it has left the Community.[91]

12.50   Under regulation 7 of the Transfrontier Shipment of Waste Regulations 1994 no one may ship waste into or out of the United Kingdom unless a certificate has been issued that certifies that the relevant competent authority is satisfied that the appropriate financial guarantee or equivalent insurance is in place in respect of the shipment. A certificate is issued on application and a determination of the application must be made within the times set out in regulation 7(4). Any person who contravenes regulation 7, or supplies false information to obtain a certificate, will commit an offence.[92]

---

[88] Transfrontier Shipment Regulation, Art. 1.3(a).
[89] *ibid.* Art. 27.
[90] D.o.E. Circular 13/94, paras 77–86.
[91] "Shipments" Regulation, Art. 27.2.
[92] S.I. 1994 No. 1137, reg. 12(4) & 12(5).

RETURN OF SHIPMENTS

Where an approved shipment of waste cannot be dealt with in accordance **12.51** with the contract or consignment note it is the duty of the competent authority of dispatch to arrange for its return to the sender unless it is satisfied that the waste can be disposed of recovered in an alternative and environmentally sound manner. Where the authority of dispatch is notified by the authority of destination or by anyone else, that the shipment cannot be properly completed, it has 90 days to ensure the notifier has it returned to the state of dispatch. A new consignment note must be raised[93] but the states of transit and dispatch should not object to a re-shipment initiated by the state of destination if this is done for the right reasons.[94]

Where illegal traffic is the responsibility of the notifier the competent **12.52** authority of dispatch should ensure the shipment is returned to its jurisdiction or, if this is impracticable, otherwise disposed of or recovered in an environmentally sound manner.[95]

The power of United Kingdom authorities to ensure the return of **12.53** shipments is contained in regulation 8 of the Transfrontier Shipment of Wastes Regulations 1994.[96] This enables the authority to serve a notice, in accordance with regulation 17, on the notifier requiring him to return the waste to a specified place by a specified time. The date specified must allow him reasonable time to comply with it, taking into account the present location of the waste.[97] If the notice is not complied with the authority can serve a further notice on him stating that it intends to act as his agent in securing the shipment's return. Non-compliance is also an offence under regulation 12(9). After the further notice is served the authority can act as his agent, and will be deemed to be his duly authorised agent acting within the scope of its authority.[98] It will be the duty of the notifier to provide the authority with such information and assistance as it may reasonably request in writing to enable it to secure the shipment's return.[99] The authority can recover its costs either against the financial guarantee or under Article 33 of the Regulation.

In a number of cases the competent authority has power under the **12.54** Regulation to make decisions. There is no right of appeal against those decisions. The possibility of providing an appeals mechanism was considered in drafting the 1994 Regulations but it was felt that the E.C. Regulation does not make specific provision for an appeals mechanism and that it would not readily accommodate an appeals process.[1]

---

[93] See D.o.E. Circular 13/94, para. 126.
[94] "Shipments" Regulation, Art. 25.
[95] *ibid.* Art. 26.2.
[96] S.I. 1994 No. 1137.
[97] *ibid.* reg. 8(2) & 8(3).
[98] *ibid.* reg. 8(4) & 8(6).
[99] *ibid.* reg. 8(5).
[1] *Hansard*, H.L. Vol. 561, col. 642.

ILLEGAL TRAFFIC

12.55   Under Article 26 a number of contraventions of the Regulation are to be considered illegal traffic. These contraventions are effecting shipments of waste;

(a) without notifying all relevant authorities as required by the Regulation;

(b) without the consent of the relevant authorities;

(c) with a consent obtained by deceit;

(d) which is not properly described in the consignment note;

(e) which results in disposal or recovery in contravention of Community or international rules;

(f) contrary to articles 14 or 16—export to a country to which exports are banned—or articles 19 or 21—imports from countries from which imports are banned.

12.56   If illegal traffic is the responsibility of the notifier its return to the state of origin must be arranged as set out in paragraph 12.44. If it is the responsibility of the consignee the state of destination must ensure that the waste is disposed of in an environmentally sound manner either by the consignee or the authority within 30 days of becoming aware of the situation, although a longer time can be agreed.[2] To implement this in the United Kingdom regulation 9 of the Transfrontier Shipment of Waste Regulations 1994 enables the authority to serve a notice on the consignee to require the wastes proper disposal in accordance with the notice and by a specified date; which should allow a reasonable time for disposal. Failure to comply with the notice will result in the authority serving a further notice stating that the authority will now deal with the waste itself. Powers of entry and inspection for this purpose are provided in regulation 9(6) which also enables the authority to remove the waste from its current site and either dispose of or recover it or arrange for this to be done. The authority can request, in writing, information or assistance from the consignee to carry out its functions and the consignee must comply with any such reasonable request. Failure to comply with the initial notice or obstruction of an officer of the authority will be an offence.[3]

12.57   If responsibility cannot be assigned to either the consignee or the notifier all competent authorities should co-operate to ensure the waste is disposed of in an environmentally sound manner.[4]

12.58   Member States are required to take appropriate legal action to prohibit and punish illegal traffic.[5] In the United Kingdom this is effected through regulation 12(1) of the Transfrontier Shipment of Waste Regulation 1994 which makes it an offence for any person to contravene the Regulation in the United Kingdom so that waste is shipped in circumstances which are deemed to be illegal traffic under Article 26.

[2] E.C. "Shipments" Regulation, Art. 26.3
[3] S.I. 1994 No. 1137, reg. 12(9) & 12(10).
[4] E.C. "Shipments" Regulation, Art. 26.4.
[5] *ibid.* Art. 26.5.

ENFORCEMENT

Under Article 30 Member States are required to take the measures **12.59** necessary to ensure that waste is shipped in accordance with the Regulation. This should include inspection of disposal and recovery sites and spot checks of shipments. The checks can be carried out at the producer's or notifier's premises, at the waste's destination, at the Community's borders and during transit. They can include the inspection of documents, confirmation of identity and of the waste.

Customs offices may be designated by a Member State through which **12.60** shipments entering or leaving the Community may be sent. If such offices are designated shipments may not be sent through any other customs offices in that state.[6] Under regulation 10 of the Transfrontier Shipment of Waste Regulations 1994 a customs officer may detain a shipment for three working days at the request of a United Kingdom competent authority. It will be an offence to intentionally obstruct a customs officer in the exercise of has powers of detention.[7]

Under Article 13 of the Basel Convention parties must provide certain **12.61** information to the Secretariat. Article 41 of the E.C. Shipments Regulation requires the annual return of information to the Secretariat with a copy for the Commission. In the United Kingdom this requirement will be implemented through regulation 16 of the Transfrontier Shipment of Waste Regulations 1994 which enables the Secretary of State to require information from competent authorities. A direction for this purpose is contained in Annex 3 of D.o.E. Circular 13/94 which requires a copy of every consignment note they issue to be sent to the Secretary.

All documents sent to or by any competent authority must be retained for **12.62** three years by the authorities, the notifier and the consignee.[8] This is considered to mean from three years after completion of the last shipment covered by the document.

In the United Kingdom illegal traffic and other breaches of the **12.63** requirements of the Regulation are criminal offences by virtue of regulation 12 of the Transfrontier Shipment of Waste Regulations 1994. Where an offence is committed by a person because of the act or default of another that other can be charged with and convicted of the offence regardless of whether the original offender is also dealt with.[9] This will enable junior company officials to be charged for offences for which they are responsible. Senior managers will be made liable under regulation 13 for offences which they have consented to, connived at or which have occurred through their neglect.

It will be a defence to a charge under the Regulations for the defendant to **12.64** show that he took all reasonable steps and exercised all due diligence to

---

[6] E.C. "Shipments" Regulation, Art. 39.
[7] S.I. 1994 No. 1137, reg. 12(10).
[8] E.C. "Shipments" Regulation, Art. 35.
[9] S.I. 1994 No. 1137, reg. 12(11).

avoid the commission of the offence.[10] Where a person is found guilty of an offence the usual penalty on summary conviction will be a fine not exceeding the statutory maximum (£2,000 in Northern Ireland) or on indictment a fine or imprisonment for up to two years. However, failure to provide the information required by Article 11 is only triable by magistrates and carries a maximum fine of level 3 on the standard scale (£400 in Northern Ireland).[11]

COSTS RECOVERY

12.65   Authorities may recover the administrative costs they incur in implementing the notification and procedure and of appropriate analyses and inspections from the notifier.[12] In the United Kingdom the Secretary of State's view is that authorities should aim for full costs recovery using the charging scheme guidance set out in the Treasury's "Fees and Charges Guide".[13] If payments are made in advance of approval there should be a refund where consent is refused.

12.66   Costs may also be recovered in respect of an authority's involvement in the return of waste or its disposal or recovery. The person charged here will be either the notifier, the consignee or the Member State of dispatch as appropriate.[14] Usually where there is a financial guarantee the costs concerned should be recoverable from that.

4. SHIPMENTS OF RADIOACTIVE WASTES

12.67   The International Atomic Energy Authority has developed a Code of Practice on the International Transboundary Movement of Radioactive Waste.[15] This has no legal force as such but has been used to develop controls on such movements. The Code of Practice establishes general principles designed to serve as guidelines in ensuring the safety of international transboundary movements of radioactive waste. It provides that such movements should take place only when they are authorised by all states involved in the movement, when all stages of the movement can be conducted in a manner consistent with international safety standards and when all states involved in the movement have the administrative and technical capacity and regulatory structure to fulfil their respective responsibilities for the movement in a manner consistent with international safety standards. The Code relies on existing relevant international standards and does not establish separate guidelines here.[16]

---

[10] E.C. "Shipments" Regulation, reg. 14(1).
[11] ibid. reg. 15.
[12] E.C. "Shipments" Regulation, Art. 33.1.
[13] D.o.E. Circular 13/94, para. 166.
[14] E.C. "Shipment" Regulation, Art. 33.2, 33.3 and 33.4.
[15] IAEA Doc. GC (XXXIV)/920, June 27, 1990.
[16] ibid. para. 6.

The Code has been dealt with in the European Community by Council 12.68
Directive 92/3/EURATOM on the supervision and control of shipments of
radioactive waste between Member States and into and out of the Community.[17] For the purposes of the directive "radioactive waste" means any
material which contains or is contaminated by radio-nuclides and for which
no use is foreseen.[18] It applies to transboundary shipments when the
quantities and concentration exceed the levels laid down in Articles 4(a) and
(b) of Directive 80/836/EURATOM.[19] The directive became effective on
January 1, 1994.[20]

Under the directive, states have established procedures by which radioac- 12.69
tive wastes being imported from, exported to or passing through their
territory must be authorised. The state of export may only release a
shipment if all the other relevant states have consented to it.[21] However, no
exports may be allowed to Antarctica, certain parties to the Lome Convention and to states which in the opinion of the authorities in the exporting
country do not have the technical, legal or administrative capacity to deal
with it.[22] If necessary wastes may have to be reshipped, particularly if the
shipment cannot be completed or there is a breach of the conditions for
shipment.[23]

In the United Kingdom the directive is implemented by the Transfrontier 12.70
Shipment of Radioactive Substances Regulations 1993.[24] Shipments come
under the control of the either the Agency in England or Wales, SEPA in
Scotland or the Alkali and Radiochemical Inspectorate in Northern Ireland.[25]
The Regulations apply to shipments between Member States and into and
out of the Community except those small, low level shipments that are
exempted under the EURATOM Directive and the return of sealed sources,
other than those containing fissile material, by the user to the supplier.[26]
Transport operations involved in a shipment must comply with Community
or national provisions and international agreements on the transport of
radioactive material.[27] However, contravention of this provision is not an
offence. Regulation 4 makes special provision for shipments involving the
Crown.

5. RESTRICTIONS ON IMPORTS AND EXPORTS

Under section 100 of the Control of Pollution Act 1974 the Secretary of 12.71
State was empowered to make regulations prohibiting or restricting the
importation of injurious substances into the United Kingdom. That section

---

[17] [1992] O.J. L35/24.
[18] Dir. 92/3/EURATOM, Art. 2.
[19] *ibid.* Art. 1 and see [1980] O.J. L246/1.
[20] *ibid.* Art. 21.1.
[21] *ibid.* Arts 7 & 12.2.
[22] *ibid.* Art. 11.
[23] *ibid.* Art. 15.
[24] S.I. 1993 No. 3031.
[25] *ibid.* reg. 1(1) and see reg. 1(2).
[26] *ibid.* reg. 3(1).
[27] *ibid.* reg. 3(2).

was replaced and extended by section 140 of the Environmental Protection Act 1990, while sections 141 and 142 of that Act give further powers to control imports and exports of waste. These new powers apply throughout the United Kingdom except that sections 140 and 142 are only applicable in respect of imports to Northern Ireland.[28] These provisions entered into force on January 1, 1991.[29]

12.72   Section 140 enables the Secretary of State to make regulations to prohibit or restrict the importation into and the landing and unloading in the United Kingdom of any specified substance[30] or article if he considers it appropriate to do so in order to prevent it from causing pollution of the environment or harming human, animal or plant life.[31] In addition regulations can also be made under this section to prohibit or restrict the use of such articles or substances for any purpose, their supply for any purpose or their storage. Any prohibition or restriction can be applied generally or to specific areas, circumstances or persons or if specified conditions are not complied with.[32] Ancillary powers may allow prescribed articles and substances used, supplied or stored in the United Kingdom despite a prohibition or restriction to be treated as waste or controlled waste and for directions to be made as to its treatment or disposal. Alternatively illegally imported material—whether in breach of regulations, other United Kingdom legislation or a Community instrument—may be required to be disposed of or re-exported.[33] Inspectors may be given powers corresponding to those under section 17 of the 1990 Act by, and other incidental or supplemental provisions may be included in, the regulations.[34] The regulations may include penalties for their contravention.[35]

12.73   An advisory committee has been established to advise the Secretary of State as to the exercise of his powers under this section.[36] He must consult this committee before making regulations; other than those enforcing E.C. or other United Kingdom legislation.[37] Once this consultation has taken place the public must be informed of the proposal and allowed to comment on the draft regulations unless the Secretary of State considers that, in respect of a particular substance or article, if this procedure is observed, there is an imminent risk that serious environmental pollution will result.[38] Any comments received on the proposal must be taken into account by the Secretary before making a decision. He may either confirm the draft on which consultation took place or do so in a modified form, but he may only make modifications if he considers that further consultations are unnecces-

[28] E.P.A. 1990, s.164(4).
[29] *ibid.* s.164(2).
[30] See *ibid.* s.140(11).
[31] *ibid.* s.140(1).
[32] *ibid.* s.140(2).
[33] *ibid.* s.140(3)(b) & 140(4).
[34] *ibid.* s.140(3)(c) & 140(3)(d).
[35] *ibid.* s.140(9) & 140(10).
[36] *ibid.* s.140(5) & Sched. 12 and S.I.s 1991 No. 1487 & 1488.
[37] *ibid.* s.140(6)(a).
[38] *ibid.* s.140(6)(b) & 140(7).

ary.[39] The Environmental Protection (Controls on Injurious Substances) Regulations 1992[40] are not concerned with imports or exports of substances.

A number of regulations have been issued under section 140. The 12.74 Environmental Protection (Controls on Injurious Substances) Regulations of 1992 and 1993[41] prohibit the supply and use of a number of substances either on their own—such as DBBT or PCP—or for certain purposes—such as cadmium in pigments. There is usually an exception for research and development uses or supply for such a use. The Regulations contain a penalty for contravention or causing or permitting a contravention of a fine of level 5 on summary conviction or on conviction on indictment of a fine or up to two years' imprisonment.

More specific regulations deal with certain types of substance. The 12.75 Environmental Protection (Non-Refillable Refrigerant Containers) Regulations 1994[42] prohibit, *inter alia*, the importation, landing or unloading of a non-refillable refrigerant container into the United Kingdom. Imports are allowed for export purposes or for research and development.

The powers under section 140 are purely for environmental protection 12.76 from harmful substances or articles. Section 141 of the 1990 Act enables the Secretary of State to make regulations prohibiting or restricting the import or export of waste—not just controlled waste—for environmental reasons and for preservation of waste disposal facilities or resources in the United Kingdom or, indeed, in any other country.[43] Different provision may be made for different types of waste or wastes of any type in different circumstances.[44] This provision allows effect to be given to the doctrine of self-sufficiency in waste disposal capability.

These regulations will be enforced by waste regulation authorities or, in 12.77 Northern Ireland, by district councils; although their normal powers in respect of the importation of waste may be extended or limited in the regulations as the Secretary of State considers appropriate.[45] The regulations may also give them power to deal with an individual consignment of waste or consignments in a series by the same person but may not enable them to control consignments or types of consignment generally. Alternatively, they may enable the Secretary of State, by direction, to transfer any powers conferred by the regulations to himself.[46]

In addition regulations under this section may prohibit imports or exports 12.78 either absolutely or only if specified conditions or procedures are not complied with. They may also impose duties that are to be complied with before, on or after any import or export by those concerned with the

[39] E.P.A. 1990, s.140(6)(c) & 140(8).
[40] S.I. 1992 No. 31.
[41] S.I. 1992 No. 31 and (No. 2) 1583 and S.I. 1993 No. 1 and (No. 2) 1643.
[42] S.I. 1994 No. 199.
[43] E.P.A. 1990, s.141(1).
[44] *ibid.* s.141(2).
[45] *ibid.* s.141(3), 141(4) & 141(7).
[46] *ibid.* s.141(5)(a).

shipment or any waste derived from it. Provision may be made for appeals to the Secretary of State from any determination made by an authority under the regulations. Inspectors of an authority may be given powers corresponding to those under section 69(3) of the 1990 Act and offences for non compliance created. Finally, a public register of information concerning imports and exports of waste may be required to be kept by the Secretary of State, waste regulation authorities and waste collection authorities.[47]

12.79   Section 142 of the Environmental Protection Act 1990 enables the Secretary of State, by regulations, to obtain information about potentially hazardous substances, products or articles. Under this provision he will be able to obtain information relating to their properties, production, distribution, importation or use or intended use and, in relation to products or articles, to their disposal as waste.[48] Regulations made under this section will be integrated with E.C. provisions and, in particular, the current proposed E.C. Regulation on "the evaluation and control of the environmental risks for existing substances.[49] They will impose requirements on importers, manufacturers or suppliers to provide specified information.

12.80   Other controls will also prevent the import of wastes, even if not specifically designed to do so. Thus the Importation of Animal Products and Poultry Products Order 1980[50] prevents the importation of such products into Great Britain without a licence. This would include the importation of any waste containing such products. Similarly the controls imposed under the Import and Export (Plant Health) (Great Britain) Order 1980[51] may also prevent the importation of waste plants, plant products or soil that may contain pests harmful to plants. Similar provisions apply in Northern Ireland.[52]

12.81   The importation of amphibole asbestos into the United Kingdom is prohibited by virtue of regulation 3 of the Asbestos (Prohibitions) Regulations 1992.[53]

---

[47] E.P.A. 1990, s.141(5)(b)–141(5)(g).
[48] *ibid.* s.142(6).
[49] [1990] O.J. C276/1.
[50] S.I. 1980 No. 14 as amended.
[51] S.I. 1980 No. 420 as amended.
[52] *e.g.*: Landing of Carcases and Animal Products Order (N.I.) 1985 S.R. 1985 No. 161.
[53] S.I. 1992 No. 3067.

13. WASTE PLANS

There are two types of waste plans; the National Waste Strategy and the recycling plans made under the Environmental Protection Act 1990 and development plans dealing with waste policies that are made under planning legislation. This chapter is concerned with the National Waste Strategy and development plans, in particular waste local plans. Waste recycling plans are considered at paragraphs 10.05 to 10.10.

Article 7 of the Framework Waste Directive[1] is concerned with waste **13.01** management planning. Its provisions are set out in paragraph 1.17. In addition Article 6 of the Hazardous Waste Directive[2] requires plans to be prepared for the management of hazardous wastes.[3] A competent authority that draws up a waste disposal strategy that does not comply with these requirements is likely to be in breach of Community law. Further, by virtue of Article 14 of the Packaging and Packaging Waste Directive,[4] waste plans drawn up under Article 7 of the Framework Waste Directive should include a specific chapter on the management of packaging and packaging waste, including measures taken for its prevention or reuse.

1. THE NATIONAL WASTE STRATEGIES

Formerly waste disposal authorities prepared and maintained waste **13.02** disposal plans under the provisions of section 2 of the Control of Pollution Act 1974. Section 2 gave way to section 50 of the Environmental Protection Act 1990 on May 31, 1991 when that section was brought into force.[5] Waste Regulation Authorities then prepared plans under that section. It was in turn repealed by the Environment Act 1995.[6] Disposal plans that were current on April 1, 1996 remained in force until the final determination of the national waste strategies or, in England and Wales, the relevant statements under it.[7]

---

[1] Dir. 91/156 [1991] O.J. L78/32.
[2] Dir. 91/689 [1991] O.J. L377/20.
[3] See para. 11.06.
[4] Dir. 94/62 [1994] O.J. L365/10.
[5] S.I. 1991 No. 1319.
[6] E.A. 1995, Sched. 22, para. 78.
[7] *ibid.* Sched. 23, paras 16 & 17.

13.03   Section 92 of the Environment Act 1995 inserted sections 44A and 44B into the Environmental Protection Act 1990. Section 44A provides for the national waste strategy for England and Wales; section 44B, that for Scotland. Both sections aim to implement the plan making requirements of the Framework Waste Directive.[8]

THE STRATEGY MAKING PROCESS

13.04   The Secretary of State is responsible for preparing the national waste strategy for England and Wales. It must contain his policies in relation to the recovery and disposal of waste in those countries; consisting either of a strategy for both countries or two or more strategies which between them relate to the whole of England and Wales.[9] He may from time to time modify the strategy.[10]

13.05   In preparing the strategy or any modification of it, the Secretary of State must consult the Environment Agency and bodies or persons representing the interests of local government and industry. He may also consult such other bodies or persons as he considers appropriate.[11]

13.06   In addition he may direct the Agency to advise him on the policies to be included in the strategy. He may also require it to carry out surveys or investigations on the kinds and quantities of waste likely to be situated in England and Wales, the facilities needed for recovering or disposing of it and any other matters he considers necessary in connection with the strategy and to report its findings to him.[12] Where he directs surveys or investigations, he must specify the matters or areas concerned and make provisions for the way in which they are to be carried out and the reporting of the findings.[13] The Agency, in carrying out such surveys or investigations must consult bodies or persons representing local planning authorities and industry and make its findings available to local planning authorities.[14]

13.07   In Scotland the national waste strategy is prepared by SEPA and may be modified by it.[15] In doing so it must consult such persons or bodies that appear to it to be representative of the interests of industry and such local authorities that appear to it to be likely to be affected by the strategy or modification. It may also consult such persons or bodies as it considers appropriate.[16] The Secretary of State for Scotland may gave SEPA directions as to the policies to be included in the strategy and requiring it to carry out similar surveys and investigations as may be required in England and Wales.[17]

[8] E.P.A. 1990, ss.44A(10) & 44B(9).
[9] E.P.A. 1990, s.44A(1) & 44A(2).
[10] *ibid.* s.44A(3).
[11] *ibid.* s.44A(5).
[12] *ibid.* s.44A(6).
[13] *ibid.* s.44A(7).
[14] *ibid.* s.44A(8).
[15] *ibid.* s.44B(1) & 44B(2).
[16] *ibid.* s.44B(4).
[17] *ibid.* s.44B(5) & 44B(7).

THE STRATEGIES

Each strategy must contain provisions relating to the type, quantity and **13.08** origin of waste to be recovered or disposed of. It must also consider any general technical requirements and any special requirements for particular wastes.[18] These will have been the subject of the surveys and investigations made by the Agencies. In addition the strategies must contain a statement as to how the objectives specified in Schedule 2A of the 1990 Act[19] will be achieved.

Schedule 2A sets out five objectives; all of which are also found in **13.09** paragraph 4 of Schedule 4 to the Waste Management Licensing Regulations 1994.[20] The first is ensuring that waste is recovered or disposed of without endangering human health and without using processes or methods which could harm the environment and, in particular, without risk to water, air, soil, plants or animals, without causing nuisance through noise or odours and without adversely affecting the countryside or places of special interest.[21]

The second objective is that of establishing an integrated and adequate **13.10** network of waste disposal installations, taking account of the best available technology (not technique[22]) not involving excessive costs. The third is to ensure that this network enables the E.C. as a whole to become. self-sufficient in waste disposal, and the Member States individually to move towards that aim, taking into account geographical circumstances or the need for specialised installations for certain types of waste. It should also enable waste to be disposed of in one of the nearest appropriate installations by means of the most appropriate methods and technologies in order to ensure a high level of protection for the environment and public health.[23]

The strategy should encourage the prevention or reduction of waste **13.11** production and its harmfulness. In particular this should be achieved by the development of clean technologies more sparing in their use of natural resources and the technical development and marketing of products designed so as to make no contribution, or to make the smallest possible contribution—by the nature of their manufacture, use or final disposal—to increasing the amount or harmfulness of waste and pollution hazards. In addition, the development of appropriate techniques for the final disposal of dangerous substances contained in waste destined for recovery should be encouraged.[24]

Finally, the strategy must also contain policies to encourage the recovery **13.12** of waste by means of recycling, reuse or reclamation or any other process with a view to extracting secondary raw materials and the use of waste as a source of energy.[25]

---

[18] E.P.A. 1990, ss.44A(4)(b) & 44B(3)(b).
[19] Added by E.A. 1995, s.92(2) & Sched. 12.
[20] S.I. 1994 No. 1056.
[21] E.P.A. 1990, Sched. 2A, para. 1.
[22] cf. E.P.A. 1990, s.7(2).
[23] E.P.A. 1990, Sched. 2A, paras 2 & 3.
[24] ibid. Sched. 2A, para. 4.
[25] ibid. Sched. 2A, para. 5.

13.13   At present no general National Waste Strategy has been issued. A national survey of waste arisings is being conducted but a final strategy is unlikely to appear before 1998.

13.14   The government has issued a White Paper entitled *Making Waste Work*.[26] This is an advisory document but local authorities should have regard to it in drawing up development plans. It sets out three main objectives—to reduce the amount of waste produced; to make the best use of waste which is produced and to choose waste management practices which minimise the risk of immediate and future environmental pollution and harm to human health. These objectives are bolstered by specific targets such as "to recover 40 per cent of municipal waste by the year 2005." In addition the Paper examines options for specific waste streams such as batteries, PCBs and tyres.

13.15   A specific strategy has been drawn up under the Act for imports and exports of waste. The *U.K. Management Plan for Exports and Imports of Waste* came into effect on June 1, 1996. It was amended with effect from January 13, 1997 in an Addendum to D.o.E. News Release 577. The plan is to be implemented by decisions of the Agency on individual shipments.

2. DEVELOPMENT PLANS

13.16   Development plans are made under the provisions of Part II of the Town and Country Planning Act 1990. There are three main types; structure plans, local plans and unitary development plans (UDPs). Structure plans in England are the responsibility of the county planning authority and set out the general policies and proposals for the development and other use of land in the plan area. They will also provide the framework for local plans. Local plans are usually prepared by district planning authorities, although counties may prepare them. They will be concerned with developing the policies and general proposals in the structure plan to a specific area and providing a detailed basis for development control. They can either concern the district's area or a specific subject matter such as waste disposal. Unitary development plans are made by planning authorities in Greater London and metropolitan counties—and, possibly, the new unitary authorities—and are combined structure and local plans for the areas of those authorities. In Wales unitary development plans are prepared by local planning authorities.

13.17   Government guidance states that

> "In drawing up development plans, planning authorities must have regard to national planning policy guidance (in P.P.G.s and M.P.G.s) as well as to any regional guidance. They must also comply with any E.C. requirements—in particular Framework Directive on Waste (75/442/EEC as amended . . .). In setting the broad framework for development in the region, regional guidance may indicate the scale of longer-term demands for land for potentially polluting development, including land for waste

---

[26] Cm. 3040.

management facilities, and any constraints on development arising from the cumulative impact of existing and proposed polluting uses. It should set out guidelines to ensure that structure plans can identify the general locations for, or criteria for the location of, particular industries or facilities that have a regional or national role (*e.g.* large municipal incinerators or specialist waste incinerators or major landfill sites.[27] The objectives for such plans are in part set out in paragraph 4(3) of Schedule 4 to the Waste Management Licensing Regulations[28] which is reproduced in paragraphs 13.11 and 13.12 above. Other guidance is given in paragraphs 2.2 to 2.5 of P.P.G. 23. In addition authorities should have regard to the White Paper *Making Waste Work*."[29]

Regional Planning Policy Guidance is the starting point for the prepara- **13.18** tion of a development plan.[30] They may deal with waste disposal matters. For example R.P.G. 10: *Regional Planning Guidance for the South West* states that

". . . Development plan policies should provide a land use framework for identifying suitable sites for disposal, storage and treatment of waste. Waste local plans should identify broad areas of search which are likely to contain sufficient new sites, or extensions to existing sites, and should set out criteria for assessing proposals . . . Development plans should make provision for alternative methods of disposal and the re-cycling of waste, including its use as a potential energy source."[31]

Structure plans (or U.D.P.s) will contain policies for waste disposal. **13.19** Guidance as to the content of Plans is also set out in P.P.G. 23 at paragraph 2.16. The Essex Structure Plan[32] sets out a number of policies including

"Sites for the deposit of waste will only be permitted where that waste emanates from sources within the county or that waste is identified as acceptable as part of the county's contribution to meeting regional disposal needs or the waste is necessary for land reclamation purposes."

Further it is said that "In considering the suitability of any site for waste disposal the County Council will take account of space already available for the deposit of waste and may refuse permission if there is adequate alternative provision.[33]

Current local plans may also include policies on environmental and waste **13.20** disposal matters. However, in England and Wales local plans made after February 10, 1992 may not contain any policies in respect of the depositing of refuse or waste materials, other than mineral waste, unless it is a plan for a National Park or for an area where such depositing is not a county matter for the purposes of Schedule 1 to the Town and Country Planning Act 1990.[34] Unitary development plans should include waste disposal policies for

---

[27] P.P.G. 23, para. 2.1.
[28] S.I. 1994 No. 1056.
[29] Cm. 3040.
[30] P.P.G. 15, paras 5–12.
[31] R.P.G. 10, paras 4.30 & 4.31.
[32] First Alteration 1987.
[33] *ibid.* Policies WD1 and WD2A.
[34] T.&C.P. Act 1990, s.36 as substituted by the Planning and Compensation Act 1991, s.27 and Sched. 4.

their areas. These should be consistent with the waste regulation authorities' disposal plans (now the waste strategy) or justify any inconsistencies.

13.21   In future therefore waste development planning in England will be in effect a function of county councils, metropolitan councils and London Boroughs. These last two authorities will deal with the topic in their unitary development plans. Counties should have dealt with it in their structure plans, but a number did not, before 1991, do so. In order to ensure that they formulate such policies Schedule 4 to the Planning and Compensation Act 1991 introduced a new section 38 to the Town and Country Planning Act 1990 which sets out provisions for waste local plans. These will be made by county councils, or the local planning authorities for a National Park (or a National Park authority when constituted).[35] Unitary authorities should include waste policies in their U.D.P.s.

13.22   Under section 38 these authorities must, within such period (if any) as the Secretary of State directs prepare a waste local plan—a plan containing waste policies—for their area or include their waste policies in their minerals local plan.[36] For these purposes "waste policies" means the authority's detailed policies and proposals[37] in respect of development which involves the depositing of refuse or waste materials other than mineral waste and of suitable disposal sites or installations for the carrying on of such development.[38] The plan should address the land-use implications of authorities' waste policies: it must consider the need for sites and facilities in particular areas, suitable locations, and the planning criteria likely to apply, including geological, hydrological and other considerations.[39] In formulating these policies the authority must have regard to such information and other considerations as the Secretary of State may prescribe or, in a particular case, direct.[40] Further, the plan should be in general conformity with the structure plan for the area.[41] Planning authorities in National Parks can either prepare waste local plans or include waste policies in their minerals local plans or their local plans.[42]

13.23   The procedures for adopting, altering or replacing a waste local plan are the same as for ordinary local plans.[43] In particular they will be subject to the public participation requirements set out in section 40 to the 1990 Act and local inquiries may be held to determine any objections to the proposals contained in a draft plan or alteration.[44] A plan will be adopted in accordance with the provisions of section 43 of the Act; although the Secretary of State has powers to direct the authority to make modifications

[35] T.&C.P. Act 1990, s.38(3).
[36] *ibid.* s.38(1) & 38(2).
[37] *ibid.* ss.36(11) and 38(6).
[38] *ibid.* s.38(1) as modified by S.I. 1994 No. 1056, Sched. 4, para. 7(3).
[39] P.P.G. Feb. 12, 1992, para. 3.14.
[40] T.&C.P. Act 1990, s.38(5).
[41] *ibid.* s.36(4) & 38(6).
[42] *ibid.* s.38(4).
[43] *ibid.* s.38(7).
[44] *ibid.* s.42 as amended by the Planning and Compensation Act 1991, Sched. 4, para. 18.

to it.[45] Alternatively the Secretary of State has power to call in the plan for his approval.[46] Once the waste local plan is adopted it will take effect from the date of adoption.[47]

Under section 46(11) the Secretary of State may make regulations to deal **13.24** with cases where provisions in a structure or local plan conflict with those of a minerals or waste local plan. The Town and Country Planning (Development Plan) Regulations 1991[48] have been made for this purpose. Regulation 36 provides that usually the waste local plan will prevail unless, in making the structure plan, the planning authority have made a statement that the waste local plan is not in conformity with it.[49]

For some time waste disposal policies in local plans will remain valid until **13.25** replaced by those in waste local plans. Local plans that were in force, or are treated as having been in force, on February 9, 1992 will be known as "saved local plans", and will continue to have effect.[50] In some cases such a plan can be transformed directly into a waste local plan if it meets all the criteria of sections 36 and 38 of the new plan.[51] Otherwise however, a new plan must be made. In drawing up the waste local plan the authority can identify certain of their policies in the old plan as "existing policies". If they do so, and an inquiry is held into the new plan, the inspector need not hear any objections that relate to a policy he is satisfied is an existing policy and in relation to which there has been no significant change in circumstances affecting it since it formed part of the saved local plan.[52]

Development plans will contain other policies that may affect the grant of **13.26** permission for a waste disposal facility. In particular these can relate to landscape and amenity in the plan area. Structure plans and local plans must now contain policies in respect of the conservation of the natural beauty and amenity of the land, the improvement of the physical environment and suitable waste disposal sites and installations.[53]

In Scotland, structure plans will be prepared by authorities within a **13.27** structure plan area designated by the Secretary of State under section 5 of the Town and Country Planning (Scotland) Act 1997. Local plans are prepared by all planning authorities. Such plans will be made in accordance with the Town and Country Planning (Structure and Local Plans) (Scotland) Regulations 1983.[54] The Planning and Compensation Act 1991 did not provide for waste local plans in Scotland, but such plans could be prepared

---

[45] T.&C.P. Act 1990, s.43(3) as amended by the Planning and Compensation Act 1991, Sched. 4, para. 19(2).
[46] *ibid.* ss.4 & 45 as amended by the Planning and Compensation Act 1991, paras. 20 & 21.
[47] *ibid.* ss.36(10) & 38(6).
[48] S.I. 1991 No. 2794.
[49] See D.o.E. Circular 18/91 (W.O. Circular 71/91) para. 29.
[50] Planning and Compensation Act 1991, Sched. 4, paras 43 & 44(1).
[51] *ibid.* Sched. 4, para. 44(2).
[52] *ibid.* para. 46.
[53] T.&C.P. Act 1990, ss.31(3) and 36(3) as substituted by the Planning and Compensation Act 1991, Sched. 4 and modified by S.I. 1994 No. 1056, Sched. 4, para. 7(1) & (2).
[54] S.I. 1983 No. 1590.

by a planning authority under section 11 of the Town and Country Planning (Scotland) Act 1997.[55]

13.28   In Northern Ireland such plans could be made by the Department of the Environment under Part III of the Planning (Northern Ireland) Order 1991[56] in accordance with the Planning (Development Plans) Regulations (Northern Ireland) 1991.[57]

3. THE EFFECT OF DEVELOPMENT PLAN POLICIES

13.29   Section 54A of the Town and Country Planning Act 1990 states

"where, in making any determination under the Planning Acts, regard is to be had to the development plan, the determination shall be made in accordance with the plan unless material considerations indicate otherwise."

This provision is considered in P.P.G. 1 where it is stated that the section

"in effect . . . introduces a presumption in favour of development which is in accordance with the development plan."[58]

13.30   The phrase "unless material considerations indicate otherwise" may give rise to some difficulty. Certainly it would cover the position where the plan was out of date or did not deal with the type of development under consideration. In addition where a plan does not accord with national or regional guidance it could also be argued that an exception should be made to the presumption.

[55] See s.11(4).
[56] S.I. 1991 No. 1220 (N.I. 11).
[57] S.I. 1991 No. 119.
[58] P.P.G.1. para. 25 and see paras 26–30.

14. WASTE DISPOSAL BY LOCAL AUTHORITIES

Under the Control of Pollution Act 1974 waste disposal authorities were responsible for both the disposal of waste and for the regulation of waste disposal activities. This dual role was severely criticised by the House of Commons Select Committee on the Environment who recommended that waste licensing and disposal should be the statutory responsibility of different bodies.[1] At the same time the government felt that the principle of competitive tendering for the provision of local services should apply to waste disposal. The result was the provisions of sections 32, 51 and Schedule 2 of the Environmental Protection Act 1990 which put these two principles into practice in England and Wales.

In Scotland the principles apply but not in the same way. Separation of 14.01 the roles of waste disposal and regulation is achieved by the administrative arrangements made under section 30(7) of the Environmental Protection Act 1990. However, there is no requirement for the establishment of local authority waste disposal companies. This is because it would be wasteful of resources to require councils in Scotland to set up such companies owing to the small and dispersed populations of many of them and the consequent problems of obtaining enough bids for the waste disposal contracts. Nevertheless, public procurement policies will apply to waste disposal operations by Scottish local authorities.

A. WASTE DISPOSAL BY AUTHORITIES IN ENGLAND AND WALES

In England a waste disposal authority will normally be a county council or 14.02 unitary authority. However, in Greater London they will be the London Waste Authorities constituted by the Waste Regulation and Disposal (Authorities) Order 1985[2] and the London Boroughs or the Common Council of the City of London, in Greater Manchester, Wigan District Council and the Greater Manchester Waste Disposal Authority, in Merseyside, the Merseyside

---

[1] H.C. Session 1988–1989 Second Report, "*Toxic Waste*", February 2, 1989.
[2] E.P.A. 1990, s.30(4) & S.I. 1985 No. 1884, Sched. 1, Parts I, II, III, IV & V.

Waste Disposal Authority and in any other metropolitan county in England, the district council. In Wales the county or county borough council will take this role[3] although regional organisations have been formed. Many of these councils were also Waste Regulation Authorities under the Act but will have been required to separate those functions.[4] The arrangements made at that time may be relevant for determining the role of the disposal authority.

14.03   Compliance with the requirements for disengagement from waste disposal activities involved two stages. The first was for those authorities who wished to continue to undertake waste disposal directly to form Local Authority Waste Disposal Companies (LAWDCs) or to make joint arrangements to this end. Otherwise waste disposal functions had to be privatised. The second was for contracts for waste disposal operations to be put out to competitive tendering.

14.04   A local authority that was a waste disposal authority before May 31, 1991, when section 32 of the Environmental Protection Act 1990 was brought into force,[5] was an "existing authority" for the purposes of that section and the transitional provisions of section 77.[6] At that date they will have had waste disposal contracts in force and operated sites and other facilities. However, after then they would be precluded from carrying out waste disposal functions by virtue of section 51(1) of the Act, to ensure a smooth transition that prohibition did not apply to an authority forming a LAWDC until the scheme came into force.[7] Otherwise it was to be the date on which the authority ceased to carry out waste disposal functions or provide waste disposal sites and equipment.[8] However, the Secretary of State may have directed that a different date should apply for a particular authority.[9]

14.05   Until the transition date an existing authority could exercise the powers in section 55(2)(a) and (b) of the Act to recycle waste or to use waste for the production of heat and power.[10] In addition, under section 48(4) it was able to object to recycling proposals of a waste collection authority when it already was engaged in recycling that waste.[11] However, even though section 14(4) of the Control of Pollution Act 1974, which required disposal authorities to ensure waste was disposed of, was still in force on the date the new schemes came into operation, an existing authority was not able to use the provisions of the 1974 Act to bypass its duties under the 1990 Act.[12]

DISENGAGEMENT FROM WASTE DISPOSAL OPERATIONS

14.06   The Environmental Protection Act 1990 required local authorities to disengage from their waste disposal operations. As much of this activity will have taken place before the main provisions of the Act entered into force,

[3] E.P.A. 1990, s.30(2) as substituted by Local Government (Wales) Act 1994, Sched. 9, para. 17.
[4] *ibid.* s.30(7) (repealed).
[5] S.I. 1991 No. 1319.
[6] *ibid.* ss.32(1) & 77(1).
[7] *ibid.* s.77(6)(a)(i).
[8] *ibid.* s.77(6)(a)(ii).
[9] *ibid.* s.77(7).
[10] *ibid.* s.77(9)(a).
[11] *ibid.* s.77(9)(b).
[12] *ibid.* s.77(6)(b), 77(7) & 77(8).

the detail of the disengagement process is now irrelevant. What is important is the position following it. An examination of that position must begin with an outline of the controls imposed by the Local Government and Housing Act 1989 on companies established by local authorities.

(a) Companies in Which Local Authorities Have Interests

Part V of the Local Government and Housing Act 1989 regulates **14.07** companies established by local authorities. The philosophy of its provisions is that a company[13] that is under the control of a local authority is an extension of the authority and should be subject to the same financial and other regimes as authorities. However, not all companies are controlled by an authority to the same degree. Part V therefore established three main categories; local authority controlled companies, local authority influenced companies and companies in which an authority have a minority interest. In addition local authority controlled companies are divided into those under the direct control of the authority and those at "arm's length". Further provisions with respect to such companies are made in the Local Authorities (Companies) Order 1995.[14]

If a local authority is a member of a company and controls the **14.08** composition of the board or if it holds more than 50 per cent in nominal value of the company's equity share capital, it will be in control of that company, or the subsidiary of such a company, for the purposes of Part V of the 1989 Act.[15] A company will also be under the control of an authority if the authority control a majority of the votes at the company's general meeting[16] or if, even though not a member of the company, it can appoint or remove a majority of the directors.[17] Finally, a company that is controlled by another company to which these provisions apply will also be considered under the control of the authority.[18] The status of a company as a "controlled company" is not static. It only applies to companies that "are for the time being" under the authority's control. Further the Secretary of State has power to direct that a particular company is not a "controlled company" for a specified time.

A company can be controlled by a local authority but detached from it if **14.09** it is an "arm's length company" in a particular financial year, as defined in section 68(6) of the 1989 Act. For the company to have this status the authority must pass a resolution before the start of each of its financial years that it will be so regarded. Eight conditions must be fulfilled throughout the year for the status to be maintained. Three concern directors. They must be appointed for a fixed term of at least two years, no director may have been removed by an ordinary resolution before his period of office expired

---

[13] As defined in s.67 of the Act.
[14] S.I. 1995 No. 849, as amended by S.I. 1996 No. 621.
[15] Local Government and Housing Act 1989, s.68(1)(a) and Companies Act 1985, s.736.
[16] ibid. s.68(1)(b) & 68(3).
[17] ibid. s.68(1)(c) & 68(4).
[18] ibid. s.68(1)(d).

(although the Secretary of State may waive this requirement[19]), and no more than 20 per cent of the directors may have been members or officers of the authority.[20] In addition the company must not have had any special treatment for tenancy arrangements with the authority[21] nor any special financial arrangements. However, the authority may loan, or guarantee, sums for the acquisition of fixed assets or for working capital, and make grants to it as long as these are arranged before the start of the company's relevant financial year.[22] No grant may be made that is connected to the company's financial results for a particular period.[23] Finally, the company must have entered into an agreement with the authority, before the relevant financial year, that it will use its best endeavours to achieve a specified positive return on its assets.[24] This latter condition will be expressed in a business strategy or market plan.

14.10   A company that is not for the time being a controlled company may be subject to local authority influence as provided by section 69 of the 1989 Act. The key criteria here are that the authority is in a "business relationship" with the authority whereby the company receives the benefits described in sections 69(3) of the Act and the authority has at least a 20 per cent interest in the company. This 20 per cent interest can be realised as either control of 20 per cent of the total voting rights of all the members entitled to vote at the company's general meeting,[25] or because at least 20 per cent of the company's directors are persons associated with the authority or because 20 per cent of the voting rights at a director's meeting are controlled by such persons.[26] The Secretary of State may issue a direction that such a company, or type of company, is not to be treated as an influenced company for a certain time or subject to certain conditions.[27] A local authority influenced company cannot be an "arms length company" because such companies may be wholly owned by the authority even if only 20 per cent of their directors are appointed by it.

14.11   Local authority controlled or influenced companies may be the subject of orders made under section 70 of the 1989 Act to regulate their activities. Different provisions may be applied to "controlled" and "arm's length" companies. Authorities must comply with such orders or risk any relevant expenditure contrary to their provisions being deemed unlawful for the purposes of Part III of the Local Government Finance Act 1982.[28] This would make those responsible for incurring the expenditure liable to repay it and members of the authority concerned with the transaction liable to disqualification from being councillors.[29] Despite the wording of these

---

[19] Local Government and Housing Act 1989, s.68(7).
[20] *ibid.* s.68(6)(a)–(c).
[21] *ibid.* s.68(6)(d).
[22] *ibid.* s.68(f) & 68(g).
[23] *ibid.* s.68(6)(h).
[24] *ibid.* s.68(6)(e).
[25] *ibid.* s.69(1)(a) & 69(5).
[26] *ibid.* s.69(1)(b) & 69(1)(c).
[27] *ibid.* s.69(2).
[28] *ibid.* s.70(2) & 70(4).
[29] Local Government Finance Act 1982, s.19.

provisions that the expenditure will be "deemed" unlawful it is submitted that the auditor would still have to apply for a declaration to this effect under section 19(1) of the 1982 Act.

The real purpose of these provisions is, however, to ensure that controlled 14.12 or influenced companies are subjected to the capital controls of Part IV of the 1989 Act. Under section 39(5) of the Act the Secretary of State may make orders for the purposes of treating anything done by a local authority controlled or influenced company as if they were done by the authority. An order would make the authority and the companies concerned members of a "local authority group". The order will apply the provisions of Part IV, with necessary modifications, to the group and regulate accounting procedures regarding dealings between the group members and changes in the capitalisation or capital structure of companies in it.[30]

A local authority company will only escape these provisions under a 14.13 direction made by the Secretary of State or if the authority has less than a 20 per cent interest in the company. Even then it will be subject to the controls of section 71 of the 1989 Act unless it is to become a "controlled" company or an "authorised company". Under section 71 a local authority will need the approval of the Secretary of State to subscribe for or acquire shares in the company; to become a member of it if the company is limited by guarantee, to nominate someone as a member, to appoint directors to it, to allow an officer to make such a nomination or appointment and to allow an officer, in the course of his employment, to become or remain a member or director of the company.[31] An officer or member appointed to an authorised company by or on behalf of the authority must be accountable to the authority.[32]

Some LAWDCs may result from joint arrangements made by local 14.14 authorities. A company so formed will be a "controlled" company, even though no one authority has the required interest, if all the authorities involved control 50 per cent or more of the relevant shares, appointments or votes. In such a case the company will be treated as being controlled by each of the authorities with an interest in it.[33] A company in which local authorities control 20 per cent or more of the relevant votes or directorships and with which more than one of those authorities have a business relationship will be regarded as "influenced", for the purposes of Part V, by each of the authorities in the group that are concerned with it.[34]

(b) Schemes for Disengagement

Two basic options were open to authorities in disengaging their waste 14.15 disposal operations. They could have made voluntary arrangements for partially or wholly transferring their sites and operations to private contractors or formed LAWDCs under the direction of the Secretary of State. As

---

[30] Local Government and Housing Act 1989, s.39(6).
[31] *ibid.* s.71(2).
[32] *ibid.* s.71(5) & 71(7).
[33] *ibid.* s.73(1).
[34] *ibid.* s.73(2) & 73(3).

these arrangements have now been made it is not intended to consider the transfer process in detail. However, those matters that continue to be of importance are dealt with in greater depth, The new arrangements were made in accordance with sections 32(2) to 32(6) and Part I of Schedule 2 to the Environmental Protection Act 1990. Further authorities had to have regard to the advice about the arrangements contained in Department of the Environment Circular 8/91.[35]

Arrangements made under the direction of the Secretary of State

14.16   Under section 32(2) of the Environmental Protection Act 1990 the Secretary of State was enabled to issue directions to Waste Disposal Authorities, or joint authorities, that were in existence on May 31, 1991[36] requiring them to form or participate in the forming of a LAWDC and to transfer their relevant undertakings and functions to that company before a specified date. Notices of direction will have been issued under paragraph 2 of Schedule 2 to the Act; although authorities had the right to request an exemption from direction on the grounds that they were establishing alternative arrangements.[37] They could also make representations about the form of the proposed direction. Where the proposal was for the joint formation of a LAWDC by several authorities, one of those authorities will have been chosen to represent the others—"the representative authority".[38]

14.17   Once the direction was issued an authority that was not allowed to make other arrangements will have formed a shell company under the Companies Act 1985. Such a company will have been wholly owned bv the authority and thus a controlled company for the purposes of the Local Government and Housing Act 1989. Before the LAWDC became effective arrangements must have been made to ensure that it was operated as an "arm's length" company by the auithority.[39]

14.18   Following the formation of the company a scheme for the transfer of relevant properties, rights and liabilities will have been drawn up under the provisions of paragraphs 6 and 7 of Schedule 2.[40] To enable the Secretary of State to approve the scheme the authority will also have drawn up a business plan for the venture. The scheme will then have been sent to the Secretary of State for his comments and tenders for the business of the company issued. Contracts were awarded and on that basis the final scheme submitted for approval, or new arrangements considered. If the scheme was approved a date on which it vested will have been set.

14.19   It is in the provisions of the transfer scheme that the advantages of forming a LAWDC become apparent. Under paragraph 9(2) of Schedule 2 assets acquired under a scheme will be treated for the purposes of capital

---

[35] W.O. Circular 24/91.
[36] E.P.A. 1990, s.32(1) & S.I. 1991 No. 1310.
[37] *ibid.* Sched. 2, para. 3 and Circular 8/91, para. 23 and Annex B.
[38] *ibid.* Sched. 2, para. 4(2) & 4(3).
[39] *ibid.* Sched. 2, para. 5.
[40] Taking into account the advice in Circular 8/91, Annex B.

gains assessment as if they were acquired on the vesting date for their then market value. LAWDCs will be able to claim capital allowances under Chapter I of Part II of the Capital Allowances Act 1990 based on these values. In addition no stamp duty was payable on a transfer of assets under a scheme.[41]

Where a waste disposal authority operated a waste disposal site it will 14.20 have granted itself deemed planning permission to do so under regulation 4 of the Town and Country Planning General Regulations 1976.[42] Such planning permissions will continue to enure for the benefit of the land concerned if transferred to a LAWDC under a scheme.[43] However, the company will have to have a waste management licence for the site.[44]

The scheme transferred the property, rights and liabilities, representing 14.21 the authority's waste disposal undertaking, to the company. Some division of assets held by the authority for one purpose may have been necessary.[45] The authority and the company will have drawn up schedules of the properties, etc., transferred to them. These may include agreements for leases or the creation of other interests in land and for the allocation of statutory responsibilities.[46] Such agreements will be construed in the manner provided by paragraph 13 and 14 of Schedule 2. The transactions between the authority and the LAWDC, as set out in the schedules, will bind third parties even if they could have otherwise prevented a particular course of action from taking place.[47] However, anyone whose property is devalued by such a transaction may recover compensation in respect of it from either the authority or company, or both; any dispute being resolved by the Lands Tribunal.[48]

The transfer scheme will have come into effect on the date directed by the 14.22 Secretary of State for it to do so. On that date the properties, rights and liabilities transferred by the scheme vested in their new owners in accordance with the scheme.[49] The company must in turn have issued the authority the securities it agreed to issue them under the scheme.[50]

Voluntary Arrangements

A waste disposal authority may have disengaged its waste disposal 14.23 operations voluntarily. Such arrangements must have been made to the satisfaction of the Secretary of State and to the extent set out in section 32(3) of the Environmental Protection Act 1990. Under the provisions of that subsection the authority must have formed its own LAWDC and

[41] D.o.E. Circular 8/91, para. D6.
[42] S.I. 1976 No. 1419 (replaced now by S.I. 1992 No. 1492).
[43] E.P.A. 1990, Sched. 2, para. 10(2).
[44] See D.o.E. Circular 8/91, paras D13 & D14.
[45] E.P.A. 1990, Sched. 2, paras 8(2) & 8(3).
[46] *ibid.* Sched. 2, paras 8(4) & 8(5).
[47] *ibid.* Sched. 15(1).
[48] *ibid.* Sched. 2, para. 15(3).
[49] *ibid.* Sched. 2, para. 6(7).
[50] *ibid.* Sched. 2, para. 6(8).

transferred to it the relevant part of its undertaking.[51] Here the company need not have been controlled by the authority but could be a joint venture in which the authority either has control, influence or a minority share. If it does retain control, the company must be an "arm's length company".[52]

14.24   Alternatively, the authority may have discontinued its waste disposal operations altogether. The discontinuance must be total. It must have disposed of its sites, plant and machinery for dealing with controlled waste to others, whether a private contractor or another LAWDC.[53] Alternatively, where an authority was reluctant to dispose of assets, it must have leased them to the new operator in a way that ensures it ceases to carry on waste disposal itself or have made arrangements to do so.[54] If arrangements have been made but not implemented in a reasonable time the Secretary of State may have issued a direction to the authority to impose a compulsory scheme on it.[55]

WASTE DISPOSAL OPERATIONS

14.25   Under the new regime the waste disposal authorities' role is limited to making arrangements for the disposal of waste collected in its area. Its new duties are set out in section 51 of the Environmental Protection Act 1990. They are to arrange for the disposal of controlled waste that is collected in its areas by the waste collection authorities and to provide civic amenity sites. However, these duties may only be performed under arrangements made with waste disposal contractors.[56]

14.26   For these purposes a "waste disposal contractor" has the meaning assigned to it by section 30(5) of the 1990 Act. The term encompasses a LAWDC or some other company, within the meaning of the Companies Act 1985, the objects of which are the collection, keeping, treatment or disposal of waste for profit. In addition a partnership or individual that does such things in the course of a business will also be a "waste disposal contractor" for these purposes.

14.27   To arrange for the disposal of waste collected by a collection authority, a disposal authority will give it a direction to deliver that waste to a specified person or place.[57] These directions will be the way in which contracts awarded under the provisions of Part II of Schedule 2 of the Act are implemented.

14.28   To facilitate waste disposal operations an authority may make certain ancillary arrangements. It may assist contractors in the provision of waste transfer stations or treatment plant and hire them plant or equipment for

[51] E.P.A. 1990, s.32(3)(a).
[52] *ibid.* s.32(8).
[53] *ibid.* s.32(3)(d).
[54] *ibid.* ss.32(3)(b) & 32(3)(c).
[55] *ibid.* s.32(5).
[56] E.P.A. 1990, s.51(1).
[57] *ibid.* s.51(4)(a).

this purpose and lease them land for such facilities.[58] It could therefore use compulsory purchase powers to acquire an interest in land for this purpose but would not be able to operate the facility itself. In addition an authority may make grants to producers of industrial or commercial waste for the provision of plant or equipment to deal with it before collection or towards a pipeline system through which it is collected by a collection authority.[59] The recycling functions of a waste disposal authority are discussed at paragraph 10.13.

A waste disposal authority will, before May 31, 1991, have operated civic **14.29** amenity sites under the provisions of section 1 of the Refuse Disposal (Amenity) Act 1978; either themselves or under agency agreements with collection authorities. That section was repealed by Schedule 16 to the Environmental Protection Act 1990. Disposal authorities now have a duty to ensure the provision of such sites under section 51 of the 1990 Act. They may only provide them through arrangements made with waste disposal contractors.[60] Agency arrangements with collection authorities, who are no longer empowered to operate such sites, have been brought to an end.[61]

A civic amenity site is a place where waste from houses in the authority's **14.30** area can be deposited. The authority's duty is to ensure that the site or sites it is responsible for are reasonably accessible to its residents, that they are open for the deposit of waste at all reasonable times, including a weekend opening, and that residents are not charged for using them. However, sites can be limited to accepting only certain types of household waste.[62] In addition the sites can be open to non-residents for the deposit of any controlled waste. The authority may make a charge for providing this latter service.[63]

The authority must arrange for the disposal of the household waste **14.31** accumulated at these sites.[64] For these purposes it may also arrange, through contractors, transfer stations or treatment centres where the waste can be kept or treated prior to disposal. It can also make available plant or equipment to contractors for these purposes, or for making waste easier to transport, and lease them land on which the waste can be kept, treated or disposed of.[65] Where waste is deposited at the site by non-residents these provisions also apply to that waste.[66] In addition the authority may make arrangements with contractors for the recycling of such waste or use, sell or otherwise dispose of it or of anything produced from it by virtue of section 55(2) of the Act. Arrangements with contractors for the operation of such sites that were not transferred to a LAWDC will be contained in management contracts.[67]

[58] E.P.A. 1990, s.51(4)(b), (c) & (d).
[59] *ibid.* s.51(4)(e)–(f).
[60] *ibid.* s.51(1).
[61] D.o.E. Circular 8/91, para. B10.
[62] E.P.A. 1990, s.51(2).
[63] *ibid.* s.51(3).
[64] *ibid.* s.51(1)(b).
[65] *ibid.* s.51(5).
[66] *ibid.* s.51(6).
[67] D.o.E. Circular 8/91, para. D8.

Control and Management of a LAWDC

14.32   As long as a waste disposal authority controls a LAWDC it must exercise that control to ensure that the company only engages in the collection, disposal, keeping or treatment of waste or activities incidental or conducive to such operations—such as consultancy work in that field.[68] However, the company will not be subject to the restrictions on activity by the authority itself that is imposed by local government legislation.[69] Companies controlled by the authority must be "arm's length companies"[70] but once control changes to influence or minority interest the company will be able to diversify its activities in accordance with its articles and memorandum of association. A LAWDC will be able to provide services to its parent authority and charge for them as long as it stays within these rules while it is controlled.[71]

14.33   The initial holdings of securities in the LAWDC will have been allocated under the transfer scheme. Nothing in the memorandum or articles of association of the company or any other regulatory document concerning it may impose restrictions on the transfer of such securities. Nor may a company pass a resolution to impose such restrictions or impose them in contracts for transfer of securities.[72] However, the Secretary of State may approve restrictions on transfer, for example to allow for employee share schemes.[73]

14.34   Where the LAWDC is an "arm's length company" only 20 per cent of its board of directors may be members or officers of the controlling authority or authorities.[74] The Secretary of State considers that at least two of the other directors should have commercial experience and that one or two of the senior managers of the company should be on the board.[75] Advice for councillors on the board as regards their duties to the council is contained in paragraph C9 of D.o.E. Circular 8/91. Local authority officers on the board will be bound by section 117 of the Local Government Act 1972 to declare their pecuniary interests, direct or indirect, to the authority.

14.35   The company will appoint auditors under section 384 of the Companies Act 1985. The auditors will have the powers set out in Chapter V of Part XI to the 1985 Act. They will produce annual reports under section 237 of the Act. The company will make an annual return to the registrar of companies under section 363 and the directors of the company will issue an annual report under section 235.

Contracts for waste disposal services

14.36   Authorities may only contract with LAWDCs or other contractors, after May 31, 1991, in accordance with the provisions of Part II of Schedule 2 to the Environmental Protection Act 1990.[76] Those provisions set out the terms

[68] E.P.A. 1990, s.32(8) and D.o.E. Circular 8/91, para. C2.
[69] *ibid.* s.32(7).
[70] *ibid.* s.32(10) and see para. 14.09.
[71] D.o.E. Circular 8/91, para. C13.
[72] *ibid.* Sched. 2, paras 23(1) & 23(2).
[73] *ibid.* para. 23(3) & D.o.E. Circular 8/91, para. C6.
[74] Local Government and Housing Act 1989, s.68(6)(c).
[75] D.o.E. Circular 8/91, para. C7.
[76] E.P.A. 1990, s.32(6) and S.I. 1991 No. 1319.

that might be included in such a contract and the procedure for tendering. The purpose of this procedure is to ensure proper competition in the award of contracts. If the procedure is not followed the disposal authority may not be properly informed and fail to identify the full cost of waste disposal. If so then its ability, in the public interest, to decide whether or not to accept any tender which has been received will be impaired.

This procedure is different from that established for other activities under 14.37 Part I of the Local Government Act 1988; which includes contracts for the collection of refuse.[77] Despite the definition of what is meant by the "collection of refuse" in Schedule 1 to the 1988 Act[78] the term "collection" is not dealt with either in that Act or the 1990 Act. As a LAWDC will collect, as well as dispose of waste, disputes may arise as to under what procedure a contract should have been put out to tender. Where the contract was dealt with under the 1988 Act the dispute may be resolved by the Secretary of State exercising his powers under sections 13 and 14 of that Act. Any dispute arising under the 1990 Act will have to be dealt with as a civil action. Breach of the tendering procedure would render the contract void.[79] However, it is only a failure to follow the tendering procedure laid down in Part II of Schedule 2 that gives rise to a judicial review action; as long as that procedure is followed, any other dispute between the parties is a private law matter.[80] In particular the authority is under a duty to act commercially, which may result in some unfairness.

In drawing up the terms of a contract under the 1990 Act concerning the 14.38 keeping, treatment or disposal of waste the authority must ensure that it avoids undue discrimination in favour of one type of contractor, for example LAWDCs, as against other types of contractor.[81] The way in which the courts will determine whether there has been undue discrimination was set out in *Attorney-General v. Wimbledon Corporation*.[82] Discrimination can be undue even if the reason for it is legitimate.[83] In *R. v. Avon County Council, ex p. Terry Adams Limited*[84] the court was concerned with claims of undue discrimination under Part II. It considered that whether or not a term is discriminatory will largely be a question of fact. Discrimination will be undue if it is unnecessary and of such effect as to destroy, with reference to the contract, any prospect of competitive bidding. The need to ensure the viability of the LAWDC does not justify discrimination in its favour. Indeed, in awarding contracts, or in inviting tenders for them, the authority may not base its decision on the fact that the contractor is, or is not, controlled by it.[85]

Consideration should be given to including terms in the contract for the 14.39 safe disposal or treatment of the waste to be dealt with under it or to

---

[77] Local Government Act 1988, s.2(2)(a).

[78] As amended by E.P.A. 1990, Sched. 15, para. 27.

[79] E.P.A. 1990, Sched. 2, para. 20(1) & D.o.E. Circular 8/91, para. A16.

[80] *Mass Energy Limited v. Birmingham City Council* [1994] Env. L.R. 298.

[81] *ibid.* Sched. 2, para. 18.

[82] [1940] Ch. 180 at 194.

[83] *South of England Electricity Board v. British Oxygen Co.* [1956] 1 W.L.R. 1069.

[84] [1994] Env. L.R. 442, C.A.

[85] E.P.A. 1990, Sched. 2, para. 21.

maximise recycling.[86] These could include conditions as to the manner in which the waste is to be disposed of or requiring pre-treatment before disposal. However, such conditions would have to avoid giving rise to undue discrimination in favour of one type of waste disposal contractor. In the *Terry Adams* case it was said that

> ". . . it is permitted for the W.D.A. to have a policy favouring, to the extent that it is reasonable to do so, disposal by incineration, incorporating waste to energy facilities. It is, however, in the interest of the W.D.A., and of the public served by the W.D.A., to have accurate information for comparing the cost of that preferred option with the cost, over a period of time, of other legitimate means of disposal. It is that information which Parliament intended that W.D.A.s should have."

In addition the contract should contain conditions to allow for shortfalls of waste because a collection authority increases the amount it recycles or to allow the disposal authority to increase the amount of waste to be disposed of by the contractor.[87]

14.40   The authority should first advertise its intention to invite tenders for waste disposal service contracts in at least two publications circulating amongst waste disposal contractors.[88] The advertisement should give a brief description of the contract work—for example disposal of household waste or operation of civic amenity sites; including the amounts of waste expected to be dealt with under the contract.[89] A detailed specification for the contract should be made available for inspection to any person, not just prospective tenderers, free of charge; while a copy of the specification must be supplied on request and on payment of the specified price.[90] The availability of the specification must be set out in the advertisement. In addition the advertisement should include a statement as to the time, which must be reasonable, within which tenderers must submit tenders.[91]

14.41   The published notice must also state that it is proposed to invite tenders for the contract and that those invitations will be sent out in accordance with paragraph 20(4) of Schedule 2 to the 1990 Act.[92] This paragraph requires the authority, after considering the initial responses, to invite at least four of those interested to tender for the contract, or if less than four responses were received, to invite all of those responding to tender. The authority should make it clear as to how tenders received will be evaluated.[93] However, as long as these requirements are complied with there is nothing to prevent the authority from allowing a particular tenderer an extension of time within which to tender, or an opportunity to improve its tender, or an opportunity to revise its tender to bring it into compliance with the specifications of the tender invitation.[94]

---

[86] E.P.A. 1990, Sched. 15, para. 19(1).
[87] D.o.E. Circular 8/91, para. A6. See also *R. v. Cardiff City Council, ex p. Gooding Investments Ltd* [1996] Env. L.R. 288.
[88] *ibid.* Sched. 2, para. 20(2).
[89] D.o.E. Circular 8/91, para. A3.
[90] E.P.A. 1990, Sched. 2, para. 20(2)(b) & (c) and (3)(b) & (c).
[91] *ibid.* paras 20(2)(d) & 20(3)(a).
[92] *ibid.* para. 20(2)(e).
[93] D.o.E. Circular 8/91, para. A8.
[94] *Mass Energy Limited v. Birmingham City Council* [1994] Env. L.R. 298 at 313.

In evaluating tenders the authority may take into account environmental **14.42** factors and the benefits of recycling as well as value for money.[95] The authority will normally be expected to award contracts to the tenderer that gives the lowest total cost for disposing of the waste, within a given set of environmental objectives.[96] It will be for the authority to judge the relative costs and environmental benefits concerned.[97] There is no restriction in these provisions as to the categories of environmental factors to which the authority may properly have regard. Providing that it acts reasonably it may have regard to its own environmental preferences.[98]

A contract that has been entered into by a disposal authority may be **14.43** varied in accordance with its terms. Where any such variation is made in respect of a contract with a LAWDC that the authority controls, the undue discrimination requirements of paragraph 18 of Schedule 2 must be observed in relation to the new terms and conditions.[99]

B. WASTE DISPOSAL BY AUTHORITIES IN SCOTLAND

In Scotland waste disposal authorities will continue to carry out waste **14.44** disposal activities under sections 53 and 56 of the Environmental Protection Act 1990. An authority's main duty will be to arrange for the disposal of waste collected by it in its capacity as a waste collection authority.[1] It can either carry out disposal activities itself or employ contractors for this purpose.

Where it carries out disposal activities an authority may operate waste **14.45** transfer stations, waste recycling plants or landfill sites or incinerators and provide plant and equipment for use there. These facilities may be situated within or ouside the area of the authority.[2] In addition it can construct and maintain pipelines for waste disposal purposes.[3] Powers to recycle waste are granted by section 56 of the 1990 Act.

The authority will now need a waste management licence to operate its **14.46** disposal facilities. Formerly it operated them pursuant to a resolution under section 54 of the Environmental Protection Act 1990. That section was repealed by the Environment Act 1995.[4] Resolutions that were in force on April 1, 1996 continued in force until an application for a licence was either granted, withdrawn or refused and either not appealed or the appeal was withdrawn or rejected. If no application was made the resolution lapsed for the site concerned on October 1, 1996.[5]

---

[95] E.P.A. 1990, Sched. 2, para. 19(2).
[96] D.o.E. Circular 8/91, para. A9.
[97] *ibid.* para. 14.
[98] *R. v. Avon County Council, ex p. Terry Adams Ltd* [1994] Env. L.R. 442, C.A.
[99] E.P.A. 1990, Sched. 2, para. 22.
[1] *ibid.* s.53(10).
[2] *ibid.* s.53(1)(a) & (b).
[3] *ibid.* s.53(2), amending s.45(7) & (10).
[4] E.A. 1995, Sched. 24.
[5] *ibid.* Sched. 23, para. 18.

14.47   Under section 58 of the Environmental Protection Act 1990 the Secretary of State for Scotland may give a waste disposal authority a direction to accept and keep, or to accept and treat or dispose of, controlled waste at specified places on specified terms. Thus he could give a direction that a particular consignment of waste should be held at a particular place pending its disposal or require an authority to dispose of it. It will be the duty of the authority to comply with such a direction.

C. WASTE DISPOSAL IN NORTHERN IRELAND

14.48   Under Article 21 of the Pollution Control and Local Government Order (Northern Ireland) 1978[6] a district council is required to arrange for the disposal of waste collected by it. In doing so it may provide transfer stations or disposal sites and plant or equipment for processing or otherwise disposing of the waste. In addition, it may provide civic amenity sites by virtue of Article 22 of the Order.

14.49   Where the council operates its own waste disposal facilities it must ensure that the provisions of Article 13 of the Order are complied with. This enables the council to operate the facilities in accordance with the conditions of a resolution passed by the council. However, if it subsequently appears to the council that the activities authorised by the resolution would cause danger to public health or serious harm to amenities then, unless the problem can be dealt with by varying the conditions of the resolution, it should discontinue those activities and rescind the resolution.

D. PUBLIC SERVICES CONTRACTS

14.50   In addition to the scheme of competitive tendering under Part II of the Environmental Protection Act 1990, W.D.A.s must also comply with the terms of the Public Services Contracts Regulations 1993[7] in letting contracts for waste disposal or collection services. Waste disposal or collection services are "Part A services contracts" in that they are included in category 16 of Part A of Schedule 1 to the Regulation—Sewerage and refuse disposal service: sanitation and similar services.

14.51   The main exeception to this will be if the contract is for what is known as a public concession. This term is not found in the Regulations but encompasses the type of contract defined in paragraph (e) of the definition of "public services contract"—

> "a contract under which a contracting authority engages a person to provide services to the public lying within its responsibility and under which the consideration given by the contracting authority consists of or includes the right to exploit the provision of the services."[8]

[6] S.I. 1978 No. 1049 (N.I. 19).
[7] S.I. 1993 No. 3228 as amended by S.I. 1995 No. 201, reg. 31.
[8] *ibid.* reg. 2(1).

Such contracts are not "public services contracts" for the purposes of the Regulations. Otherwise the Regulations will normally apply to any contract for waste disposal or collection which has an estimated value (net of VAT) of 200,000 ECU or more.[9]

[9] S.I. 1993 No. 3228, reg. 7(1).

15. UNAUTHORISED HANDLING OF WASTE

Section 33 of the Environmental Protection Act 1990 is concerned with the unauthorised or harmful depositing, treatment or disposal of controlled waste. It makes such activities an offence, subject to various exemptions and defences. It is a similar provision to section 3 of the Control of Pollution Act 1974 which it replaces but there are enough differences between the two for any cases decided under the 1974 Act to be treated with caution. Further the incorporation of Annex IIA and IIB of the framework Directive into section 33(1) by the Waste Management Licensing Regulations 1994[1] also makes earlier cases suspect.

UNAUTHORISED HANDLING OF CONTROLLED WASTE

15.01 Section 33(1)(a) states that a person shall not

> "deposit controlled waste, or knowingly cause or knowingly permit controlled waste to be deposited in or on any land unless a waste management licence authorising the deposit is in force and the deposit is in accordance with the licence."

"Land" for these purposes includes land covered by water where that land is above the low water mark of ordinary spring tides.[2] The definition extends beyond open land to include land forming part of premises.[3] A deposit made by injection into underground strata could be made into both land and "controlled waters" as defined by section 104(l)(d) of the Water Resources Act 1991.

15.02 The deposit of waste includes both permanent and temporary deposit on land. In *R. v. Metropolitan Stipendary Magistrate and Ors, ex p. London Waste Regulation Authority*[4] the divisional court, determining that an earlier decision of that court—*Leigh Land Reclamation Ltd v. Walsall Metropolitan Borough*

---

[1] S.I. 1994 No. 1056, Sched. 4, para. 9(3)–9(5).
[2] E.P.A. 1990, s.29(8) and see Interpretation Act 1978, Sched. 1.
[3] *Gotech Industrial and Environmental Services v. Friel* [1995] S.C.C.R. 22.
[4] [1993] 3 All E.R. 113.

Council[5]—was wrongly decided, held that the provision applied when waste was deposited at its final resting place or when it was deposited temporarily, for example at a waste transfer station. The word "deposit" is to be construed, unless the context otherwise requires, in a broad sense. It can include a continuing state of affairs such as the failure to cover deposited waste contrary to licence conditions.[5a]

A deposit of waste will normally be made by someone like the driver of a **15.03** vehicle carrying it who physically places it onto the land. However, the driver's employers can also be vicariously liable for his actions. To avoid technical arguments on vicarious liability, section 33(5) provides that where Directive waste is unlawfully deposited from a vehicle the person who controls or who is in a position to control the use of the vehicle will be treated for the purposes of section 33(1)(a) as knowingly causing that deposit, whether or not he gave any instructions for that to be done. For these purposes it will make no difference if the driver is an employee or an independent contractor.

The phrase "knowingly cause or knowingly permit to be deposited" is **15.04** intended to catch the person ordering the deposit or accepting it illegally. Under section 3 of the 1974 Act the offence was to "cause or knowingly permit . . ." The addition of "knowingly" before "cause" will make a difference to the way the courts deal with this provision. In *Alphacell Ltd v. Woodward*[6] it was held that knowledge was not a necessary ingredient of an offence of "causing" water pollution so that such an offence was one of strict liability. However, "knowingly cause" does imply some knowledge on the part of a defendant. Now it would have to be shown that the defendant not only caused the deposit—*i.e.* by giving instructions for it to be done—but also that he knew that the waste was controlled waste but not that the site where it was deposited did not have a waste management licence or that the deposit was not in accordance with its conditions.[7] This is not an undue burden on the prosecution. They will have to prove the basic ingredients of the offence—the giving of instructions for the deposit. Then, if the defendant chooses not to give evidence of his absence of knowledge, and there are no circumstances which suggest his ignorance, the court may properly convict him without direct evidence as to knowledge. If he does raise the absence of knowledge then the court will have to be sure that he did have the requisite knowledge before they can convict; section 101 of the Magistrates' Courts Act 1980 having no relevance here.[8]

To "knowingly permit" an offence it would seem the same applies. In **15.05** *Ashcroft v. Cambro Waste Products Ltd*[9] it was held that the offence was made out if the defendant knew of the deposit of controlled waste; the prosecution did not have to go on to show that he knew it was in contravention of a

---

[5] [1991] 3 J.Env.L. 281.
[5a] *Thames Waste Management Ltd v. Surrey County Council* [1977] Env. L.R. 148.
[6] [1972] 2 All E.R. 475.
[7] *Shanks & McEwan (Teeside) Ltd v. Environment Agency, The Times,* January 24, 1997.
[8] *Westminster City Council v. Croyalgrange Ltd* [1986] 1 W.L.R. 674.
[9] [1981] 3 All E.R. 703.

waste disposal licence. Further, a person who was reckless as to the position could be convicted here.[10]

15.06   Section 33(l)(b) forbids anyone to treat, keep or dispose of controlled waste, or knowingly cause or knowingly permit it to be done, in or on any land or by means of any mobile plant[11] except under and in accordance with a waste management licence. Waste is "treated" when it is subjected to any process, including making it re-usable or reclaiming substances from it.[12] Waste can be kept in storage while a person can keep something even though he derives no personal benefit from it and it is also in the possession of someone else.[13] The "disposal" of waste includes its disposal by way of deposit in or on land.[14] "Land" here means treated, etc., on the surface of the land or any structure set into the surface.[15] The effect of this paragraph is to control the business side of waste disposal. Paragraph (a) is more concerned with "fly-tipping". Paragraph (b) would control, amongst other things, scrap yards, chemical transfer stations and mobile incinerators.

15.07   Section 33(l)(c) prohibits anyone to treat, keep or dispose of controlled waste in a manner likely to cause pollution of the environment or harm to human health. Thus it is the *manner* in which the waste is dealt with that is important. This applies even if the defendant is operating in accordance with the conditions of a licence. The environment here means land, water or air.[16] It is polluted for these purposes by the escape or release, into any environmental medium, from land on which controlled waste is treated, kept or deposited or from fixed—or mobile[17]—plant by means of which it is treated or disposed of, of substances (whether natural or artificial; solid, liquid or gaseous[18]) or articles constituting or resulting from waste and capable of causing harm to man or any other living organisms supported by the environment. No actual harm has to be shown. The release or escape merely has to be capable, by reason of the quantity or concentrations involved, of causing harm.[19] Thus if emissions from a waste incinerator "harm" the marine environment the offence will be made out. "Harm" here means harm to the health of living organisms or other interference with the ecological systems of which they form part. Thus air pollution that damages lichen or seepage from a site that kills water insects are covered. As far as harm to man is concerned this includes offence to any of his senses or harm to his property. This is particularly important in that smells from licensed waste disposal operations may not be easy to control. "Harmless" in this context means an escape or release that does not have such effects.[20]

---

[10] *Wilson v. Bird* (1962) 106 S.J. 880.

[11] Plant designed to move or be moved whether on roads or other land—E.P.A. 1990, s.29(9).

[12] E.P.A. 1990, s.29(6).

[13] *D.P.P. v. Turton, The Guardian*, June 8, 1988.

[14] E.P.A. 1990, s.29(6).

[15] *ibid.* s.29(8).

[16] *ibid.* s.29(2).

[17] *ibid.* s.29(4).

[18] *ibid.* s.29(11).

[19] *ibid.* s.29(3).

[20] *ibid.* s.29(5).

Under the 1974 Act there was some doubt as to whether the breach of a **15.08** condition of a waste disposal licence was an offence in itself. This has been clarified by section 33(6) of the Environmental Protection Act 1990 which states that anyone contravening section 33(1) or any condition of a waste management licence commits an offence. As "any condition" is specified this means that the contravention of each condition is an offence rather than the contravention of four conditions just being one offence. This provision overturns the *dicta* in *Leigh Land Reclamation Ltd & Ors v. Walsall MBC*[21] that not every breach of a condition amounts to an offence.

A person committing an offence under section 33(6) will be liable on **15.09** summary conviction to imprisonment for up to six months or a fine not exceeding £20,000 or both or, on indictment to imprisonment for up to two years or a fine or both.[22] However, if the case involves special waste the penalty on indictment is increased to up to five years imprisonment or a fine or both.[23]

Under section 157(1) of the Environmental Protection Act 1990 where a **15.10** body corporate has committed an offence under the Act which is proved to have been committed with the consent or connivance or through the neglect[24] of any director, manager or similar officer of that body or of any person purporting to act in that capacity, he will also be guilty of the offence and liable to be proceeded against and punished accordingly. Connivance is essentially culpable acquiescence in a course of conduct that is reasonably likely to lead to the offence being committed. In the context of adultery, for example, it has been held that for connivance to be established there must be an element of encouragement and it may lapse in time or by the lack of any causal connection between it and the act complained of.[25] If the affairs of the body corporate are managed by its members then a member who has management functions may be liable in the same way as a director.[26]

The purpose of this type of provision is to fix liability only on those who **15.11** are in a real position of authority. On those who have the power and responsibility to decide corporate policy—the "directing minds" of the company. Thus someone who has the title of "general manager" or "assistant manager" but no responsibility for actually governing the company will not be liable under this provision.[27]

This does not mean that junior managers can escape liability. If they **15.12** actually knowingly caused or knowingly permitted the unlawful deposit they can be charged with an offence under section 33(1) in the same way as the company. Alternatively they may be aiders, abettors, counsellors or procurers of the offence.

---

[21] [1991] 3 J.Env.L. 281.
[22] E.P.A. 1990, s.33(8).
[23] *ibid.* s.33(9).
[24] See *Hirschler v. Birch* (1987) 151 J.P. 396.
[25] *Godfrey v. Godfrey* [1965] A.C. 444.
[26] E.P.A. 1990, s.157(2).
[27] *R. v. Boal* [1992] 2 W.L.R. 890 and *Woodhouse v. Walsall Metropolitan Borough Council* [1994] Env. L.R. 30.

15.13   In addition section 158 of the Act provides that where an offence under Part II has been committed by one person but is due to the act or default of some other person, that other person can be charged with and convicted of the offence whether or not proceedings for it are taken against the first defendant

EXEMPTIONS

15.14   The Secretary of State has a discretion to make exemptions from the provisions of section 33(1)(a), (b) or (c) by virtue of section 33(3) of the Environmental Protection Act 1990. In doing so he should have regard to the expediency of removing from control any deposits that are so insignificant or of such a temporary nature that they can be excluded, innocuous methods of treatment or disposal and activities covered by other controls.[28] This discretion has to be read in the light of Article 11 of the Framework Waste Directive which restricts the scope of exemptions. The first point is that exemptions can only be applied to "establishments or undertakings". This term is not defined in the Directive or the Regulations, but is considered to mean organisations as opposed to private individuals.[29] Secondly, exemptions are restricted to such organisations that either carry out their own waste disposal at the place of production or are engaged in waste recovery. In addition any exemptions must be subject to general conditions as to the amounts and types of waste covered by them, the way in which the activity is carried on, and cannot result in a breach of the requirement for prevention of environmental harm imposed by Article 4. Additional exemptions are imposed in the United Kingdom regulations to deal with activities that are not covered by the Directive.[30] Exceptions were made under regulation 9 of the Controlled Waste Regulations[31] but these are no longer relevant.

15.15   A wide range of exemptions is provided by regulation 17 of, and Schedule 3 to the Waste Management Licensing Regulations 1994.[32] These apply only to "ordinary" deposits prohibited under section 33(1)(a) and (b). A deposit that is likely to cause environmental harm or harm to human health in contravention of section 33(1)(c) cannot benefit from the exemptions contained in Schedule 3.[33] The Schedule lists a number of activities as "exempt activities". Deposits in the course of these activities will be exempt if any relevant condition imposed either in regulation 17 or Schedule 3 is complied with. However, where the activity is operated by an establishment or undertaking, the exemption only applies if the type and quantity of waste submitted to the activity, and the method of disposal or recovery, are consistent with the environmental and public protection objectives laid down

---

[28] E.P.A. 1990, s.33(4).
[29] D.o.E. Circular 11/94, Annex 5, para. 5.10.
[30] *ibid.*
[31] S.I. 1992 No. 588.
[32] S.I. 1994 No. 1056.
[33] *ibid.* reg. 17(1).

by the Framework Waste Directive.[34] None of these partial exemptions apply where the waste concerned is special waste unless it expressly applied to such waste in a particular exemption.[35] Where exempt waste is deposited at a licensed site the licence conditions do not apply to such a deposit.[35a]

The prohibitions in section 33(1) do not extend to household waste from a   **15.16** domestic property that is treated, kept or disposed of within the curtilage of the dwelling by, or with the permission of, its occupier.[36] This exemption does not cover the treating, keeping or disposal of asbestos, clinical waste or mineral or synthetic oils or grease as these are excluded from the definition of household waste.[37]

DEFENCES

Under section 33(7) of the Environmental Protection Act 1990 three   **15.17** defences are provided to charges of unauthorised handling of waste. It is for the defendant to prove, on the balance of probabilities, that one of them applies to his case.

It will be a defence to show that he took all reasonable precautions and   **15.18** exercised all due diligence to avoid the commission of the offence.[38] There have been a number of decisions on the way this defence should be applied; although none of them lay down any general principle. Each case will rest on its own facts. In particular a court will look at the size of the business involved to determine the extent of the precautions they should have taken as against the risks the precautions are supposed to avoid. The risks here are environmental pollution or harm to amenity, while the precautions will be similar to those set out in the code of practice on the duty of care under section 34 of the 1990 Act. To establish this defence it will have to be shown that a system had been created that could rationally be said to be so designed that the commission of the offence would be avoided.[39] This "system" can have two bases. It can be an inquiry to a responsible person, such as an agency, as to whether a particular course of action would constitute an offence. As long as all relevant information is given to the authority then, particularly for a small producer, it is submitted that the defence will be made out[40]; although a defendant will not be allowed to shift responsibility for his actions to the authority.[41] However, another firm, whether producer or disposer, may have to implement some form of sampling regime. This will depend on the type of waste involved; if there is a risk of pollution from it then sampling will be more necessary than from

---

[34] S.I. 1994 No. 1056, reg. 17(4).
[35] S.I. 1994 No. 1056, reg. 17(3).
[35a] *London Waste Regulation Authority v. Drinkwater Sabey Ltd* [1997] Env. L.R. 137.
[36] E.P.A. 1990, s.33(2).
[37] Controlled Waste Regulations 1992, S.I. 1992 No. 588, reg. 3(1).
[38] E.P.A. 1990, s.33(7)(a).
[39] *Tesco Ltd v. Nattrass* [1972] A.C. 153 *per* Lord Morris at 180F.
[40] See *Riley v. Webb & Ors* (1987) 151 J.P. 372.
[41] *Taylor v. Lawrence Fraser (Bristol) Ltd* (1977) 121 S.J. 757.

innocuous materials.[42] The scope of the regime will again depend on the waste involved and its origins but a court will expect it to be more than an occasional check unless the standard of the waste is always the same.[43] In addition where it is not possible to check each drum, etc., a defendant will probably have to have verified contents with the originator in accordance with the duty of care.[44] If a company have established such a system which fails because of the actions of an employee then, as long as they can show that the system was being operated properly and that they had exercised all due diligence in relation to it, they will succeed in making out their defence.[45] Merely establishing a system will not be enough.

15.19   Under section 33(7)(b) it is also a defence for a person to show that he acted under instructions from his employer and neither knew nor had reason to suppose that what he was doing was in contravention of section 33(1). "Reason to suppose" here is likely to be interpreted by the courts as "reason to suspect".[46]

15.20   Finally it will be a defence to show that the acts alleged to constitute the contravention were done in an emergency in order to avoid danger to the public and that, as soon as reasonably practicable after they were done, particulars of them were furnished to the relevant waste regulation authority.[47] In *Larchbank v. British Petrol*[48] an "emergency" was considered to be "A condition of things causing a reasonable apprehension of the near approach of danger." However, if the emergency was caused as a result of the defendant's own negligence the defence may not be available while the measures taken to deal with it must be proportionate to the risks involved.[49]

15.21   Section 33(5) of the Act provides that it will not be a defence for the controller of a vehicle from which controlled waste is deposited to plead ignorance. He will be treated as having "knowingly" caused the deposit whether or not he gave any instructions for this to be done. Thus the only defence in these circumstances will be for the controller to be able to show that he actively forbade waste to be dumped from the vehicle. A notice in a vehicle cab to this effect might be sufficient but each case will turn on its own facts.

REMOVAL OF UNAUTHORISED DEPOSITS

15.22   Under section 59 of the Environmental Protection Act 1990 where directive waste is deposited in or on any land[50] in the area of one of the Agencies or of a waste collection authority in contravention of section 33(1)

---

[42] *Hurley v. Martinez & Co. Ltd* (1990) 154 J.P. 821.

[43] *Rotherham MBC v. Raysun (U.K.) Ltd, The Times,* April 27, 1988.

[44] *Amos v. Melcon (Frozen Foods) Ltd* (1985) 149 J.P. 712.

[45] *Tesco Ltd v. Nattrass* [1972] A.C. 153.

[46] *Kent County Council v. Rackham,* February 4, 1991 Q.B.D. (Unreported).

[47] E.P.A. 1990, s.33(7)(c).

[48] [1943] A.C. 299 *per* Lord Atkin at 304.

[49] *Perka et al v. The Queen* [1985] 13 D.L.R. (4th) 1 (CAN).

[50] See E.P.A. 1990, s.29(8).

that Agency or authority may require the removal and cleaning up of that land. This enables action to be taken not only in respect of fly-tipped waste but also waste that has been deposited in breach of licence conditions. However, it is submitted that the latter power should only be exercised by an Agency.

The power is exercised by the service of a "removal notice" on the **15.23** occupier of the land. Such a notice should be served in accordance with section 160 of the Act. The notice may require the removal of the waste within a period that must be not less than 21 days from the date the notice was served or the taking of specified steps with a view to eliminating or reducing the consequences of the deposit or both.[51] As the Act provides for steps to be specified, a notice merely requiring the elimination or reduction of the consequences of the deposit is unlikely to be valid.[52]

The person served with the notice has 21 days from the date of service to **15.24** appeal against it to a magistrates' court or, in Scotland, to the sheriff by way of summary application.[53] This right of appeal must be set out in the notice.[54] The notice will have no effect pending a decision on the appeal.[55] In determining such an appeal the court must quash the notice if it is satisfied that the appellant neither deposited, nor knowingly caused or knowingly permitted the deposit of, the waste on the land or if there is a material defect in the notice. Otherwise it may modify the notice or dismiss the appeal[56] or extend time for compliance with it.[57] There would be a material defect here if the deposit was not prohibited under section 33(1), or if it was prohibited but the occupier would have a defence under section 33(7). Either party to the proceedings may appeal to the Crown Court against the magistrates' decision—or in Scotland from the sheriff to the Court of Session.[58] If such an appeal is brought then the notice will remain suspended until the determination of the appeal.[59] It will be the duty of the Agency or the waste collection authority to give effect to the decision of the court finally determining the matter.[60]

If a person served with a remedial notice fails, without reasonable excuse, **15.25** to comply with it, he will be guilty of an offence and liable on summary conviction to a fine not exceeding level 5 on the standard scale. In addition he will be liable to a further fine of up to one fifth of level 5 for each day that the notice remains not complied with after conviction and before the relevant authority exercises powers to fulfil its requirements.[61] Following the conviction the relevant authority will be able to carry out the requirements

---

[51] E.P.A. 1990, s.59(1).
[52] *Sterling Homes v. Birmingham City Council* [1996] Env. L.R. 121.
[53] E.P.A. 1990, s.59(2).
[54] *ibid.* s.73(4).
[55] *ibid.* s.59(4).
[56] *ibid.* s.59(3).
[57] *ibid.* s.59(4).
[58] *ibid.* s.73(1) & 73(2).
[59] *ibid.* s.73(3).
[60] *ibid.* s.73(5).
[61] *ibid.* s.59(5).

of the notice itself and recover from the defendant any expenses reasonably incurred in so doing.[62]

15.26   In addition to their powers to serve removal notices, relevant authorities may also take remedial action in respect of unlawful deposits under section 59(7) of the Act. This power can be exercised where immediate steps are necessary in order to remove or prevent pollution or harm to human health or to eliminate or reduce the consequences of the deposit or both. It can also be used where there is no occupier of the land in question or any occupier of it was not responsible for the deposit in question.

15.27   Where the Agency or a waste collection authority take remedial action under section 59(7) it will be entitled to recover the costs of doing so and of disposing of the waste concerned.[63] Such waste will become the property of the authority removing it.[64] The costs recoverable by collection authorities may include a sum reflecting their establishment charges,[65] the Agencies may charge for the service rendered by virtue of section 43 of the Environment Act 1995. Charges may be recovered from the occupier of the relevant land unless he proves that he neither made nor knowingly caused or knowingly permitted the deposit. In any case they may be recovered from the person who did make or knowingly cause or knowingly permit it. However, recovery is limited to those costs necessarily incurred in taking the relevant action.[66]

NORTHERN IRELAND

15.28   In Northern Ireland the offence of unlicensed disposal of waste is provided in Article 5 of the Pollution Control and Local Government Order (Northern Ireland) 1978.[67] This should be read in conjunction with the Waste Collection and Disposal Regulations (Northern Ireland) 1992.[68]

---

[62] E.P.A. 1990, s.59(6).
[63] ibid. s.59(8).
[64] ibid. s.59(9).
[65] Local Government Act 1974, s.36 and Local Government (Scotland) Act 1947, s.193.
[66] E.P.A. 1990, s.59(8).
[67] S.I. 1978 No. 1049 (N.I. 19).
[68] S.R. 1992 No. 254.

16. PARTIAL EXEMPTIONS FROM CONTROL

This chapter is entitled "Partial Exemptions from Control" to reinforce the point that none of the exemptions contained in Schedule 3 to the Waste Management Licensing Regulations 1994[1] give total immunity from action. They do not apply if:

(1) an offence is committed under section 33(1)(c),
(2) the waste concerned is special waste—unless a particular exemption specifically covers that waste,
(3) if the activity concerned is conducted in a way that contravenes Article 4 of the Framework Waste Directive.

These, and other restrictions on the scope of the "exemptions", should make those covered by them wary of relaxing in the belief that they have a total immunity. They will not need a waste management licence but that does not mean that they are exempt from control. In addition if the activity is carried on by an establishment or undertaking it will have to be registered under regulation 18 of the 1994 Regulations.

STATUTORY EXEMPTIONS

16.01 The Secretary of State has a discretion to make exemptions from the provisions of section 33(1)(a), (b) or (c) by virtue of section 33(3) of the Environmental Protection Act 1990. In doing so he should have regard to the expediency of removing from control any deposits that are so insignificant or of such a temporary nature that they can be excluded, innocuous methods of treatment or disposal and activities covered by other controls.[2] This discretion has to be read in the light of Article 11 of the Framework Waste Directive which restricts the scope of exemptions. The first point is that exemptions can only be applied to "establishments or undertakings". This term is not defined in the Directive or the Regulations, but is considered to mean organisations as opposed to private individuals.[3] Secondly exemptions are restricted to such organisations that either carry out

---

[1] S.I. 1994 No. 1056.
[2] E.P.A. 1990, s.33(4).
[3] D.o.E. Circular 11/94, Annex 5, para. 5.10.

their own waste disposal at the place of production or are engaged in waste recovery. In addition any exemptions must be subject to general conditions as to the amounts and types of waste covered by them, the way in which the activity is carried on, and cannot result in a breach of the requirement for prevention of environmental harm imposed by Article 4. Additional exemptions are imposed in the 1994 Regulations to deal with activities that are not covered by the Directive.[4]

16.02   The prohibitions on unlawful dealing with waste in section 33(1) do not extend to household waste from a domestic property that is treated, kept or disposed of within the curtilage of the dwelling by, or with the permission of its occupier.[5] This exemption does not cover the treating, keeping or disposal of asbestos, clinical waste or mineral or synthetic oils or grease as these are excluded from the definition of household waste.[6]

16.03   In Scotland controlled waste that is deposited in or on land occupied by the waste disposal authority for the area or is kept, treated or disposed of in or on such land or by means of mobile plant operated by them, will be exempt from the provisions of section 33(1)(a) and (b).[7] This exemption will only apply if conditions specified by SEPA as to the use or operation of the site or plant by the disposal authority are complied with. If the site or plant is run by anyone else they must have the consent of the disposal authority and operate it in accordance with any conditions to which that consent is subject.[8] It should be noted that there is no exemption for the "environmental harm" provision in section 33(1)(c) here.

UNDER THE WASTE MANAGEMENT LICENSING REGULATIONS 1994

16.04   A wide range of exemptions is provided by regulation 17 of, and Schedule 3 to, the Waste Management Licensing Regulations 1994.[9] These apply only to "ordinary" deposits prohibited under section 33(1)(a) and (b). A deposit that is likely to cause environmental harm or harm to human health in contravention of section 33(1)(c) cannot benefit from the exemptions contained in Schedule 3.[10] The Schedule lists a number of activities as "exempt activities". Deposits in the course of these activities will be exempt if any relevant condition imposed either in regulation 17 or Schedule 3 is complied with. However, where the activity is operated by an establishment or undertaking, the exemption only applies if the type and quantity of waste submitted to the activity, and the method of disposal or recovery, are consistent with the environmental and public protection objectives laid down by the Framework Waste Directive.[11] None of these partial exemptions

---

[4] D.o.E. Circular 11/94, Annex 5, para. 5.10.
[5] *ibid.* s.33(2).
[6] Controlled Waste Regulations 1992, S.I. 1992 No. 588, reg. 3(1).
[7] E.P.A. 1990, s.54(1).
[8] *ibid.* s.54(3).
[9] S.I. 1994 No. 1056.
[10] *ibid.* reg. 17(1).
[11] *ibid.* reg. 17(4) and see D.o.E. Circular 11/94, Annex 5, para. 5.34/5.

apply where the waste concerned is special waste unless it expressly applied to such waste in a particular exemption. Nor can an establishment or undertaking disposing of special waste at the place of production benefit from these exemptions.[12]

In addition to the activities exempted under regulation 17, regulation 16 **16.05** excludes activities controlled by other regulatory regimes from waste licensing. This exclusion applies to offences under section 33(1)(a), (b) and (c) and cases involving special waste. Activities involving the deposit in or on land, the recovery or disposal of waste and designated for central control under Part I of the Environmental Protection Act 1990 will be excluded if authorised under that regime,[13] as will incinerators disposing of waste that are authorised by local authorities by virtue of section 5.1 of Schedule 1 to the Prescribed Process and Substances Regulations 1991.[14] These exclusions only apply in relation to releases to the air or water. They do not cover an operation that involves the final disposal of waste by deposit in or on land.[15]

Regulation 16 excludes activities involving the discharge of liquid waste **16.06** under a consent granted by virtue of Chapter II of Part III of the Water Resources Act 1991 or Part II of the Control of Pollution Act 1974. The exclusions under the 1974 Act will now mainly apply in Scotland. Disposals at sea authorised under the Food and Environment Protection Act 1985 are also excluded from the waste management licensing regime.[16]

The specific exemptions under Schedule 3 are set out below. An exempt **16.07** activity will still have to be registered with the waste regulation authority or another specified body by virtue of regulation 18. Registration is dealt with at the end of this chapter. However, the provisions of regulation 18(10) which sets out the appropriate registration authority for each activity are incorporated in the text concerning a particular exemption. In addition the text will also state whether the exemption only applies if the activity is carried on by or with the consent of the occupier of the land on which it takes place or if the operator is otherwise entitled to conduct his activity on that land as provided by regulation 17(2).

A condition of an exemption will often be that the waste concerned is **16.08** kept in "secure storage" or is stored in a "secure container, lagoon or place". A container, lagoon or place is secure for these purposes if all reasonable precautions are taken to ensure that the waste kept in it cannot escape from it and members of the public are unable to gain access to the waste. "Secure storage" means storage in a secure container, lagoon or place.[17] To meet the requirements of a "secure storage" exemption the operator may have to show a detailed and considered level of attention by way of precaution,

---

[12] S.I. 1994 No. 1056, reg. 17(3) & 17(3A) added by S.I. 1996 No. 972, Sched. 3.
[13] S.I. 1994 No. 1056, reg. 16(1)(9a), as amended by S.I. 1995 No. 288, reg. 3(4).
[14] S.I. 1991 No. 472, amended by S.I. 1991 No. 836, S.I. 1992 No. 614, S.I. 1993 Nos. 1749 & 2405 and S.I. 1994 No. 1271.
[15] S.I. 1994 No. 1056, reg. 16(2).
[16] *ibid.* reg. 15(1)(b) & (c).
[17] *ibid.* reg. 17(5).

which it itself takes account of a number of possibilities whereby escape of waste can occur and from many different possible causes.[18]

ACTIVITIES CONTROLLED BY LOCAL AUTHORITIES UNDER PART I OF THE ENVIRONMENTAL PROTECTION ACT 1990

16.09   The use of waste glass in the course of an operation authorised by an authority under Part 1 of the Environmental Protection Act 1990 will be exempt from licensing if the total quantity of waste glass used in the process does not exceed 600,000 tonnes in any period of twelve months. This exemption will also cover the storage of the glass that is intended to be used in the authorised process.[19] The registration authority for this exemption will be the local enforcing authority under Part I of the 1990 Act.

16.10   The burning of certain types of waste as fuel is exempted by paragraph 3 of Schedule 3 if the burning is authorised under Part I of the Environmental Protection Act 1990. The materials that can be burned here are straw, poultry litter or wood, waste oil[20]—including oil which is special waste[21]—or solid fuel that has been manufactured from waste by a process involving the application of heat. The exemption will cover the secure storage of these wastes—except waste oil—for such burning and the feeding of it into the furnace, etc., in which it is to be burned. It will not matter if the waste is stored at a place outside the site where it is to be burned.[22] Storage of any waste oil for these purposes is only covered if the waste remains at the place at which it is produced and is not kept for more than twelve months.[23]

16.11   The burning of waste tyres is also dealt with by paragraph 3. This operation is exempted if it is, or forms part of, a process under Part B of section 1.3 of Schedule 1 to the 1991 Prescribed Processes and Substances Regulations. The exemption extends to the shredding and feeding of tyres into the appliance in which they are to be burned. The storage of tyres for this process is only covered if they are kept in a secure place, they are stored separately, none of them are kept for more than twelve months and the number kept at any one time does not exceed 1,000. The body with whom these operations should be registered for the purposes of regulation 18(10) of the 1994 Regulations is the local enforcement authority under Part I of the 1990 Act.

16.12   Concrete crushing is exempted from waste management licensing by paragraph 24 of Schedule 3 if the operation is authorised under Part I of the 1990 Act as a process falling into Part B(c) of Section 3.4 of Schedule 1 to

---

[18] *North Yorkshire County Council v. Boyne* [1997] Env. L.R. 91, Newman J. at 98.
[19] S.I. 1994 No. 1056, Sched. 3, para. 1.
[20] As defined in reg. 1(3).
[21] Inserted by S.I. 1996 No. 972, Sched. 3.
[22] D.o.E. Circular 11/94, Annex 5, para. 5.51.
[23] S.I. 1994 No. 1056, Sched. 3, para. 3(c) as amended by S.I. 1996 No. 972, Sched. 3.

the 1991 Regulations. For these purposes concrete crushing includes the size reduction of waste concrete, bricks or tiles. However, if the operation takes place elsewhere than the site at which the waste is produced the exemption only applies if it is done to recover or reuse the waste. The exemption covers on-site storage of waste to be crushed as long as no more than 20,000 tonnes of waste is stored at any one time. The operation should be registered with the local enforcing authority under Part I of the 1990 Act.

BURNING AS FUEL IN PROCESSES NOT CONTROLLED BY PART I

Paragraph 5 of Schedule 3 exempts the burning of waste as a fuel in an appliance if that appliance has a net rated thermal input—the rate at which fuel can be burned at the maximum continuous rating of the appliance multiplied by the net calorific value of the fuel and expressed as megawatts thermal—of less than 0.4 megawatts or, if it is used together with other appliances (whether or not it is operated simultaneously with them[24]), the aggregate net rated thermal input of all the appliances is less than 0.4 megawatts. The lower limit of control under Part B(e) of section 1.3 of Schedule 1 to the 1991 Regulations is 0.4 megawatts. The exemption also covers the secure storage of waste intended to be submitted to such burning. **16.13**

The burning of waste oil[25] as a fuel in the engine of an aircraft, hovercraft, mechanically propelled vehicle, railway locomotive, ship or other vessel is exempted by paragraph 6 of Schedule 3 as long as the total amount of such oil burned in any one engine does not exceed 2,500 litres an hour. The exemption also covers the storage, in a secure container, of waste oil intended to be so used. The registration authority for the processes covered in these two paragraphs will be the relevant Agency. **16.14**

DISPOSAL OF WASTE BY BURNING

The disposal of waste by burning it in an incinerator is exempted from licensing by paragraph 29 of Schedule 3 to the 1994 Regulations. For this provision to apply the incinerator concerned must be an "exempt incinerator" for the purposes of section 5.1 of Schedule 1 to the 1991 Regulations. Such incinerators are those on premises where there is plant designed to incinerate waste, including animal remains, at a rate of not more than 50kgs an hour, the weight of the waste being determined by its weight as fed into the incinerator. However, incinerators employed to incinerate clinical waste, sewage sludge, sewage screenings or municipal waste are not "exempt" for these purposes.[26] In addition the exemption under the 1994 Regulations only **16.15**

[24] S.I. 1994 No. 1056 as amended by S.I. 1995 No. 288, reg. 3(12).
[25] As defined in reg. 1(3).
[26] S.I. 1991 No. 472, as amended by S.I. 1992 No. 614.

applies if the waste is incinerated at the place where it is produced by the person producing it. The exemption extends to the secure storage of waste awaiting incineration at the production site.

16.16   The burning of wood, bark or other plant matter can be carried out on open land without a licence under paragraph 30 of Schedule 3. The exemption only applies to waste that is produced on land which is the operational land of a railway, light railway, tramway, internal drainage board, or the Environment Agency, or which is a forest, woodland, park, garden, verge, landscaped area, sports ground, recreation ground, church-yard or cemetery, or is produced as a result of demolition work. The fire must burn on the land on which the waste was produced and no more than 10 tonnes may be burnt in any period of 24 hours. If the fire is built by an establishment or undertaking it may only burn its own waste. The exemption extends to the storage of the waste before it is burned as long as it is stored on the land where it is to be burned. The registration authority for these operations is the relevant Agency.

USE OF WASTE FOR AGRICULTURE OR ECOLOGICAL IMPROVEMENT

16.17   Paragraph 7 of Schedule 3 to the 1994 Regulations gives a number of exemptions in relation to spreading of wastes on land used for agriculture. For these purposes "agriculture" has the same meaning as in section 109(3) of the Agriculture Act 1947 or section 86(3) of the Agriculture (Scotland) Act 1948. In these sections it is widely defined so as to include horticulture, fruit growing, seed growing, dairy farming, livestock breeding and keeping, the use of land as grazing land, meadow land, osier land, market gardens and nursery grounds, and the use of land for woodlands where that use is ancillary to the farming of land for other agricultural purposes. None of these exemptions apply unless the activity is carried on by or with the consent of the occupier of the relevant land or the operator is otherwise entitled to carry it on on that land.[27]

16.18   The wastes allowed to be spread under this exemption are listed in Table 2 to the Regulations.

TABLE 2

PART I

Waste soil or compost.
Waste wood, bark or other plant matter.

PART II

Waste food, drink or materials used in or resulting from the preparation of food and drink.

---

[27] S.I. 1994 No. 1056, reg. 17(2).

Blood and gut contents from abattoirs.

Waste lime.

Lime sludge from cement manufacture or gas processing.

Waste gypsum.

Paper waste sludge, waste paper and de-inked paper pulp.

Dredgings from any inland waters.[28]

Textile waste.

Septic tank sludge.

Sludge from biological treatment plants.

Waste hair and effluent sludge from a tannery.

Paragraph 7 allows any of the wastes in Table 2 to be spread on land **16.19** which is used for agriculture. In addition the wastes in Part I of Table 2 may be spread on the operational land of a railway, light railway, internal drainage board[29] or the Environment Agency or on land which is a forest, woodland, park, garden, verge, landscaped area, sports ground, recreation ground, churchyard or a cemetery.[30] However, this exemption is subject to three conditions. First no more than 250 tonnes of waste—5,000 tonnes of dredging wastes—can be spread per hectare in any period of twelve months. Secondly, the operation must result in benefit to agriculture or ecological improvement; a requirement that would probably exclude land raising operations. It is advisable to have an expert opinion that the operation will result in these benefits.[31]

Finally, where the waste is to be spread by an establishment or **16.20** undertaking on agricultural land the relevant Agency must be provided with the name, address, telephone number and fax number (if any) of the body concerned, a description of the waste, including the process from which it arises, where the waste will be stored before it is spread, an estimate of the amount to be spread and the time and place of the operation. In the case of a single spreading these details should be provided in advance of the operation. Where there is to be regular or frequent spreading of wastes of a similar composition the particulars should be furnished at six monthly intervals or, if the waste on a particular occasion is going to be different to that usually spread a separate notification must be made. For regular spreadings the particulars must include an estimate of the total quantities to be spread over the six months period and the frequency of spreadings.[32]

The exemption extends to the storage, at the place where it is to be **16.21** spread, of any Table 2 waste—other than septic tank sludge (residual sludge from septic tanks and similar installations for the treatment of sewage,[33] so excluding waste from cesspits)—which is intended to be spread under the

---

[28] As defined in reg. 1(3).

[29] See para. 7(9).

[30] S.I. 1994 No. 1056, Sched. 3, para. 7(1) & 7(2).

[31] See D.o.E. Circular 11/94, Annex 5, para. 5.74.

[32] S.I. 1994 No. 1056, Sched. 3, para. 7(3) & (4) and see D.o.E. Circular 11/94, Annex 5, para. 5.77.

[33] As defined in the Sludge (Use in Agriculture) Regulations 1989, reg. 2(1)—S.I. 1989 No. 1263 as amended by S.I. 1990 No. 880.

exemption.[34] Waste in liquid form may only be stored in a secure container or lagoon that holds no more than 500 tonnes of such waste. More than one such container may be used.[35] Septic tank sludge to be spread on land used for agriculture may be stored in a secure container or lagoon, without limit as to the amount, while watered sludge can be stored in a secure place.[36]

16.22   An establishment or undertaking carrying out the spreading of waste on lands used for agriculture need not follow the registration procedure in regulation 18(1) to (6) and so will not be guilty of the offence of failing to register. However, if the relevant Agency receives notification of the operation pursuant to the conditions of the exemption it should enter the particulars provided on its register.[37] Establishments, etc., spreading wastes for ecological improvement will have to register with the relevant Agency.

16.23   The spreading of sludge and septic tank sludge[38] on agricultural land is controlled by the Sludge (Use in Agriculture) Regulations 1989.[38a] "Sludge" here means the residual sludge from sewage plants treating domestic or urban waste waters and from other plants treating similar types of waste waters. Paragraph 8 of the 1994 Regulations exempts the storage of such matter as long as it is in a secure container or lagoon and is on the agricultural land on which it is to be spread in accordance with the 1989 Regulations.

16.24   The spreading of such sludge on land which is not agricultural land[39] is only exempted if it will result in ecological improvement and the concentrations of metals, etc., listed in column 1 of the soil table set out in Schedule 2 to the 1989 Regulations does not exceed the limits specified in column 2.[40] The exemption is extended to the storage of sludges for this purpose in a secure container or lagoon—or if it is dewatered sludge, in a secure place—on the land on which it is to be spread.[41]

16.25   Paragraph 9 of Schedule 3 is concerned with land reclamation. It enables the spreading of waste consisting of soil, rock, ash or sludge[42] or of wastes arising from the dredging of inland waters or from construction or demolition work on any land in connection with its reclamation or improvement. However, the exemption cannot be used to dispose of waste at a site designed or adapted as a landfill site.[43] To achieve this exemption the operation must satisfy three conditions. The land concerned must be such that by reason of industrial or other development it is incapable of

---

[34] S.I. 1994 No. 1056, Sched. 3, para. 7(5) and see D.o.E. Circular 11/94, Annex 5, para. 5.79.
[35] *ibid.* para. 7(6).
[36] *ibid.* para. 7(7).
[37] S.I. 1994 No. 1056, reg. 18(7).
[38] As defined in the 1989 Regulations—see para. 16.20 above.
[38a] S.I. 1989 No. 1263 as amended by S.I.s 1990 No. 880 and S.I. 1996 No. 593.
[39] As defined in the 1989 Regulations as opposed to the 1994 Regulations.
[40] S.I. 1994 No. 1056, Sched. 3, para. 8(2).
[41] *ibid.* para. 8(3).
[42] As defined in the 1989 Regulations—see para. 16.22 above.
[43] S.I. 1994 No. 1056, Sched. 3, para. 9(3).

beneficial use without treatment. This would again rule out landraising operations on agricultural land. It must also be carried out in accordance with a planning permission for the reclamation or improvement of the land and result in benefit to agriculture or ecological improvement. Finally, no more than 20,000 cubic metres per hectare of such waste may be spread on the land. The exemption extends to the storage of such waste at the place where it is to be spread.[44] The exemption only applies if the activity is carried on by or with the consent of the occupier of the relevant land or the operator is otherwise entitled to carry it on on that land.[45] The operation must be registered with the relevant Agency if carried on by an establishment or undertaking.

DREDGING WASTES

In addition to the exemptions for spreading dredging wastes from inland 16.26 waters[46] on land contained in paragraphs 7 and 9, paragraph 25 of Schedule 3 enables the deposit of such wastes on banks and towpaths. The exemption also covers the deposit of waste from clearing plant matter from such waters; although care should be taken in such an operation not to commit an offence under section 90(2) of the Water Resources Act 1991 or section 49(1)(b) of the Control of Pollution Act 1974 (Scot). Such deposits can either be made on the banks or towpaths of the waters where the dredging or clearing takes place or on any other bank or towpath if doing so will result in a benefit to agriculture or ecological improvement. No more than 50 tonnes of waste a day may be deposited per metre of the bank or towpath along which it is being deposited and if the operation is conducted by an establishment or undertaking then deposit at the dredging or clearing site can only be done if the waste is their waste.[47] Unlike other exemptions this one does not extend to waste stored or deposited in a container or lagoon.[48]

Dredgings or cleared plant matter can be treated by screening or de- 16.27 watering. Such operations are also covered by paragraph 25 as long as the treatment is done prior to the use of the matter in one of the activities exempted by paragraphs 25, 7 or 9(1) of Schedule 3. The exemption will cover such treatment either on the banks or towpath where the dredging or clearing has taken place or at the site where the waste is to be spread in reliance on the relevant exemption. As far as paragraph 9(1) is concerned that only covers dredgings from inland waters, not cleared plant matter.[49] None of these exemptions apply unless the activity is carried on by or with the consent of the occupier of the relevant land or the operator is otherwise entitled to carry it on on that land.[50] The operation should in all cases be

[44] S.I. 1994 No. 1056, Sched. 3, para. 9(2).
[45] *ibid.* reg. 17(2).
[46] As defined in reg. 1(3).
[47] S.I. 1994 No. 1056, Sched. 3, para. 25(2) & 25(4).
[48] *ibid.* para. 25(3).
[49] *ibid.* para. 25(5).
[50] S.I. 1994 No. 1056, reg. 17(2).

registered with the relevant Agency if carried on by an establishment or undertaking.

SEWAGE RECOVERY AND DISPOSAL OPERATIONS

16.28   Paragraph 10(1) of Schedule 3 exempts a recovery operation carried on within the curtilage of a sewage treatment works in relation to sludge or septic tank waste brought onto the site from another sewage treatment works if the total quantity of imported waste in any period of twelve months does not exceed 10,000 cubic metres. This exemption will also cover the storage of such waste at the site it is to be treated.[51] The storage of such waste at the exporting site will also be exempt by virtue of paragraph 41 of Schedule 3.

16.29   This provision needs to be read with regulation 7(1) of the Controlled Waste Regulations 1992.[52] Under that regulation sewage, sludge or septic tank sludge which is treated, kept or disposed of—otherwise than by means of a mobile plant—within the curtilage of a sewage treatment works as an integral part of the operation of those works is not to be regarded as controlled waste. Similarly sludge or septic tank sludge that is supplied or used in accordance with the Sludge (Use in Agriculture) Regulations 1989[53] is not controlled waste. There is no special exemption for waste arising from boreholes or excavations carried out in connection with sewerage operations.[54]

16.30   The discharge of waste onto the track of a railway from a sanitary convenience or sink forming part of a passenger train is exempted as long as the discharge does not exceed 25 litres.[55] Further, paragraph 32 enables the burial of sanitary waste on sites such as fairgrounds or caravan sites. The exemption covers waste from a sanitary convenience that is equipped with a removable receptacle only. No more than five cubic metres of such waste may be buried on the premises within any period of twelve months. These operations should be registered with the relevant Agency.

STORAGE OF WASTE PENDING RECOVERY

16.31   Under paragraph 17 of Schedule 3[56] the wastes set out in Table 4 to the Regulations may be stored in a secure place on any premises if the total quantity of the kind of waste so stored at any time does not exceed the

[51] See D.o.E. Circular 11/94, Annex 5, para. 5.98.
[52] S.I. 1992 No. 588.
[53] S.I. 1989 No. 1263 as amended by S.I. 1990 No. 880.
[54] D.o.E. Circular 11/94, Annex 5, para. 5.204.
[55] S.I. 1994 No. 1056, Sched. 3, para. 31.
[56] As amended by S.I. 1996 No. 1279.

quantity limit set in the Table. The waste must be intended to be reused or to be subjected to the handling activities set out in paragraph 11 or to a recovery operation. Each kind of listed waste stored on the premises must be stored separately[57] and no waste so stored may be kept on the premises for longer than twelve months. An operation conducted by an establishment or undertaking should be registered with the relevant Agency. It can only be carried on by or with the consent of the occupier of the relevant premises or if the operator is otherwise entitled to carry it on on those premises.[58]

TABLE 4

Kind of waste	Maximum total quantity
Waste paper or cardboard	15,000 tonnes
Waste textiles	1,000 tonnes
Waste plastics	500 tonnes
Waste glass	5,000 tonnes
Waste steel cans, aluminium cans or aluminium foil	500 tonnes
Waste food or drink cartons	500 tonnes
Waste articles which are to be used for construction work which are capable of being so used in their existing state	100 tonnes
Solvents (including solvents which are special waste)	5 cubic metres
Refrigerants and halons (including refrigerants and halons which are special waste)	18 tonnes
Tyres	1,000 tyres
Waste mammalian protein	60,000 tonnes
Waste mammalian tallow	45,000 tonnes

The wastes listed in Table 4—except waste solvents, refrigerants or halons and waste mammalian protein or tallow—and waste oil may also be stored in secure containers under the provisions of paragraph 18 of Schedule 3. This exemption is mainly aimed at bottle banks, etc., but covers any such storage. Waste oil is included in the exemption even if it is special waste.[59] 16.32

A number of conditions are imposed on this exemption. The storage capacity of the container or containers must not exceed 400 cubic metres in total and no more than 20 containers may be used on the premises where the storage takes place. Where waste oil is involved the total capacity of the container or containers must not exceed 3 cubic metres and arrangements must be made to prevent the oil escaping into the ground or a drain. Each 16.33

[57] D.o.E. Circular 11/94, Annex 5, para. 5.137.
[58] S.I. 1994 No. 1056, reg. 17(2).
[59] ibid. Sched. 3, para. 18(2)(b) as amended by S.I. 1996 No. 972, Sched. 3.

kind of waste listed in Table 4, or the waste oil, must be kept separately and no waste may be stored on the premises for longer than twelve months. The waste must be intended for reuse, or for use in a handling operation exempted under paragraph 11 or any other recovery activity. The operation should be registered with the relevant Agency. It can only be carried on by or with the consent of the occupier of the relevant premises or if the operator is otherwise entitled to carry it on on those premises.[60] Further the person storing the waste must either be the owner of the container or have the owner's consent.[61]

PREPARATORY TREATMENT OF WASTE

16.34   Table 3 to the Regulations lists certain waste handling operations that can be carried out in respect of certain types of wastes. These activities are exempt from licensing under paragraph 11 of Schedule 3 if carried on with a view to the recovery or reuse of that waste—either by the person carrying on the activity or by someone else—and that the volume limit which is set in tonnes per week, is not exceeded in respect of each material. Weights are based on the weight of the materials before they are treated.[62] The exemption only applies if the activity is carried on by or with the consent of the occupier of the relevant land or the operator is otherwise entitled to carry it out on that land.[63] An activity conducted by an establishment or undertaking must be registered with the relevant Agency.

TABLE 3

Kind of waste	Activities	Limit (tonnes per week)
Waste paper or cardboard	Baling, sorting or shredding	3,000
Waste textiles	Baling, sorting or shredding	100
Waste plastic	Baling, sorting, shredding, densifying or washing	100
Waste glass	Sorting, crushing or washing	1,000
Waste steel cans, aluminium cans or aluminium foil	Sorting, crushing, pulverising, shredding, compacting or baling	100
Waste food or drink cartons	Sorting, crushing, pulverising, shredding, compacting or baling	100

[60] S.I. 1994 No. 1056, reg. 17(2).
[61] *ibid.* Sched. 3, para. 18(1)(g).
[62] See D.o.E. Circular 11/94, Annex 5, para. 5.104.
[63] S.I. 1994 No. 1056, reg. 17(2).

Paragraph 21 of Schedule 3 exempts the chipping, shredding, cutting or 16.35 pulverising of waste plant matter—including wood or bark—or sorting and baling of sawdust or wood shavings on any premises. The activity must be carried on for the purposes of the recovery or reuse of these substances. No more than 1,000 tonnes of such waste can be dealt with on the premises in any period of seven days.[64] The exemption extends to the storage of such wastes before, during and after processing on the premises where the activity takes place as long as no more than 1,000 tonnes are stored at any one time.

Paragraph 4 of Schedule 3 exempts the cleaning, washing, spraying or 16.36 coating of waste packaging or containers so that it or they can be reused. The total quantity of such waste so dealt with at any place may not exceed 1,000 tonnes in any period of seven days. This exemption is aimed at the reuse of whole containers, not parts of them.[65] The exemption covers the storage of such waste pending its treatment at the treatment site if no more than 1,000 tonnes is stored at any one time and, where metal containers for the transport or storage of chemicals are dealt with, no more than one tonne of such containers are dealt with in any seven day period.

The exemption only applies if the activity is carried on by or with the 16.37 consent of the occupier of the relevant land or the operator is otherwise entitled to carry it on on that land.[66] Further, while usually the operation will have to be registered with the relevant Agency, where the coating or spraying of metal containers as, or as part of, a process within Part B of section 6.5 of the 1991 Regulations is carried on the registration authority will be the local enforcement authority who authorise the process under Part I of the 1990 Act.[67]

The composting of biodegradable wastes is exempted under paragraph 12 16.38 of Schedule 3. For the purposes of this paragraph "composting" includes any other biological transformation process that results in materials which may be spread on land for the benefit of agriculture or ecological improvement. The exemption covers such activities either at the place where the waste is produced or at any other place occupied by the producer, or the place where the compost is to be used or any other place occupied by the user. Only 1,000 cubic metres of material may be composted at the relevant place at any one time, but this is extended to 10,000 cubic metres where the compost is for use in the cultivation of mushrooms. The exemption also covers the storage of biodegradable waste to be composted either at the place where the waste is produced or where it is to be composted. Composting for the cultivation of mushrooms is an activity that must be authorised under Part I of the 1990 Act[68] thus activities relating to use or storage for this purpose should be registered with the local enforcement authority under Part I.[69] All other activities conducted by an establishment or undertaking should be registered with the relevant Agency.

[64] See D.o.E. Circular 11/94, Annex 5, para. 5.162.
[65] See D.o.E. Circular 11/94, Annex 5, para. 5.58.
[66] S.I. 1994 No. 1056, reg. 17(2).
[67] *ibid.* reg. 18(10)(ii).
[68] S.I. 1991 No. 472 (as amended) Sched. 1, section 6.9. Part B "exempt process" (xi).
[69] S.I. 1994 No. 1056, reg. 18(10)(a)(iii).

16.39   The laundering or otherwise cleaning of waste textiles with a view to their recovery or reuse is exempted from licensing by paragraph 20 of Schedule 3. This exemption extends to the storage of waste textiles at the place where they are to be laundered or cleansed. After cleansing it is likely that the materials would no longer be waste. The operation must be registered with the relevant Agency.

16.40   Waste arising at a water treatment works from the treatment operation can be treated within the curtilage of those works without a licence as long as the total amount dealt with in any twelve month period does not exceed 10,000 cubic metres. This exemption extends to the storage of such waste at the works before it is treated.[70]

REUSE OF WASTE

16.41   A general exemption for activities involving the reuse of waste items, if they can be put to beneficial use without further treatment, is granted by paragraph 15 of Schedule 3. The exemption extends to the storage of the waste pending its reuse. The use or storage of the waste must not involve its disposal. Further if the activity is covered by the exemptions contained in paragraphs 7, 8, 9, 19 or 25 of Schedule 3, or would be covered but for any condition or limitation to which one of those exemptions is subject, then the exemption under paragraph 15 is inapplicable.[71] The exemption is intended to deal with salvaged items that can be put to beneficial use without repair or other treatment or materials that can be reused.[72] An activity conducted by an establishment or undertaking should be registered with the relevant Agency and can only be carried on by or with the consent of the occupier of the relevant land or if the operator is otherwise entitled to carry it on on that land.[73]

16.42   Construction operations can store and use certain types of waste without a licence if the works are "relevant works" for the purposes of paragraph 19 of Schedule 3. "Relevant works" here means construction work including the deposit of waste on land in connection with the provision of recreational facilities on that land or the construction, maintenance or improvement of a building, highway, railway, airport, dock or other transport facility on that land. This wide definition is extended by regulation 1(3) to include the repair, alteration or improvement of existing works, while "work" includes preparatory work. However, it does not go so far as to allow any deposit of waste not needed for these operations or any work involving land reclamation.[74]

16.43   Wastes arising from demolition or construction or tunnelling or other excavations or which consists of ash, slag, clinker, rock, wood or gypsum can

---

[70] S.I. 1994 No. 1056, Sched 3, para. 10(2) & (3).
[71] S.I. 1994 No. 1056, Sched. 3, para. 15(3).
[72] See D.o.E. Circular 11/94, Annex 5, para. 5.122.
[73] S.I. 1994 No. 1056, reg. 17(2).
[74] *ibid.* Sched. 3, para. 19(4) and see D.o.E. Circular 11/94, Annex 5, para. 5.159.

be stored at a construction site if it is suitable for use in the "relevant work" that will be carried on at the site. Any waste imported onto the site may not be stored there for longer than three months before operations start. These wastes can be used in the operations as long as they are suitable for those purposes.[75] Road planings can be stored anywhere for use in "relevant works" if no more than 50,000 tonnes are so stored and they do not stay at the storage site for more than three months.[76] The exemption only applies if the activity is carried on by or with the consent of the occupier of the relevant land or the operator is otherwise entitled to carry it on on that land.[77] The storage or use should be registered with the relevant Agency.

MANUFACTURING FROM WASTE MATERIALS

Wastes consisting of ash, slag, clinker, rock, wood bark, paper, straw, or 16.44 gypsum and waste which arises from demolition or construction work or tunnelling or other excavations can be used in certain manufacturing processes under paragraph 13 of Schedule 3. It can be used in making timber products, straw board, plasterboard, bricks, roadstone or aggregate if the activity is carried on by, or with the consent of, the occupier of the relevant land or the operator is otherwise entitled to carry it on on that land.[78] In addition it can be used for the manufacture of soil substitutes subject to that condition and if the operation is carried out at the place where either the waste is produced or where the product is to be applied on the land. Further, no more than 500 tonnes of material may be manufactured on any day.

Waste soil or rock can be treated without a licence if it is to be spread on 16.45 land under the exemptions provided by paragraphs 7 or 9 of Schedule 3. For this exemption to apply the material must be treated at the place where the waste is produced or where the product is to be spread, with the consent of the occupier or through a right the operator has to work on that land. In addition the total amount of material treated at that place may not exceed 100 tonnes in any one day.[79]

The storage of wastes to be submitted to any of these manufacturing or 16.46 treatment processes is also permitted if the waste is stored at the place where the activity is to be carried on. In addition no more than 20,000 tonnes of waste should be stored at the site, although for road planings this limit is increased to 50,000 tonnes if it is to be used for the manufacture of roadstone. In all cases these operations should be registered with the relevant Agency.

The manufacture of finished goods from waste metals, plastic, glass, 16.47 ceramics, rubber, textiles, wood, paper or cardboard is exempted under

[75] S.I. 1994 No. 1056, Sched. 3, para. 19(1) & 19(2).
[76] ibid. Sched. 3, para. 19(3).
[77] S.I. 1994 No. 1056, reg. 17(2).
[78] S.I. 1994 No. 1056, reg. 17(2).
[79] ibid. Sched. 3, para. 13(3).

paragraph 14 of Schedule 3.[80] This exemption extends to the storage of such wastes for the purposes of manufacturing if the waste is stored at the place it will be used and the total amount of any particular kind of waste stored there at any time does not exceed 15,000 tonnes. The operation must be carried on by, or with the consent of, the occupier of the relevant land unless the operator is otherwise entitled to carry it on on that land.[81] In addition the activity should be registered with the relevant Agency.

RECOVERY OF WASTE MATERIALS

16.48   The recovery, at any premises, of silver from waste produced in connection with printing or photographic processing is exempted from licensing under paragraph 22 of Schedule 3. However, no more than 50,000 litres of such waste may be dealt with on those premises in any day. The exemption extends to the storage of such waste on those premises pending the recovery operation.

16.49   Paragraph 23 of Schedule 3 exempts from licensing the keeping or treatment of animal by-products in accordance with the Animal By-Products Order 1992.[82] Animal by-products are materials arising from the slaughter of animals such as meat, offals, hides, wools, feathers, etc. The term is defined in article 3(1) of the Order to mean any carcase or part of any animal or product of animal origin not intended for direct human consumption but does not include animal excreta or catering waste or meat cooked at a knacker's yard for use as food for animals whose flesh is not intended for human consumption. The Minister approves rendering premises under article 8 of the Order, while the appropriate Minister compiles a register of premises collecting or using animal by-products for petfood, technical or pharmaceutical products under article 9. A register is also compiled under article 10, by the appropriate Minister, for premises used for the feeding of animal by-products to zoo, circus or fur animals, recognised packs of hounds and maggot farming for fishing bait. The appropriate registration authority for the purposes of regulation 18 of the 1994 Regulations in respect of such activities will be the Minister or the appropriate Minister.[83] Where the activity is carried on at a knacker's yard licensed under section 1 of the Slaughterhouses Act 1974 or section 6 of the Slaughter of Animals (Scotland) Act 1980 the registration authority will be the relevant local authority.[84]

16.50   The processing of waste food as defined in article 2(1) of the Diseases of Animals (Waste Food) Order 1973[85] will be licensed under article 7 of the Order, while places holding such matter to be fed to livestock or poultry will

[80] See D.o.E. Circular 11/94, Annex 5, para. 5.119.
[81] S.I. 1994 No. 1056, reg. 17(2).
[82] S.I. 1992 No. 3303.
[83] S.I. 1994 No. 1056, reg. 18(10)(c)(i) & 18(10)(c)(ii).
[84] ibid. reg. 18(10)(c)(iii).
[85] S.I. 1973 No. 1936, as amended by S.I. 1987 No. 232.

be licensed under article 8. Such licensed activities are exempted from waste management licensing by virtue of paragraph 16 of Schedule 3 to the 1994 Regulations. The exemption only applies if the activity is conducted in accordance with the conditions and requirements of the 1973 Order. The operation should be registered, for the purposes of regulation 18 of the 1994 Regulations, with the issuing authority responsible for granting the relevant licence.[86]

RECOVERY OR DISPOSAL AS AN INTEGRAL PART OF THE PRODUCTION PROCESS

The recovery or disposal of waste at the place where it is produced, as an 16.51 integral part of the process that produces it, is exempted from licensing under paragraph 26 of Schedule 3. This covers cases where recovery on site is involved because the waste is reincorporated in the production process, such as scrap materials being added to the raw material. It also deals with final disposal in the process and treatment as part of that process.[87] However, the final deposit of waste on land is excluded from the exemption. The exemption also extends to the storage at the place where it is produced of waste to be so recovered or disposed of. The operation should be registered with the reelvant Agency.

BALING, COMPACTING, ETC.

The baling, compacting, crushing, shredding or pulverising of waste, at 16.52 the place where it is produced is exempted from licensing by paragraph 27 of Schedule 3. The exemption also covers the storage of waste, at the place where it is to be produced, that is to be so dealt with. The activity should be registered with the relevant Agency.

STORAGE OF RETURNED GOODS

Goods returned to a manufacturer, distributor or retailer will usually be 16.53 considered as waste by the person returning them. Even if they are returned for repair they would still normally be regarded as waste if they are incapable of use without repair. However, the Department of the Environment considers that they will cease to be waste at the moment when the intention is formed to repair them. Thus no licensing exemption is necessary for the storage or repair of returned goods.[88]

The storage of waste returned goods, and the secure storage of goods that 16.54 are special waste, by their manufacturer, distributor or retailer is exempted

---

[86] S.I. 1994 No. 1056, reg. 18(10)(b).
[87] See D.o.E. Circular 11/94, Annex 5, para. 5.179.
[88] See D.o.E. Circular 11/94, Annex 5, para. 5.184.

from licensing by virtue of paragraph 28 of Schedule 3 pending their recovery or disposal.[89] They can only be stored for one month under this exemption. An activity carried on by an establishment or undertaking should be registered with the relevant Agency.

DISPOSAL OPERATIONS

16.55   The keeping or deposit of waste consisting of excavated material arising from peatworking is exempted from licensing under paragraph 33 of Schedule 3 if the material is kept or disposed of at the place where the operation takes place. If the activity is carried on by an establishment or undertaking the waste must belong to that body. The operation should be registered with the relevant Agency.

16.56   Paragraph 34 of Schedule 3 exempts the keeping or deposit of spent railway ballast on land at the place where it is produced if the land is the operational land of a railway, light railway or tramway. "Operational land" for these purposes bears the same definition as in sections 263 and 264 of the Town and Country Planning Act 1990[90] which the D.o.E. interprets to mean that the land must be clearly used for the purposes of the railway, etc.[91] Only 10 tonnes of spent ballast may be so kept or deposited for each metre of track from which it derives. If the operation is carried on by an establishment or undertaking the waste must belong to that body and the activity be registered with the relevant Agency.

16.57   Waste from prospecting for minerals, whether from a borehole or other activity, will be exempted from licensing under paragraph 35 of Schedule 3 if it is deposited in or on the land at the place where it is excavated. In addition no more than 45,000 cubic metres per hectare may be deposited in any twelve month period. This exemption only applied if the operation is permitted development under either the General Permitted Development Order 1995 or the General Permitted Development (Scotland) Order 1992 and the conditions subject to which the development is permitted are observed. The relevant permissions under those orders relate to the drilling of boreholes, the carrying out of seismic surveys and making other excavations for the purposes of mineral explorations.

16.58   The burial of dead domestic pets in the garden of the domestic property where the pet lived is exempted under paragraph 37 of Schedule 3. However, the exemption does not apply if a dead pet may prove hazardous to anyone who may come into contact with it or if the pet died elsewhere than the premises and the burial is carried on by an establishment or undertaking. Further, the burial must take place by or with the consent of the occupier of the land on which the pet is buried or if the person burying the pet is otherwise entitled to do so on that land.[92] It is only in the unlikely

---

[89] S.I. 1994 No. 1056, Sched. 3, para. 28 as substituted by S.I. 1996 No. 972, Sched. 3.
[90] In Scotland T.&C.P.(S.) Act 1997, ss.215 & 216.
[91] See D.o.E. Circular 11/94, Annex 5, para. 5.196.
[92] S.I. 1994 No. 1056, reg. 17(2).

event that the burial of a pet that died in the home is carried out by an establishment or undertaking that the activity will have to be registered with the relevant Agency.

WASTE FROM SHIPS

Garbage landed from ships—a vessel of any type whatsoever operating in the marine environment including submersibles and a structure which is a fixed or floating platform—can be stored for seven days without a licence under paragraph 36(1) of Schedule 3. The storage of such waste, which can include special waste, must be at reception facilities provided in a harbour area under the Merchant Shipping (Reception Facilities for Garbage) Regulations 1988[93] and be incidental to the collection and transport of the waste. The amount of garbage so stored may not exceed 20 cubic metres for each ship from which garbage has been landed **16.59**

Tank washings—waste residues from the tanks, other than fuel tanks, or holds of a ship or waste arising from their cleaning—may also be temporarily stored without a licence.[94] The waste, which can include special waste, must be stored at reception facilities provided within a harbour area in accordance with the Prevention of Pollution (Reception Facilities) Order 1984.[95] The storage must also be incidental to the collection and transport of the waste. Where the washings consist of dirty ballast the amount so stored may not exceed 30 per cent of the total deadweight of the ships from which such washings have been landed. Where they consist of waste mixtures containing oil the amount so stored may not exceed 1 per cent of the total deadweight of the ships from which they have been landed. The activity should also be registered with the relevant Agency. **16.60**

SAMPLES OF WASTE

The deposit or storage of samples of waste, including that of special waste, pending or during testing or analysis can be done under paragraph 38 of Schedule 3 without a licence if they are deposited or stored at the place where they are or will be tested or analysed. The actual testing and analysis is neither a disposal nor a recovery operation and so need not be licensed.[96] Only samples taken for certain purposes may benefit from this exemption. It only applies to samples taken: **16.61**

(a) in the exercise of any power under the Radioactive Substances Act 1993, the Sewerage (Scotland) Act 1968, the Control of Pollution Act 1974, the Environmental Protection Act 1990, the Water Industry Act 1991 or the Water Resources Act 1991;

---

[93] S.I. 1988 No. 2293.
[94] S.I. 1994 No. 1056, Sched. 3, para. 36(2).
[95] S.I. 1984 No. 862.
[96] D.o.E. Circular 11/94, Annex 5, para. 5.220.

(b) by or on behalf of the holder of a waste management licence in pursuance of the conditions of that licence;

(c) by or on behalf of a person carrying on an activity exempted by Schedule 3 or regulation 16(1) of the 1994 Regulations;

(d) by or on behalf of the owner or occupier of the land from which the samples are taken;

(e) by or on behalf of a person to whom the duty of care under section 34 of the Environmental Protection Act 1990 applies and in connection with that duty;

(f) for the purposes of research.

If the activity is conducted by an establishment or undertaking it must be registered with the relevant Agency.

STORAGE OF WASTES PENDING THEIR DISPOSAL

16.62   Waste medicines, including those that are special wastes, which have been returned to a pharmacy by households or individuals may be stored securely there pending their disposal there or elsewhere without a licence by virtue of paragraph 39(1) of Schedule 3. The waste medicine need not be returned to the pharmacy at which it was bought.[97] The total quantity of such wastes at the pharmacy must not exceed five cubic metres at any time and any waste medicine returned to the pharmacy may not remain there for longer than six months.

16.63   Medical, nursing or veterinary practices may store waste, including special waste, produced in carrying on their practice at their premises without a licence under paragraph 39(2) of Schedule 3. The total quantity of such waste at the premises at any time may not exceed five cubic metres and no such waste may be stored there for longer than three months. These storage activities must be registered with the relevant Agency if the practice is operated by an establishment or undertaking.

16.64   The temporary storage of scrap rails on the operational land of a railway, light railway or tramway is exempted from licensing under paragraph 40(2) of Schedule 3. The storage does not have to be at the place where the waste was produced. However, it must be incidental to the collection or transport of the waste and no more than 10 tonnes may be kept in any one place at any time.

16.65   Any non-liquid waste may be stored at a place other than where it is produced without a licence under paragraph 40(1) of Schedule 3. This exemption is aimed at storage incidental[97a] to the collection and transport of the waste. Storage in connection with another purpose, such as in a waste transfer station, at a scrap metal yard or vehicle dismantling sites, cannot benefit from this paragraph; storage at such places being specifically

---

[97] D.o.E. Circular 11/94, Annex 5, para. 5.223.
[97a] See *North Yorkshire County Council v. Boyne* [1997] Env. L.R. 91.

excluded.[98] The waste must be stored in a secure container or containers and cannot exceed 50 cubic metres in total at any one time. The person carrying on the activity must also either be the owner of the containers or have the owner's consent for such storage. In addition the storage must be carried on by or with the consent of the occupier of the land on which it takes place or the storer must otherwise be entitled to use the land for the storage.[99] No waste may be kept in storage for longer than three months. Where waste is stored by an establishment or undertaking it must be registered with the relevant Agency.

Waste, either solid or liquid, may also be stored temporarily without a **16.66** licence on the site where it was produced under paragraph 41 of Schedule 3. However, scrap metal yards or vehicle dismantling sites cannot benefit from this exemption.[1] There is no set time limit on the storage, nor on the volume that can be stored but after a while it is likely that the onus would be on the storer to show that the storage was indeed temporary. There are special rules for special wastes. They may not be stored for more than twelve months. Liquid wastes must be stored in a secure container or containers and the maximum volume allowed at any one time is 23,000 litres. Other special wastes should be stored securely; if in secure container the total volume is limited to 80 cubic metres at any one time or if in a secure place a lower level of only 50 cubic metres at any one time is allowed. In addition the storage must be carried on by or with the consent of the occupier of the land on which it takes place or the storer must otherwise be entitled to use the land for the storage.[2] Where waste is stored by an establishment or undertaking it must be registered with the relevant Agency.

SCRAP METAL OPERATIONS

"Scrap metal" is defined in section 9(2) of the Scrap Metal Dealers Act **16.67** 1964 to include any old metal and any broken, worn out, defaced or partly manufactured articles made wholly or partly of metal, and any metallic wastes, and also includes old, broken, worn out or defaced tooltips or dies made of any of the materials commonly known as hard metal or of cement or sintered metallic carbides. Waste motor vehicles are not scrap metal for these purposes.[3] The exemptions in respect of some of these operations are discussed in D.o.E. Circular 6/95.[4]

A general exemption for scrap metal operations was granted by paragraph **16.68** 42 of Schedule 3 as there was difficulty in resolving the way in which such operations should be controlled.[5] Paragraph 42 therefore provided a transi-

---

[98] S.I. 1994 No. 1056, Sched. 3, para. 40(1)(c) & 40(1A) as added by S.I. 1995 No. 288, reg. 3(13).
[99] *ibid.* reg. 17(2).
[1] Para. 41(1A) added by S.I. 1995 No. 288, reg. 3(13).
[2] *ibid.* reg. 17(2).
[3] See D.o.E. Circular 11/94, Annex 5, para. 5.242.
[4] W.O. Circular 25.95, Scottish Office Env. Dept. 8/95.
[5] See D.o.E. Circular 11/94, Annex 5, para. 5.242.

tional exemption for the treatment, keeping or disposal by any person at any premises of waste, including special waste, that consists of scrap metal or waste motor vehicles which are to be dismantled. The terms of the exemption were that the operator had to have been carrying on the activity at those premises before April 1, 1995[6] and have applied for a waste disposal licence before that date to authorise his activity and the application was pending on that date. The exemption ceases either on the date the licence applied for is granted or if it is refused, the date on which the period for appealing expires without an appeal being made or on which any appeal is withdrawn or finally determined.

BURNING OF METAL WASTES

16.69   The operation of scrap metal furnaces with a designed holding capacity of less than 25 tonnes is exempted from licensing by virtue of paragraph 2 of Schedule 3. However, the furnace must be authorised under either paragraphs (a), (b) or (d) of Part B of Section 2.1 (iron and steel) of Schedule 1 to the Prescribed Processes and Substances Regulations 1991[7] or paragraphs (a), (b) or (e) of Part B of Section 2.2 (non-ferrous metals) of that Schedule. This requires authorisation by the local enforcing authority under Part I of the Environmental Protection Act 1990 and the activity must also be registered with that authority under regulation 18 of the 1994 Regulations.

16.70   The exemption extends to the loading and unloading of such furnaces in connection with their operation and to the storage of scrap metal at the site of the furnace which is intended to be used in it. However, the exemption in respect of storage does not apply if the furnace is on premises used for carrying on business as a scrap metal dealer or, in Scotland, as a metal dealer.[8]

16.71   Heating certain metals to remove grease, oil or any other non-metallic contaminants is exempted from licensing by paragraph 44 of Schedule 3.[9] However, it does not apply to the removal by heat of plastic or rubber covering from scrap cable[10] or of any asbestos contaminant. The appliance or appliances in which the metal is heated must have in aggregate a net rated thermal input of less than 0.2 megawatts.[11]

16.72   The metals that can be heated under this exemption are iron, steel or any ferrous alloy[12] unless the process concerned is related to one described in any of paragraphs (a) to (h) or (j) to (m) of Part A or paragraphs (a) to (c) or (e) or (f) of Part B to Section 2.1 of Schedule 1 to the 1991 Regulations. In

---

[6] S.I. 1994 No. 1056, Sched. 3, para. 42(1)(a) as amended by S.I. 1995 No. 288, reg. 3(14).

[7] S.I. 1991 No. 472 as amended.

[8] For definition see Scrap Metal Dealers Act 1964, s.9(1) or in Scotland Civic Government (Scotland) Act 1982, s.37(2).

[9] Added by S.I. 1995 No. 288, reg. 3(16).

[10] "Cable burning" is an offence under s.33 of the Clean Air Act 1993 unless authorised under Part I of the E.P.A. 1990.

[11] See para. 16.12.

[12] As defined in para. 44(7).

addition the heating of non-ferrous metals or alloys is exempted unless the process concerned is one related to a process described in any of paragraphs (a) to (g) or (i) to (k) of Part A of Section 2.2 of that Schedule. These processes would be authorised under the 1991 Act and thus there is no need for dual control. The exemption also extends to the secure storage of waste before it is processed if the waste, or the containers in which it is stored, is stored on an impermeable pavement provided with a sealed drainage system.[13]

SCRAP METAL YARDS AND VEHICLE DISMANTLING SITES

The operations set out in Table 4A to the 1994 Regulations may be 16.73 carried on at a secure place designed or adapted as a scrap metal yard or vehicle dismantling site in relation to the kinds of wastes specified in the table without a licence if they comply with the provisions of paragraph 45 of schedule 3.[14] In particular the quantity limits in relation to the kinds of wastes specified must not be exceeded. Thus if a dismantling site processes more than 40 vehicles in any period of seven days it will need to be licensed. The operation must be carried on with a view to the recovery of the waste. In addition any plant or equipment used in carrying out the activity must be maintained in reasonable working order.[15]

TABLE 4A

Kind of Waste	Activities	Seven day limit
Ferrous metals or ferrous alloys in metallic non-dispersible form (but not turnings, shavings or chippings of those metals or alloys)	Sorting; grading; baling; shearing by manual feed; compacting; crushing; cutting by hand-held equipment	8,000 tonnes
The following non-ferrous metals, namely copper, aluminium, nickel, lead, tin, tungsten, cobalt, molybdenum, vanadium, chromium, titanium, zirconium, manganese or zinc, or non-ferrous alloys, in metallic non-dispersible form, of any of those metals (but not turnings, shavings or chippings of those metals or alloys)	Sorting; grading; baling; shearing by manual feed; compacting; crushing; cutting by hand-held equipment	400 tonnes

[13] See D.o.E. Circular 6/95, Annex 1, para. 1.27.
[14] Added by S.I. 1995 No. 288, reg. 3(16).
[15] See also para. 16.75 below.

Kind of Waste	Activities	Seven day limit
Turnings, shavings or chippings of any of the metals or alloys listed in either of the above categories	Sorting; grading; baling; shearing by manual feed; compacting; crushing; cutting by hand-held equipment	300 tonnes
Motor vehicles (including any substance which is special waste and which forms part of, or is contained in, a vehicle and was necessary for the normal operation of the vehicle)	Dismantling, rebuilding, restoring or reconditioning, but, in relation to lead acid batteries, only their removal from motor vehicles	40 vehicles
Lead acid motor vehicle batteries (including those whose contents are special waste), whether or not forming part of, or contained in, a motor vehicle	Sorting (including removal from motor vehicles)	20 tonnes

16.74   The storage of the wastes listed in Table 4B is allowed at a secure place designed or adapted for use as a scrap metal yard or vehicle dismantling site. The storage must be for the purposes of an operation in Table 4A in relation to that kind of waste or be otherwise for a recovery or recycling activity licensed under Part II of the 1990 Act or authorised under Part I of that Act.[16] The total quantity of any waste so stored must not exceed the quantities in relation to that type of waste. Note that the quantities of vehicles allowed to be stored depends on the type of surface on which they are kept. No waste may be stored at the site or yard for longer than twelve months.

TABLE 4B

Kind of waste	Maximum total quantity
Ferrous metals or ferrous alloys in metallic non-dispersible form (but not turnings, shavings or chippings of those metals or alloys)	50,000 tonnes
The following non-ferrous metals, namely copper, aluminium, nickel, lead, tin, tungsten, cobalt, molybdenum, vanadium, chromium, titanium, zirconium, manganese or zinc, or non-ferrous alloys, in metallic non-dispersible form, of any of those metals (but not turnings, shavings or chippings of those metals or alloys)	1,500 tonnes

[16] S.I. 1994 No. 1056, Sched. 3, para. 45(2) as added by S.I. 1995 No. 288.

226

Kind of waste	Maximum total quantity
Turnings, shavings or chippings of any of the metals or alloys listed in either of the above categories	1,000 tonnes
Motor vehicles (including any substance which is special waste and which forms part of, or is contained in, a vehicle and was necessary for the normal operation of the vehicle);	
where any such vehicle is stored on a hardstanding which is not an impermeable pavement;	100 vehicles
where all such vehicles are stored on an impermeable pavement:	1,000 vehicles
Lead acid motor vehicle batteries (including those whose contents are special waste) whether or not forming part of, or contained in, a motor vehicle	40 tonnes

Where waste is stored each kind of waste stored must either be stored 16.75 separately or be kept in separate containers. However, if a load of more than one kind of waste is delivered to the yard it may be kept unseparated pending sorting for up to two months.[17] Liquid wastes or waste motor vehicle batteries must be stored in a secure container. Waste motor vehicles from which all fluids have been drained should either be stored on a hardstanding, in which case only 100 can be stored, or on an impermeable pavement, where the limit rises to 1,000. Vehicles that have not been drained of all fluids cannot take advantage of the exemption. All other wastes, or the containers they are in, must be stored on an impermeable pavement provided with a sealed drainage system. No pile or stack of waste may exceed five metres in height.

Operations in the yard or site must also be carried on on an impermeable 16.76 pavement provided with a sealed drainage system.[18] This means a drainage system with impermeable components that does not leak and which will ensure that no liquid will run off the pavement otherwise than through the system and that, unless lawfully discharged into a sewer or controlled waters, all liquids entering the system are collected in a sealed sump.[19] In addition the operations or storage of non-scrap waste must be carried on by or with the consent of the occupier of the land where it is done or the operator must be otherwise entitled to use the land for that purpose.[20]

To ensure these provisions are complied with the person responsible for 16.77 the management of the yard or site—not necessarily the site manager—

[17] S.I. 1994 No. 1056, para. 45(2)(d).
[18] ibid. para. 45(1)(c).
[19] ibid. para. 45(7) and see D.o.E. Circular 6/95, para. 1.27.
[20] ibid. reg. 17(2) as amended by S.I. 1995 No. 288, reg. 3(6).

must have established an administrative system to ensure that the waste accepted is of a kind listed in Tables 4A or 4B and that the quantity restrictions imposed in those tables are not breached. A monthly audit must be carried out to confirm compliance with the terms and conditions of the exemption.[21] The records required to be kept under the Regulations should show, for each month, the total quantity of each kind of waste recovered at the site during the month. In addition the records required to be kept by holders of a waste management licence must also be kept by the operator.[22]

16.78   Each year the operator should send the relevant Agency details of the total quantity of each kind of waste recovered over the last year. He should also send an up to date plan of the site showing the site's boundaries, the locations where the relevant activity is carried on, the locations and specifications of any hardstandings or impermeable pavements and drainage systems and the location of secure containers.[23] These details should be accompanied by the annual fee of £150.[24]

16.79   An initial fee of £300 is payable when the establishment or undertaking running the site registers with the relevant Agency in accordance with regulation 18 of the 1994 Regulations. Failure to register carries an enhanced penalty on summary conviction of a fine not exceeding level 2 on the standard scale.[25] Registration must be effected by a notice in writing from the operator of the site giving the details of the site as set out in paragraph 16.77. The fee is payable in respect of each site the operator registers.[26] The annual fee of £100 becomes payable following the service of a notice to pay on the registered operator. The notice should be served not later than one month before the anniversary of the date when the operator sent his registration details to the authority. It should specify the amount due, the method of payment, the date of the anniversary and that payment is either due on that date or, if later, a month after the date of the notice. The notice should also state the effect of payment not being made at the due time.[27]

16.80   Failure to make payment within two months of the due date will lead to loss of the exemption. If the fee is due in respect of one or more sites then all those sites lose their exemption. If it is paid in respect of some but not all the registered sites then those sites in respect of which it is unpaid will be de-registered.[28] The effect of this will be to make the operation of the site unlawful unless it has a waste management licence. Any further operations at it would be an offence under section 33(1) of the Environmental Protection Act 1990. The Agency concerned must give the operator written notice of the removal of any of his sites from the register.

---

[21] S.I. 1994 No. 1056, Sched. 3, para. 45(3)(a).
[22] *ibid*. Sched. 4, para. 14(3) as added by S.I. 1995 No. 288, reg. 3(21).
[23] *ibid*. Sched. 3, para. 45(3)(b) & 45(3)(c).
[24] *ibid*. para. 45(3)(2) as amended by S.I. 1996 No. 634, reg. 2(7).
[25] S.I. 1994 No. 1056, reg. 18(6) as amended by S.I. 1995 No. 288, reg. 3(11)(a).
[26] *ibid*. reg. 18(4A) as added by S.I. 1995 No. 288, reg. 3(10).
[27] *ibid*. Sched. 3, para. 45(4) as added by S.I. 1995 No. 288, reg. 3(10) and see D.o.E. Circular 6/95, paras. 1.35–1.38.
[28] *ibid*. reg. 18(4B) as inserted by S.I. 1995 No. 288, reg. 3(10).

The controls on scrap metal yards and vehicle dismantling sites are **16.81** designed to be largely self-regulating. The relevant Agency should inspect the site within two months of receiving the notice requesting registration. Thereafter it should inspect at least once a year but it is not required to inspect more often.[29]

The exemption granted under paragraph 45 extends to the temporary **16.82** storage of "non-scrap waste" at a site pending its collection.[30] To benefit from this exemption the non-scrap waste must not be of any type specified in Table 4B and must have been delivered to the site as part of a consignment of which at least 70 per cent by weight was waste motor vehicles or 95 per cent by weight was any other type of waste described in Table 4B. The non-scrap waste delivered in this way should be capable of being sorted from the Table 4A waste by sorting or hand dismantling. This provision is intended to stop the site being used as a waste transfer station under cover of this exemption.

In addition the non-scrap waste can only remain at the site for three **16.83** months. While it is there it should be stored in a secure place. If it consists of liquid wastes it should be kept in a secure container. The waste, or the container or containers, must be stored on an impermeable pavement which is provided with a sealed drainage system.

REGISTRATION OF EXEMPTIONS

It will be an offence for an establishment or undertaking to carry on an **16.84** exempt activity involving the recovery or disposal of waste—other than the spreading of waste on land used for agriculture under paragraph 7(3)(c) of Schedule 3[31]—without being registered with the appropriate registration authority.[32] An "establishment or undertaking" means any organisation and an individual carrying on a business as a sole trader. It does not encompass an individual in his or her private capacity.[33] A person guilty of this offence will be liable on summary conviction to a fine not exceeding level 2 on the standard scale.[34]

Most activities should have been registered on December 31, 1994, but **16.85** for scrap metal yards or vehicle dismantling sites the date was December 30, 1995.[35] However, exempt activities in Scotland that operate under a resolution made by virtue of section 54 of the Environmental Protection Act 1990 and in accordance with the conditions of such a resolution need not be registered.[36]

---

[29] S.I. 1994 No. 1056, Sched. 4, para. 13(3)–13(5) as added by S.I. 1995 No. 288, reg. 3(18).

[30] *ibid.* Sched. 3, para. 45(5) as added by S.I. 1995 No. 288, reg. 3(16).

[31] But the details supplied under that paragraph should be entered—see D.o.E. Circular 11/94, Annex 6, para. 6.8 & 6.9.

[32] S.I. 1994 No. 1056, reg. 18(1) & 18(7).

[33] D.o.E. Circular 11/94, para. 6.19.

[34] S.I. 1994 No. 1056, reg. 18(6).

[35] *ibid.* regs. 18(1) & 18(1A) added by S.I. 1995 No. 288, reg. 3(8).

[36] *ibid.* reg. 18(1B) added *ibid.*

16.86   Each registration authority must establish and maintain an exemption register for the activities in respect of which it is the authority.[37] The register must contain the name and address of the establishment or undertaking concerned, the activity that constitutes the exempt activity and the place where it is carried on.[38] As the singular includes the plural, several places where an activity is carried on can be registered in one entry.[39] There does not have to be a formal application for registration. The authority must enter the relevant particulars if it receives notice of them in writing or if it otherwise becomes aware of them[40]. Authorities who authorise or otherwise control activities under regulation 18(10)(a), (b) or (c) will be deemed to be aware of them and so should register them without more ado.[41] For other activities which have to be registered with the relevant Agency it would be wise to formally apply for registration. Registration remains valid indefinitely.[42]

16.87   Each registration authority must ensure that its register is available to the public, free of charge, at all reasonable hours. It must also allow copying of entries for a reasonable amount.[43] Registers may be kept in any form.[44]

NORTHERN IRELAND

16.88   In Northern Ireland exemptions from licensing requirements are contained in the Waste Collection and Disposal Regulations (Northern Ireland) 1992.[45]

---

[37] S.I. 1994 No. 1056, reg. 18(2).
[38] *ibid.* reg. 18(3).
[39] D.o.E. Circular 11/94, Annex 6, para. 6.4.
[40] S.I. 1994 No. 1056, reg. 18(4).
[41] *ibid.* reg. 18(5).
[42] D.o.E. Circular 11/94, Annex 6, para. 6.5.
[43] S.I. 1994 No. 1056, reg. 18(8).
[44] *ibid.* reg. 18(9).
[45] S.R. 1992 No. 254.

17. PLANNING PERMISSION FOR WASTE MANAGEMENT FACILITIES

This chapter is concerned with the requirement for planning permission for the development of land for waste management purposes, the considerations on granting permission and conditions that can be attached to it. In England and Wales the appropriate legislation is the Town and Country Planning Act 1990 while for Scotland it is mainly the Town and Country Planning (Scotland) Act 1997. In addition procedures for granting permission and variations in the scope of development control are provided in England and Wales by the General Permitted Development Order 1995[1] and the General Development Procedure Order 1995,[2] in Scotland the Orders are the General Permitted Development (Scotland) Order 1992[3] and the General Development Procedure (Scotland) Order 1992,[4] In this chapter a reference in square brackets is to the Scottish legislation.

The relationship between planning control and waste management 17.01 licensing is set out in Waste Management Papers No. 4 and 26 and, for England, in P.P.G. No. 23—"Planning and Pollution Control". The relationship is close but planning controls are concerned with the impact of the development on land and planning permission enures for the benefit of the land. Waste management licences, on the other hand, are more concerned with the day to day control of the licensed activities and only operate during the life of the site.[5] The fact that planning permission has been granted for a site should not inhibit the relevant Agency in refusing to grant a licence for it if this course of action is justified.[6]

There is no definition of "waste" in the Planning Acts or General 17.02 Permitted Development Orders. However, the term was considered for planning purposes in *R. v. Rotherham Metropolitan Borough Council, ex p.*

---

[1] S.I. 1995 No. 418 as amended.
[2] S.I. 1995 No. 419.
[3] S.I. 1992 No. 223 as amended.
[4] S.I. 1992 No. 224.
[5] W.M.P. No. 4, paras 3.6–3.10.
[6] *Gateshead Metropolitan Borough Council v. Secretary of State for the Environment* [1995] Env. L.R. 37.

Rankin[7] where the judge felt that while other decisions of the courts on the meaning of "waste" under the Control of Pollution Act 1974 were useful, a court determining planning matters is not bound by them. Nevertheless, planning decisions have followed the general view that the issue is to be looked at from the point of view of the person disposing of the material in question.[8]

THE REQUIREMENT FOR PERMISSION

17.03   Planning permission is required for all "development" unless excepted by the legislation. "Development" for these purposes means the carrying out of building, engineering, mining or other operations in, on, over or under land,—operational development—or the making of any material change in the use of any buildings or other land.[9] Thus if it is proposed to develop an engineered landfill site on land previously used for agriculture, planning permission will be required for the change of use of the land to industrial/waste disposal and for the engineering of the landfill.

17.04   Generally the deposit of waste on land will be a change of use of the land.[10] This is confirmed by section 55(3)(b) of the Town and Country Planning Act 1990 [26(3)(b), 1997 Act] which states that, "for the avoidance of doubt" the deposit of refuse or waste materials on land involves a material change in its use, notwithstanding that the land is in a site already used for that purpose, if the superficial area of the deposit is extended or if its height is raised and exceeds the level of the land adjoining the site. It may also be a change of use if different materials are tipped from those previously deposited. Thus in *Alexandra Transport Co. Ltd v. Secretary of State for Scotland*[11] it was held that to deposit household waste in a quarry that had previously been used to deposit quarry waste was a change of use.[12]

17.05   The question as to whether a waste disposal activity involves a change of use or an operation can be important. The tipping of waste material on land in itself is unlikely to be "operational" development even if connected with other construction works. Rather it is a change of use of land.[13] However, if the primary intention behind the activity is operational, such as to raise the level of land to facilitate drainage and not to provide a last resting place for waste then it will be an "engineering operation" rather than a "change of use".[14] Thus in all cases regard should be had to the primary intention of the tipping operation, although the primary intention may not be the overt one put forward by the developer.[15]

---

[7] (1990) 2 J.Env.L 250.
[8] See Planning Decision at [1989] J.P.L. 379.
[9] T.&C.P. Act 1990, s.55(1) [26(1): 1997 Act].
[10] *Bilboe v. Secretary of State for the Environment* (1980) 39 P. & C.R. 495.
[11] (1974) 27 P. & C.R. 252.
[12] But see *R. v. Derbyshire County Council, ex p. North Derby District Council* (1979) 77 L.G.R. 389.
[13] Planning Appeal *Re: Morwood Farm* Ref. T/APP/C/90/H3700/3/P6.
[14] *Northavon District Council v. Secretary of State for the Environment* (1980) 40 P. & C.R. 332.
[15] Planning Appeal *Re: land at Tygarw*, Swansea W.O. Ref. 84/1754.

Waste disposal developments are not specified in any of the use classes of   17.06
the Use Classes Order 1987.[16] Thus they will be *sui generis* uses of land
rather than being able to take advantage of the benefits the Order confers.
Any change of use for waste disposal or management purposes will require
permission. For example in one appeal concerning the change of a derelict
barn to a chemical waste transfer station the inspector considered that the
new use would be *sui generis* use reasonably related to an industrial use.[17]

Even if a deposit of waste requires planning permission, no formal   17.07
application will be necessary if it can take the benefit of the deemed
planning permissions set out in Schedule 2 to the General Permitted
Development Order 1995 [Sched. 1, 1992 Order]. Three of these permissions only concern sites that were used for the deposit of waste on July 1,
1948. Thus Part 8 Class D of Schedule 2 gives permission for wastes, other
than mining wastes, from an industrial process to be deposited at such a
site.[18] Part 21, Class B[19] is concerned with colliery waste tips—although
these will now require retrospective permission or approval by virtue of
conditions B1 to that Class—while Class A deals with deposits of mineral
waste on mining land. Part 12, Class B allows local authorities to continue
to use refuse sites in existence in 1948 even if the area or height of the
deposit is extended; although this will be largely irrelevant now. This
provision is not extended to Scotland.

Class A of Part 6 of the Order permits   17.08

> "the carrying out on agricultural land comprised in an agricultural unit of . . . any
> excavation or engineering operations reasonably necessary for the purposes of
> agriculture within that unit."[20]

This permission was previously available to farmers to allow them to use
waste to infill gullies on their land or raise it above flood levels. However, if
the development involves the deposit of waste materials on or under the land
that waste can only originate from the agricultural unit itself; except
hardcore that is promptly used for building works or the creation of a hard
surface.[21] This limitation on the rights of farmers was introduced in 1985 so
that earlier cases on this aspect of Part 6 are of little direct relevance now.
However, if the imported material is not "waste" then the limitation does
not apply.[22]

APPLYING FOR PERMISSION

In England waste disposal is a "county matter" for the purposes of Town   17.09
and Country Planning legislation. This means that county councils, metropolitan district and London borough councils will be the relevant planning

---

[16] S.I. 1987 No. 764 [S.I. 1989 No. 147 (SCOT)].
[17] Planning Appeal *Re: land at Garmondsway*, Bishop Middleham Ref: T/APP/Y1300/
A/89/121886/P7.
[18] [S.I. 1992 No. 223, Sched. 1, Class 26.]
[19] *ibid.* [Sched. 1, classes 63 and 64.]
[20] *ibid.* [Sched. 1, class 18].
[21] G.P.D.O. 1995, Sched. 2, Part 6, para. A.2(1)(c).
[22] See planning decision at [1988] J.P.L. 663.

authorities for determining applications[23]; unless unitary authorities have been established in particular areas. "County matters" are defined in paragraph 1 of Schedule 1 to the Town and Country Planning Act 1990 which enables regulations to be made to prescribe other uses or operations not contained in that paragraph.[24] Under previous legislation the Town and Country Planning (Prescription of County Matters) Regulations 1980[25] were made. These state that the use of land or the carrying out of operations in or on land for the deposit of waste materials will be a county matter. In addition the erection of any building, plant or machinery designed to be used wholly or mainly to treat, store, process or dispose of refuse or waste materials will also be dealt with by county planning authorities.[26] If there is a mixture of county and other matters in the application the test is to look at the predominant purpose of the application.[27] In Wales and Scotland and English unitary authorities such applications will be dealt with by the local planning authority.

17.10   Where the matter is a "county matter" an application for planning permission will be made to the county planning authority directly[28] instead of to the local planning authority. Otherwise it will be made to the local planning authority. An application should be made on a form provided by the authority and in, England and Wales, in accordance with the Town and Country Planning (Applications) Regulations 1988.[29] The authority can require the applicant to provide further details of his proposals.[30] The application must be accompanied by the fee payable under the Town and Country Planning (Fees for Applications and Deemed Applications) Regulations 1989.[31] The appropriate fee will be determined in accordance with Schedule 1 to the Regulations.

17.11   Before granting planning permission the authority must consult with other relevant organisations in accordance with articles 10 and 11 of the General Development Procedure Order {Art 15 G.D.P.O. (SCOT)}. A County Council determining an application must consult the district planning authority for the area in which the development will take place[32] and any relevant parish or community council.[33] In addition a waste disposal application should be the subject of consultation with the relevant Agency.[34]

17.12   Applications for developments listed in Schedule 1 or Schedule 2 of the Town and Country Planning (Assessment of Environmental Effects) Regu-

---

[23] T.&C.P. Act 1990, s.1.
[24] ibid. Sched. 1, para. 1(j).
[25] S.I. 1980 No. 2010 and see Planning (Consequential Provisions) Act 1990, s.2.
[26] ibid. reg. 2.
[27] R. v. Berkshire County Council, ex p. Wokingham District Council [1996] Env. L.R. 71.
[28] See S.I. 1995 No. 419, art. 5(1)(b).
[29] S.I. 1988 No. 1812 [S.I. 1992 No. 224].
[30] ibid. reg. 4 [ibid. reg. 8(1)].
[31] S.I. 1989 No. 193 as amended by S.I. 1991 No. 2735 [S.I. 1990 No. 563 as amended by S.I. 1990 No. 2474 & S.I. 1991 No. 2765.
[32] G.D.P.O. 1995, art. 12
[33] T.&C.P. Act 1990, Sched. 1, para. 8; G.D.P.O. 1995, art. 13.
[34] G.D.P.O. 1995, art. 10(1), Table (r) [S.I. 1992 No. 224, art. 15(h)(vii)].

lations 1988[35] must be made the subject of an environmental impact assessment. The assessment takes the form of an "environmental statement" which must contain the information specified in Schedule 3 to the Regulations. Before planning permission is granted there must be wider consultation than for ordinary applications and the statement must be advertised. For these purposes the statement must be accompanied by a non-technical summary of its contents. Copies of the statement itself should be made available at a reasonable charge. The provisions of the Regulations are explained in Department of the Environment Circular 15/88.

Schedule 1 to the Regulations sets out a list of projects for which 17.13 environmental assessment is mandatory. These include a "waste disposal installation for the incineration or chemical treatment of special waste" and "the carrying out of operations whereby land is filled with special waste, or the change of use of land (where a material change) to use for the deposit of such waste".[36] A "deposit" here would probably cover the change of use of a building to a chemical waste transfer station. Schedule 2 projects include "an installation for the deposit of controlled waste . . . not being an installation falling within Schedule 1".[37] Assessment is only required for such projects when the development is not exempted from one and is such that it would be likely to have a significant effect on the environment by virtue of factors such as its nature, size or location.[38] It is considered that proposals for development, including landfills, concerning the transfer, treatment or disposal of household or commercial wastes with a capacity of 75,000 tonnes a year may well be candidates for assessment even though considerations concerning hazardous wastes may not arise. Otherwise, except in sensitive areas, assessment for sites taking smaller tonnages or inert wastes, is not thought to be necessary.[39]

DETERMINATION OF THE APPLICATION

Section 70 of the Town and Country Planning Act 1990 [37: 1997 Act] 17.14 sets out the general principles to be observed in the determination of planning applications. In dealing with an application an authority is to have regard to the provisions of the development plan, so far as they are material to the application, and to any other material considerations.[40] However, under section 54A of the 1990 Act [25: 1997 Act] where, in making any determination under the planning Acts, regard is to be had to the development plan, the determination shall be made in accordance with the plan unless material considerations indicate otherwise. The "development plan", for these purposes will be structure and local plans and adopted or approved alterations to them.[41] The status of development plans is consid-

---

[35] S.I. 1988 No. 119 [S.I. 1988 No. 1221].
[36] *ibid.* Sched. 1, para. 9 [*ibid.* Sched. 1, para. 9].
[37] *ibid.* Sched. 2, para. 11(c) [*ibid.* Sched. 2, para. 11(c)]
[38] *ibid.* reg. 2(1) [*ibid.* regs. 4 & 6(1)(a)] and see Circular 15/88, paras 18–33.
[39] D.o.E. Circular 15/88, App. A, para. 23.
[40] T.&C.P. Act 1990, s.70(2) [37(1): 1997 Act].
[41] *ibid.* ss.54(1) & 336(1) [277(1): 1997 Act].

ered in chapter 13. This chapter is concerned with "other material considerations".

17.15   "Material considerations" are not defined in the Planning Acts. However, in *Stringer v. Minister for Housing and Local Government*[42] Cooke J. said that,

> "In principle it seems to me that any consideration which relates to the use and development of land is capable of being a planning consideration. Whether a particular consideration falling within that broad class is material in any given case will depend on the circumstances."

In addition an exception may be made to the rule that the consideration has to relate to the use and development of land where the decision may cause personal hardship or difficulty to a business of value to the community.[43] However, while in waste applications some authorities have sought to raise matters relating to the record of the applicant it is submitted that these are not proper planning considerations

17.16   The effect on the landscape and amenity of the area will always be material. Normally it will be considered that waste tipping will have an adverse effect on landscape and, particularly where the site is in an area designated as an Area of Outstanding Natural Beauty or Local Landscape Area, permission should be refused.[44] However, a proposal can be considered acceptable from the point of view of visual intrusion and resulting change in landscape character.[45] The final appearance of the landform may be important here; one appeal against deemed refusal being dismissed on the grounds that "the landform proposed here would be out of keeping and detract from the appearance of the local landscape".[46] If the operations would improve the appearance of a site then their temporary unsightly effect may be ignored.[47] As far as landraising is concerned, P.P.G. 23 states that while this may be appropriate, the site must be "designed to blend in with the surrounding landscape".[48]

17.17   Nature conservation interests will also be taken into account. If the proposed site incorporates a Site of Special Scientific Interest then permission is likely to be refused,[49] and even where the nature conservation interests are not protected by statute they may still be significant enough to result in refusal.[50] Protection of species is also important, so that an application will be refused if it is likely to have an extinguishing effect on the population of the species in the locality.[51] Guidance as to development in protected areas is set out, for England, in P.P.G. 9.

17.18   Special consideration applies where the proposed development is in the Green Belt In P.P.G. 2 only certain types of development are considered to

---

[42] [1971] 1 All E.R. 65 at 77.
[43] *Great Portland Estates plc v. Westminster City Council* [1985] A.C. 661.
[44] Planning Appeal *Re: Fulmer Chase Farm* Ref: T/APP/C/90/A0400/1/P6.
[45] *ibid. Re: Trench Lane, Winterbourne* Ref: T/APP/F0100/A/90/173176/P5.
[46] *ibid. Re: Raglington Farm, Shedfield* Ref: T/APP/21700/A/90/167353/P3.
[47] *ibid. Re: Chesterhill, Swarland* Ref: T/APP/R2900/A/94/240896/P5.
[48] P.P.G. 23, para. 5.10.
[49] Planning Appeal *Re: Halnaker Chalk Pit, Chichester* Ref: T/APP/P3800/A/91/174947/P5.
[50] *ibid. Re: Strokins Farm, Kingsclere* Ref: T/APP/21700/A/91/175721/P7.
[51] *ibid. Re: Tai Cwplau Farm, Rhigos* W.O. Ref: APP51–29NP.

be appropriate development in the Belt. A waste disposal site will be usually inappropriate development in such an area, even though it is a temporary facility that can be restored to give a high quality landscape.[52] However, there may be cases where, despite Green Belt policies, tipping in the Belt will be justified on the grounds of need and lack of suitable alternatives.[53] In another appeal concerning landraising and the construction of a Ski Centre in the Green Belt the inspector concluded, that,

(1) the proposed landraising, being inappropriate, would by definition be harmful to Green Belt purposes;
(2) the acceptability of the landraising would depend on the balance between the harm (Green Belt, visual impact, strategic policy, potential pollution) and the benefits (need for waste disposal and for a Ski Centre);
(3) the proposed, Ski Centre, being appropriate, would, in principle, be in conformity with Green Belt policy;
(4) the acceptability of the Ski Centre would depend on the balance between the harm (visual and environmental) and the need for such a facility."

In the end he decided that "the potential environmental harm attributable to both proposals must weigh heavily against them. In the case of the landraising proposal this harm would be additional to the harm inherent in inappropriate development in the Green Belt."[54]

Residential amenity will also be material. The effect of a waste disposal 17.19 site in the vicinity of houses can include the noise of operations, smells, windblown rubbish or dust, visual intrusion and the effects of traffic. These factors may not be sufficiently harmful in themselves, but the Secretary of State will take into account their cumulative effect on amenity.[55] However, permission may be granted, even where houses bound part of the proposed site, if there is a need for tipping space and the operations will be of short duration.[56] In addition the problem of landfill gas must also be taken into account. Special consideration must be given to sites within 250 metres of existing residential development.[57] Even a remote risk of gas migration is not acceptable in a location close to housing.[58]

Traffic issues centre around the ability of the roads in the area of the 17.20 proposed development to cope with the extra heavy vehicles that will use them, the number of extra vehicle trips a day that will be involved and the access to the site. In one appeal the inspector considered that while the principal road involved was residential with no footpaths and with an old

---

[52] Planning Appeal *Re: New Pump Farm, Warley* Ref: T/APP/C/90/H1515/000021/P6.
[53] *ibid. Re: land between Moss House Lane and Vicars Hall Lane, Boothstown* Ref: T/APP/U4230/A/89/144538/P5.
[54] *ibid. Re: White's Pit, Poole* Ref: APP/R1200/A/94/236953 and APP/T1220/A/94/237075.
[55] *ibid. Aveley 3 Claypit, Aveley* Ref: APP/B5480/A/93/431730.
[56] *ibid. Re: Former railway land, Bootle* Ref: T/APP/M4320/A/90/149229/P5.
[57] D.o.E. Circular 17/89, para. 11.
[58] Planning Appeal—*Lancashire County Council v. Ashburn Brothers* (1993) 8 P.A.D. 651.

people's home and an ambulance station along its length, the use of that road by an anticipated eight return trips a day would not be detrimental to highway safety.[59] On the other hand where the proposed access road had poor alignment, very restricted width, general setting and a relatively quiet character with extensive use by equestrian interests it was considered so unsuitable to HGV traffic as to justify refusal.[60]

17.21 The possibility of pollution from the development will also be taken into account. In particular the dangers from landfill gas and leachate will be addressed. If there is not enough information about the potential levels of landfill gas from the site when completed permission may be refused on that ground alone.[61] Otherwise, while these are matters to be considered, they will often be controlled by conditions or under planning agreements. But if underground levels require protection from leachates or gases and an agreement, rather than conditions, is required, permission may be refused if such an agreement is not forthcoming.[62] The interface between planning and pollution controls is discussed in P.P.G. 23, particularly paragraphs 1.31–1.37.

17.22 The deposit of waste on land may affect its drainage or make other land more susceptible to flooding. Permission for tipping on land that forms part of a floodplain may be refused on the grounds that even though it would have no direct effect on flood levels it would remove part of the plain and so increase the danger of floods.[63] However, where practicable, these issues can be dealt with by a drainage scheme.[64]

17.23 In cases where permission might otherwise be refused on environmental grounds, the need for disposal facilities may result in it being granted. The need for disposal facilities varies over time, with the spaces available for the category of waste concerned in the application being filled and new ones created. The waste disposal plan—or national waste strategy—and local plans will be of assistance but are not conclusive as the position can change rapidly. Further, it is important to note that a failure to demonstrate need for a development is not in itself a reason for refusal.[65] An argument of the need for waste facilities may have to be balanced with the aim of sustainable development. One inspector has said that

> ". . . the aim of encouraging recycling, waste reduction and energy recovery is unlikely to be furthered by a comparative abundance of landfill capacity, so that a cautious approach to adding landfill capacity is justified on the basis of the policy for sustainable development."[66]

17.24 Finally, the site's restoration, aftercare management and after-use will also be material considerations. If the final restoration profile of the site,

[59] Planning Appeal—*Ludlow Farm, Warminster* Ref: T/APP/W3900/A/90/152844/P2.
[60] *ibid. Re: Land at and contiguous with Streat Sandpit, Streat* Ref: T/APP/C/88/F1400/000001/P6.
[61] *ibid. Re: Erith Quarry, Erith* Ref: APP/D5120/A/89/130168.
[62] *ibid. Re: land near Blaenavon and Pwll-du, Gwent* Ref: P33/989.
[63] *ibid. Re: Withy Pool, Charlton* Ref: T/APP/F1800/A/89/123665/P5.
[64] *ibid. Re: Upper Trelyn, Blackwood* W.O. Ref: P32/422.
[65] *ibid. Re: Patterson's Pit, Colney Heath* Ref: T/APP/M1900/A/89/132266/P3.
[66] *ibid. Re: Pond Farm, Lingfield* Ref: T/APP/B3600/A/94/233149/P4.

whatever features of nature and conservation were introduced, would leave a visually incongruous feature, refusal is likely.[67] Restoration of a site to allow agricultural use, or to improve existing agricultural use, may be a significant factor in favour of an application but will depend on the scale of the operations.

As far as waste transfer stations are concerned, these are an appropriate 17.25 use on an industrial estate. They will be allowed providing they do not undermine the estate's satisfactory development and do not cause adverse effects for local residents. These effects can largely be dealt with by conditions.[68] They are unlikely to be appropriate in a rural area, particularly if there are public footpaths or recreational grounds nearby.[69]

PLANNING CONDITIONS AND AGREEMENTS

A planning authority may grant permission unconditionally or subject to 17.26 such conditions as they think fit.[70] While this is a wide power any condition, to be valid, must fairly and reasonably relate to the permitted development. An authority cannot use their powers for an ulterior object, however desirable it may seem in the public interest. Further, a condition must not be so unreasonable that no reasonable planning authority, properly advised, could have imposed it.[71]

Guidance as to conditions that can be imposed is given in Department of 17.27 the Environment Circular 11/95. It suggests that while normally matters that are controlled by other legislation should not be the subject of planning conditions there are exceptions to this general rule. In particular where other controls are also available a planning condition may be needed when the considerations material to the exercise of the two systems of control are substantially different, since it might be unwise in these circumstances to rely on the alternative control being exercised in the manner or to the degree needed to secure planning objectives. Conditions may also be needed to deal with circumstances for which a concurrent control is unavailable such as to secure the restoration of a waste disposal site.[72]

Part I of Schedule 5 to the Town and Country Planning Act 1990 was 17.28 amended by the Planning and Compensation Act 1991 to apply provisions for aftercare conditions, that formerly only concerned minerals permissions, to permissions for the depositing of refuse or waste materials.[73] Similar amendments are made in Schedule 3 to the 1997 Act. Nevertheless, permissions for landfill sites that were granted before the provisions of the 1991 Act were brought into force will normally have aftercare conditions

---

[67] Planning Appeal—*Re: Lower Loxley, Uttoxeter* Ref: T/APP/N3400/A/93/232284/P5.
[68] *ibid. Norfolk County Council v. Green (E.) & Son* (1994) 9 P.A.D. 79.
[69] *ibid. Re: Keepers Cottage, Aldermaston* Ref: T/APP/C/93/U0330/628029–31.
[70] T.&C.P. Act. 1990, s.70(1) [37(1)(a): 1997 Act].
[71] *Newbury D.C. v. Secretary of State for the Environment* [1981] A.C. 578.
[72] D.o.E. Circular 11/95, Annex, para. 23.
[73] T.&C.P. Act. 1990, Sched. 5, para 2(1)(a) [1997 Act, Sched. 3].

imposed on them. The new powers will strengthen the scope and effect of such conditions. They are discussed in more detail at paragraph 20.16 *et seq.*

17.29 Planning conditions will control a wide range of matters. In particular they will specify the types of waste allowed on the site, for example stating that the waste to be deposited shall be soil, subsoil or clean rubble. They may also require the boundaries of the operations and the final levels of the deposit to be marked out before operations commence and specify the total volume of wastes that can be deposited. If waste is to be stored on the site prior to being landfilled the areas for storage may be set out.[74]

17.30 A condition may require a scheme to be submitted to the authority before operations commence showing how the site is to be operated and tipped and that the site is operated in accordance with that scheme. More detailed requirements can provide for a scheme for the landscaping of the site while it is operational, for controlling vehicle movements, for the provision of bunds around certain areas, site fencing and wheel cleaning facilities. The hours in which the site may be open will be specified. Pollution controls may also be imposed, in particular to reduce noise and dust and to prevent water pollution. A provision that if operations at the site cease for a specified period the site must be restored in accordance with a scheme agreed with the authority may also be inserted.

ENFORCEMENT

17.31 For the purposes of the Town and Country Planning Acts the carrying out of development without the required planning permission, or failing to comply with any condition or limitation subject to which planning permission has been granted, constitutes a breach of planning control.[75] Guidance to planning authorities on the policy issues relating to enforcement of planning control is contained in Planning Policy Guidance Note 18. The way in which enforcement powers should be used is set out in Department of the Environment Circular 21/91.

17.32 Where a planning authority considers that there has been a breach of planning control in respect of any land they may serve a "planning contravention notice" on the owner or occupier of the land or the person who carried out the unauthorised development.[76] This notice requires the person served to give specified information to the authority about the use of the land and his interest in it. It will be an offence to fail to comply with such a notice within 21 days of being served with it or to make a false statement in response to it.[77] Notices concerning waste management developments may be served by both county and district councils in England; although the district should consult the county before serving a notice in such cases.[78]

---

[74] See P.P.G. 23, para. 5.13.
[75] T.&C.P. Act 1990, s.171A(1) [s.123A: 1997 Act].
[76] *ibid.* s.171C [s.125: 1997 Act].
[77] *ibid.* s.171D [s.126: 1997 Act].
[78] *ibid.* Sched. 1, para. 11(1) & 11(3).

The principal power an authority has is to serve an enforcement notice 17.33
under section 172 of the 1990 Act [127: 1997 Act]. In the past problems
have been encountered in the drafting of such notices so that notices would
not be valid if the right breach of control was not specified. Now a notice
must specify the breach of planning control as set out in section 171A(1)
[127: 1997 Act]; either unauthorised development or failure to comply with
a condition. Further, in specifying the matters that constitute the alleged
breach the notice will be sufficient if it enables anyone on whom a copy is
served to know what those matters are.[79]

The notice must also specify what must be done to remedy the breach. 17.34
Where the breach involves the unauthorised deposit of waste the notice may
require the contour of the deposit to be modified by altering the gradient or
gradients of its sides.[80] Where the breach is unauthorised tipping the notice
may require the cessation of all tipping operations forthwith and the removal
of the waste from the land within a specified period, with site restoration to
be carried out by a further period. The use of "forthwith" in this context has
been criticised[81] and it is suggested that a clear time should be stated in such
a notice; such as within seven days from the receipt of this notice. Other
requirements can include the removal of plant or equipment such as skips.

An enforcement notice may be varied or withdrawn by the planning 17.35
authority; although withdrawal does not affect their power to serve a further
notice.[82] Anyone with an interest in the land or the relevant occupier may
appeal to the Secretary of State against the notice. The grounds of such an
appeal can include, amongst other things, that planning permission should
be granted for the development in question or that the steps required to be
taken by the notice are excessive or that the time in which compliance is
required is too short.[83] Otherwise, if no appeal is made, or it is dismissed, the
requirements of the notice must be complied with within the specified time.
If it is not the authority can enter on the land to carry out any necessary
works to secure compliance and recover their expenses in doing so from the
owner of the land.[84] Further non-compliance is an offence under section 179
of the 1990 Act [136: 1997 Act] punishable on summary conviction by a
fine of up to £20,000 or on indictment by a fine.

In addition to an enforcement notice the authority may also serve a stop 17.36
notice in respect of a breach of planning control.[85] This will require the
unauthorised activity to cease before the time limit set out in an enforcement
notice. This in effect imposes an immediate ban on the unauthorised activity.
There is no right of appeal against such a notice as it depends on the
enforcement notice to which it relates. If an appeal against the enforcement
notice is successful the stop notice lapses. Thus an authority can use their

---

[79] T.&C.P. Act 1990, s.173(1) & 173(2) [s.128: 1997 Act].
[80] ibid. s.173(5)(d) [s.128: 1997 Act].
[81] Planning Appeal Re: Fulmer Chase Farm, Fulmer Ref: T/APP/C/90/A0400/1/P6.
[82] T.&C.P. Act 1990, s.173A [s.129: 1997 Act].
[83] ibid. s.174 [s.130: 1997 Act].
[84] ibid. s.178 [s.135: 1997 Act].
[85] ibid. s.183 [s.140: 1997 Act].

powers to serve a stop notice to bring into effect the provisions of an enforcement notice that has been suspended during an appeal. However, service of a stop notice may give rise to payment of compensation under section 186 of the 1990 Act [143: 1997 Act].

17.37　Planning conditions can be enforced under section 187A of the 1990 Act [145: 1997 Act]. This involves the service of a breach of condition notice and, if that notice is not complied with the person served with it will be guilty of an offence. In addition an authority can apply to the High Court or a county court for an injunction—or for interdict in the Court of Session or the sheriffs court—to restrain any breach, or apprehended breach, of planning control.[86] However, this jurisdiction should be exercised by the courts sparingly and with great caution and there must be something more than a mere infringement of the criminal law; although a deliberate and flagrant flouting of it suffices. Further, a court should consider whether the offender will be likely to continue his objectionable activities unless he is restrained by injunction.[87]

PLANNING IN NORTHERN IRELAND

17.38　Planning control is a function of the Department of Environment of Northern Ireland under the Planning (Northern Ireland) Order 1991.[88] Article 11 of the Order defines development in the same way as section 55 of the Town and Country Planning Act 1990, while article 11(3)(b) states that the deposit of refuse or waste material on land involves a material change of use thereof, even if the land is already used for waste disposal purposes, if it enlarges an existing deposit.

17.39　Planning permission will normally be required for the carrying out of any development of land.[89] The exceptions to this requirement are set out in the Planning (General Development) Order (N.I.) 1993.[90] An application for planning permission will be made under article 20 of the 1991 Order. Any environmental assessment must be furnished in accordance with the Planning (Assessment of Environmental Effects) Regulations (N.I.) 1989.[91] The application must be publicised in accordance with the requirements of article 21 of the 1991 Order and the General Development Order. For some waste disposal facilities the special procedure for major planning applications set out in article 31 of the 1991 Order may be applied. After all the relevant procedures have been followed the Department will determine the application having regard, where relevant, to any development plan and to any other material considerations. It may grant the application conditionally or unconditionally or may refuse it.[92] Appeals are made under article 32 of the 1991 Order.

---

[86] T.&C.P. Act 1990, s.187B [s.146: 1997 Act].
[87] See City of London Corporation v. Bovis Construction Ltd [1988] 86 L.G.R. 660.
[88] S.I. 1991 No. 1220 (N.I. 11).
[89] ibid. art. 12.
[90] S.I. 1993 No. 278.
[91] S.I. 1989 No. 20.
[92] S.I. 1991 No. 1220 (N.I. 11), arts 25(1) & 27.

Planning controls are enforced under the provisions of Part VI of the 1991 Order. A breach of planning control is defined in article 68 of the Order in similar terms to that in section 171A of the Town and Country Planning Act 1990. Enforcement notices will be served in accordance with article 68 and can require the contouring of waste deposits.[93] Appeals are made under article 69 while failure to comply with a notice will be an offence under article 72. 17.40

---

[93] S.I. 1991 No. 1220 (N.I. 11), art. 68(11).

18. LICENCES FOR WASTE FACILITIES

18.01   Formerly waste disposal sites were licensed under Part I of the Control of Pollution Act 1974. From May 1, 1994, when the licensing provisions of Part II of the Environmental Protection Act 1990 were brought into force, any waste disposal licences that were then in force—"existing disposal licences"[1]—will, on and after that day, be treated as waste management licences under the 1990 Act until they expire or cease to have effect. They will be subject to the controls imposed by the 1990 Act and, in particular, may only be surrendered or transferred in accordance with sections 39 and 40 of that Act.[2] These licences, if "current or recently current", must be shown on the public register.[3]

18.02   Waste management licences are granted and supervised under the provisions of Part II of the Environmental Protection Act 1990 as supplemented by the Waste Management Licensing Regulations 1994.[4] In particular Schedule 4 to those Regulations brings practice in Great Britain into line with the requirements of the E.C. "Waste" Directive. In Northern Ireland licences are granted under the 1978 Order, as supplemented by the 1992 Regulations.

1. GRANT OF LICENCES

18.03   The Environmental Protection Act 1990 is concerned with "waste management licences". A waste management licence is one granted by a waste regulation authority that authorises the treatment, keeping or disposal of any specified description of controlled waste in or on specified land—a "site licence"—or the treatment or disposal of such waste by means of specified mobile plant—"a mobile plant licence".[5]

18.04   Mobile plant here means plant that is designed to move or be moved, whether on roads or other lands; although regulations may prescribe what is

---

[1] E.P.A. 1990, s.77(1).
[2] *ibid.* s.77(2).
[3] S.I. 1994 No. 1056, reg. 10(1)(a) and see para. 19.31.
[4] S.I. 1994 No. 1056.
[5] E.P.A. 1990, ss.35(1) & 35(12).

or is not to be treated as mobile plant for these purposes.[6] Under regulation 12 of the Waste Management Licensing Regulations 1994[7] mobile plant is further defined as plant designed to move, or being capable of moving, from place to place with a view to being used at each such place. Plant prescribed for this purpose are, an incinerator exempted from the controls of Part I of the 1990 Act by section 5.1 of Schedule 1 to the Environmental Protection (Prescribed Processes and Substances) Regulations 1991,[8] plant for the recovery of waste oil from electrical equipment by way of filtration or heat treatment, plant for the destruction by dechlorination of PCBs, plant for the vitrification of waste, plant that treats clinical waste by microwave and plant for the treatment of waste soil.

A site licence will be granted to the occupier of the relevant land.[9] An **18.05** "occupier" is not defined for these purposes but will be someone with some degree of control associated with, and arising from, presence or use of the land. Thus there can be more than one occupier of a piece of land, although the 1990 Act implies that only one person can be—"the person who is in occupation". A mobile plant licence will be granted to the person who operates the plant in question.[10] This is presumably the person in control of the business in which the plant is used.

Advice can be sought from an authority as to whether a licence is or is not **18.06** required for a particular operation. The authority cannot charge for this advice.[11] Where the authority maintain a licence is required they are not making an administrative decision that is subject to judicial review. Rather they are giving advice which an operator is free to ignore at the risk of prosecution.[12]

An application for a licence will be made to the relevant waste regulation **18.07** authority—in effect the Agency or SEPA. For a site licence this will be the Agency in whose area the land is situated while, for a mobile plant licence, it will be the Agency in whose area the operator has his principal place of business.[13] In Northern Ireland applications are made to the district council under article 7 of the 1978 Order.[14]

An application must be made on a form provided by the Agency[15]; **18.08** although the Regulations merely require that the application be made in writing.[16] The form must be accompanied by such information as the authority reasonably requires. Guidance is given in Waste Management

---

[6] E.P.A. 1990, s.29(9) & 29(10).
[7] S.I. 1994 No. 1056 as amended by S.I.s 1995 No. 288 and 1996 No. 634.
[8] S.I. 1991 No. 472 as amended.
[9] E.P.A. 1990, s.35(2)(a).
[10] *ibid.* s.35(2)(b).
[11] *McCarthy & Stone (Developments) v. Richmond L.B.C.* [1992] 2 A.C. 48.
[12] *R. v. London Waste Regulation Authority, ex p. Specialist Waste Management Ltd* [1989] C.O.D. 288.
[13] E.P.A. 1990, s.36(1).
[14] S.I. 1978 No. 1049 (N.I. 19).
[15] E.P.A. 1990, as amended by E.A. 1995, Sched. 22, para. 68(2).
[16] S.I. 1994 No. 1056, reg. 2(1).

Paper No. 4 as to what information is necessary for the determination of an application.[17] If the applicant fails to provide any required information the authority may either refuse to process his application or to do so until the information is provided.[18] The application must be accompanied by the fee payable in respect of it under a charging scheme made under section 41 of the Environment Act 1995. Fees are detailed in the Waste Management Licensing (Fees and Charges) Scheme 1996/7 which has been issued by the Department of the Environment.[19] Details of the application should be entered on the public register held by the authority.[20]

18.09   Although not required by the legislation, a Working Plan should accompany the application. The Working Plan should consist of detailed design and operational statements to explain how the facility is to be developed, operated, restored and completed so as to enable the authority to draft the licence.[21] Special guidance for scrap metal sites is contained in "Licensing of Metal Recycling Sites"—W.M.P. No. 4A. In addition the Agencies, in licensing sites will have regard to other Waste Management Papers and in particular W.M.P. 26B "Landfill Design, Construction and Operational Practice".

18.10   A licence may not be issued for a use of land for which planning permission is required under Planning Acts unless such permission is in force in relation to that use or unless the planning authority has issued a certificate of lawfulness of that existing use in respect of it[22] under section 191 of the Town and Country Planning Act 1990 or section 150 of the Town and Country Planning (Scotland) Act 1997. In some cases an established use certificate under the former provisions of those Acts will also suffice. If no permission is required because the site has historic use rights then there will be no need for an actual permission to be in force because none is "required".[23-24] The purpose of this provision is to ensure that consideration has been given to the impact on land use of the operation of the facility, including matters such as access, effect on the local environment and amenity and planning policies. P.P.G. 23 gives advice as to the relationship between planning and pollution controls.

18.11   Subject to the planning and consultation requirements of the Act, an authority may not reject an application that has been properly made if it is satisfied that the applicant is a fit and proper person[25] to hold a licence unless rejection is necessary to prevent pollution of the environment, harm to human health or serious detriment to local amenities. However, this last

---

[17] W.M.P. No. 4, para. 2.9.

[18] E.P.A. 1990, s.36(1A) as added by E.A. 1995, Sched. 22, para. 68(2).

[19] Available from Waste Policy Division, Room A2.22, D.o.E. Romney House, 43 Marsham Street, London SW1P 3PY.

[20] S.I. 1994 No. 1056, reg. 10(1)(b).

[21] W.M.P. No. 4, para. 2.10.

[22] E.P.A. 1990, s.36(2).

[23-24] *Berridge Incinerators Ltd v. Nottinghamshire County Council*, April 14, 1987, Q.B.D. (unreported).

[25] See paras 18.17 *et seq.*

factor is excluded where planning permission is in force in respect of the use to which the land will be put under the licence.[26] This requires an actual permission rather than "historic rights". Where an application is made in respect of a site without an actual permission the "competent" authority— here the relevant planning authority—should take the "specified actions" as defined by paragraph 1 of Schedule 4 to the 1994 Regulations which in effect requires them to go through the planning processes relevant to the application and to consider a discontinuance order.[27]

"Pollution of the environment" is defined in section 29(3) of the Act to **18.12** mean pollution of any environmental medium from the escape or release of substances or articles comprising or resulting from the waste from the land on or in which, or the fixed or mobile,[28] plant by means of which, it was treated, kept or deposited. Any such escape or release must also be capable of causing harm to man or any other living organisms supported by the environment. "Harm" for these purposes means harm to the health of living organisms or other interference with the ecological systems of which they form part and, in the case of man, includes offence to any of his senses or harm to his property.[29] The provisions of the "Groundwater" directive must also be observed.[30] Rejection must be "necessary" to prevent one of these problems from materialising. Thus if a potential problem can be dealt with by condition, rejection of the application will not be "necessary"

Consultations in England and Wales will take place with the appropriate **18.13** planning authority (as defined in section 36(11)) and the Health and Safety Executive.[31] If the land to be used has been notified as a site of special scientific interest the Nature Conservancy Council for England or the Countryside Council for Wales must also be consulted.[32] 28 days must be allowed for these bodies to make representations about the application or such longer period as may be agreed in writing between the authority and the relevant body.[33] Any representations made must be considered by the waste regulation authority.[34] A copy of any representation or decision must be entered on the register.[35] Consultation in the special case of where off site works may be required is dealt with at paragraph 18.48.

In Scotland the position is much the same, the planning authority **18.14** consulted being the council constituted under section 2 of the Local Government, etc., (Scotland) Act 1994. The Nature Conservancy Council for Scotland will be consulted about operations on sites of special scientific interest.[36] Consultees have 28 days, or such longer period as may be agreed

---

[26] E.P.A. 1990, s.36(3).
[27] S.I. 1994 No. 1056, Sched. 4, para. 9(7) & para. 3, Table 5.
[28] E.P.A. 1990, s.29(4).
[29] *ibid.* s.29(5).
[30] See paras 18–32–18.36.
[31] E.P.A. 1990, s.36(4)(a) as amended by E.A. 1995, Sched. 22, para. 68(3)(a).
[32] *ibid.* s.36(7).
[33] *ibid.* s.36(10) as amended by E.A. 1995, Sched. 22, para. 68(6).
[34] *ibid.* s.36(4)(b) & 36(7)(b).
[35] S.I. 1994 No. 1056, reg. 10(1)(b)(ii) & (iii).
[36] *ibid.* s.36(7).

in writing, to make representations about the application.[37] Copies of decisions and representations must again be entered on the register.[38] Consultation in the special case of where off site works may be required is dealt with at paragraph 18.48.

18.15   If, within four months of the date the authority received an application, or such longer period as may be agreed in writing, the authority has neither granted the licence nor notified the applicant that it has rejected the application, the authority will be deemed to have rejected it.[39] The applicant will have a right to appeal to the Secretary of State against that rejection.[40] Notice of such an appeal must be given within six months of the date of the deemed rejection.[41] However, these provisions will not apply if the authority has not received relevant information and is not proceeding with the application under section 36(1A). If, under section 36(1A), it is waiting for information before dealing with the application, the four months' period starts to run on the day it receives that information.[42]

18.16   The authority can formally reject the application by a notice of rejection. A copy of the notice must be entered on the register.[43] An appeal against the rejection can be made within six months of the date of rejection.[44]

18.17   Otherwise the authority will grant the licence. The scheme of the legislation seems to be that licences should be granted unless there are good reasons for refusal. Certainly on an appeal against refusal it is for the authority to justify its decision. A licence must be genuine. It will be an offence to, with intent to deceive, forge or use a licence or to make or have in one's possession a document so closely resembling a licence as to be likely to deceive. A person found guilty of such an offence will be liable on summary conviction to a fine not exceeding the statutory maximum or, on indictment, to a fine or up to two year's imprisonment or to both.[45]

2. "FIT AND PROPER PERSON"

18.18   Under section 74(2) of the 1990 Act an Agency should determine whether or not a person is a fit and proper person to hold a licence in the context of the activities that are or are to be licensed and compliance with licence conditions. However, this provision does not apply to sites in Scotland operated under section 54 of the Act because no "fit and proper person" determination is required by the authorisation process.[46]

---

[37] S.I. 1994 No. 1056, s.36(10).
[38] S.I. 1994 No. 1056, reg. 10(1)(b)(ii) & (iii).
[39] E.P.A. 1990, s.36(9).
[40] ibid. s.43(1)(a).
[41] S.I. 1994 No. 1056, reg. 7(1).
[42] E.P.A. 1990, s.36(9A) as added by E.A. 1995, Sched. 22, para. 68(5).
[43] ibid. reg. 10(1)(b)(iv).
[44] E.P.A. 1990, s.43(1) & S.I. 1994 No. 1056, reg. 7(1)(a).
[45] ibid. s.35(7B) & 35(7C) as added by E.A. 1995, Sched. 22, para. 66(2).
[46] See E.P.A. 1990, s.74(1).

An authority will determine whether an applicant is a "fit and proper 18.19
person" for these purposes in accordance with section 74 of the Act. The
first criteria is that the applicant, or another relevant person, has been
convicted of a relevant environmental offence.[47] The second is that the
management of the activities to which the licence relates will not be in the
hands of a technically competent person—someone with the qualifications
and experience required by regulations made under section 74(6).[48] Finally,
the authority can determine that someone is not a fit and proper person if he
has not made and either has no intention of making, or is in no position to
make, adequate financial arrangements to discharge any obligations that
may arise from the licence.[49] This means that finance should be available to
comply with aftercare conditions in a planning permission or to deal with
problems such as emissions of landfill gas.

(a) RELEVANT OFFENCES

"Relevant offences" for the purposes of section 74(3)(a) are set out in 18.20
regulation 3 of the Waste Management Licensing Regulations 1994. They
are offences under any of the following enactments;

(a) section 22 of the Public Health (Scotland) Act 1897;
(b) section 95(1) of the Public Health Act 1936;
(c) sections 3, 5(6), 16(4), 18(2), 31(1), 34(5), 78, 92(6) or 93(3) of
the Control of Pollution Act 1974;
(d) section 2 of the Refuse Disposal (Amenity) Act 1978;
(e) the Control of Pollution (Special Waste) Regulations 1980;
(f) section 9(1) of the Food and Environment Protection Act 1985;
(g) the Transfrontier Shipment of Hazardous Waste Regulations
1988;
(h) the Merchant Shipping (Prevention of Pollution by Garbage)
Regulations 1988;
(i) sections 1, 5, 6(9) or 7(3) of the Control of Pollution (Amendment) Act 1989;
(j) sections 107, 118(4), or 175(1) of the Water Act 1989;
(k) section 23(1), 33, 34(6), 44, 47(6), 57(5), 59(5), 63(2), 69(9),
70(4), 71(3) or 80(4) of the 1990 Act;
(l) sections 85, 202, or 206 of the Water Resources Act 1991;
(m) section 33 of the Clean Air Act 1993;
(n) regulation 12 of the Transfrontier Shipment of Waste Regulations
1994[50];
(o) the Special Waste Regulations 1996.[51]

An offence under paragraph 15(1), (3), (4) or (5) of Schedule 5 to the
Finance Act 1996 which concerns the landfill tax will also be a relevant
offence for these purposes.[52]

---

[47] See E.P.A. 1990, s.74(3)(a) and see s.74(7) and paras 9.10–9.14.
[48] *ibid.* s.74(3)(b).
[49] *ibid.* s.74(3)(c).
[50] Added by S.I. 1994 No. 1137, reg.19.
[51] Added by S.I. 1996 No. 972, Sched. 3.
[52] S.I. 1997 No. 351, reg. 2.

18.21   It will be noted that some of these enactments have been repealed. Offences committed under them will still however count towards a "fit and proper person" assessment. Under the Rehabilitation of Offenders Act 1974 offences committed by indiviauals can become spent after a time—for example if dealt with by a fine they become spent after five years. Thus after five years have passed an individual need not notify the authority that he has committed the offence and the authority cannot rely on it in a "fit and proper person" assessment. The Department of Environment considers that while companies should disclose any offences, those that are spent should be disregarded.[53]

18.22   Despite a relevant conviction an authority can still consider a person to be a fit and proper person to hold a licence.[54] This gives authorities a discretion in determining such questions. The discretion must be exercised in accordance with the principles of administrative law. Guidance as to the way in which authorities should approach these matters is set out in paragraphs 3.17–3.35 of Waste Management Paper No. 4. Generally it is considered that it is for the applicant, etc., to satisfy the authority that he is a fit and proper person; either because there were mitigating circumstances in relation to the offences or that he has taken all the steps that he reasonably can to ensure that there is no repetition.

(b) TECHNICAL COMPETENCE

18.23   The requirement of section 74(3)(b) that the management of the licensed activities should be in the hands of a technically competent person refers to the management of the day to day activities at the site rather than the management of the company. There can be more than one technically competent person in respect of a site.[55] Usually the authority will ask the licensee to provide a list of the technically competent persons in respect of the site and to keep it informed of changes as appropriate.

18.24   There are two ways in which a person can qualify as being technically competent; either formal qualification or "grandfather rights". Formal qualifications are achieved through the certificates awarded by the Waste Management Industry Training and Advisory Board (WAMITAB). The relevant certificates for particular operations are set out in Table 1 to regulation 4 of the 1994 Regulations and that table is reproduced below.

[53] Waste Management Paper No. 4, paras 3.31–3.33.
[54] E.P.A. 1990, s.74(4).
[55] W.M.P. No. 4, paras 3.37–3.39.

TABLE 1

Type of facility	Relevant certificate of technical competence
A landfill site which receives special waste.	Managing landfill operations: special waste (level 4).
A landfill site which receives bio-degradable waste or which for some other reason requires substantial engineering works to protect the environment but which in either case does not receive any special waste.	1. Managing landfill operations: bio-degradable waste (level 4); or 2. Managing landfill operations: special waste (level 4).
Any other type of landfill site with a total capacity exceeding 50,000 cubic metres.	1. Landfill operations: inert waste (level 3); or 2. Managing landfill operations: bio-degradable waste (level 4); or 3. Managing landfill operations: special waste (level 4).
A site on which waste is burned in an incinerator designed to incinerate waste at a rate of more than 50 kilograms per hour but less than 1 tonne per hour.	Managing incinerator operations: special waste (level 4).
A waste treatment plant where special waste is subjected to a chemical or physical process.	Managing treatment operations: special waste (level 4).
A waste treatment plant where waste is subjected to a chemical or physical process and none of the waste is special waste.	1. Treatment operations: inert waste (level 3); or 2. Managing treatment operations: special waste (level 4).
A transfer station where — (a) biodegradable, clinical or special waste is dealt with; and (b) the total quantity of waste at the station at any time exceeds 5 cubic metres.	Managing transfer operations: special waste (level 4).
A transfer station where — (a) no biodegradable, clinical or special waste is dealt with; and (b) the total quantity of waste at the station at any time exceeds 50 cubic metres.	1. Transfer operations: inert waste (level 3); or 2. Managing transfer operations: special waste (level 4).
A civic amenity site.	Civic amenity site operations (level 3).

251

The qualification in respect of incinerators does not apply to those incinerators operated under IPC controls; in which case the qualification should only be concerned with the waste handling aspects of the operation.[56] Qualifications in respect of scrap metal sites and vehicle dismantling operations are excluded from Table 1[57] while nothing is said about small sites taking only inert waste or small transfer stations. For these sites the authority should make its own assessment of competence, following the guidance set out in paragraphs 3.56–3.66 of Waste Management Paper No. 4.

18.25   "Grandfather rights"—or transitional provisions—are contained in regulation 5 of the 1994 Regulations. They allow anyone who had managed a waste facility at any time in the 12 months before August 10, 1994, and who had applied to WAMITAB for a certificate before that date, to be treated as if they were competent to run the relevant type or types of facility they had experience in until August 10, 1999. Anyone who was over 55 on August 10, 1994 and had at least five years experience in running a facility in the previous ten years will be considered technically competent to manage the type of facility of which he had experience until August 10, 1999 regardless of whether some of the waste he dealt with is now "special waste".[58] A "facility" for these purposes is one either licensed or authorised under the Control of Pollution Act 1974 or the Environmental Protection Act 1990. For managers of waste recovery or disposal processes subject to the integrated pollution control regime of Part I of the Environmental Protection Act 1990 should have applied by July 10, 1995[60]; except for those concerned with the biological or physiochemical pre-treatment of waste for which the application date is March 31, 1996.[61] Operators of facilities for which licensing was deferred by virtue of paragraph 43 of the 1994 Regulations should have applied by July 31, 1995.[62]

18.26   Managers of sites concerned with special waste as defined by the old 1980 Regulations are governed by the original provisions of regulations 4 and 5. However, where, before September 1, 1996, they dealt with wastes that are newly special by virtue of the Special Wastes Regulations 1996[63] then regulations 4 and 5 have effect as amended by regulation 20 of the 1996 Regulations. This allows them to remain technically competent to August 10, 2000 as long as they applied to WAMITAB for a certificate of technical competence at Level 4 in respect of special waste before March 1, 1997.

18.27   For managers in England and Wales who had operated local authority facilities, and are not treated as competent by any other provision of regulations 4 or 5, the time is extended to October 1, 2001; those over 55 being regarded as competent until October 1, 2006.[64] Those in Scotland

[56] W.M.P. No. 4, para. 3.49.
[57] S.I. 1994 No. 1056, reg. 4(2) but see S.I. 1995 No. 288, reg. 4(3) & 4(4).
[58] S.I. 1994 No. 1056, reg. 5(2) and S.Is. 1997 No. 251, reg. 2 & 257 reg. 2.
[60] S.I. 1995 No. 288, reg. 4(1) & 4(2).
[61] ibid. as amended by S.I. 1995 No. 1950.
[62] ibid. reg. 4(3) but see reg. 4(4) for exceptions.
[63] S.I. 1996 No. 972.
[64] 1994 Regulations, reg. 5(4)–5(7) as added by S.I. 1996 No. 634.

who were managing facilities pursuant to a resolution under section 54 (repealed) of the 1990 Act have rights to the same dates.[65] In both cases applications for certificates were made before October 1, 1996. WAMITAB has established a procedure for verifying claims of experience and will issue an exemption certificate to those it considers qualify for one. The Agencies may also determine such matters themselves.[66]

(c) FINANCIAL PROVISION

18.28 Guidance as to the financial provision to be made by a licensee in accordance with section 74(3)(c) of the Act is set out in paragraphs 3.68–3.122 of Waste Management Paper No. 4. The financial provision made should depend on the site's potential capability to cause environmental harm which turns on the ability to ensure compliance with conditions. Thus before provision can be made the licensee needs to know, through a draft copy of the licence, the conditions with which he will have to comply.[67]

18.29 The first stage in meeting this requirement is for the authority to assess whether the applicant is of sufficient financial standing to meet his obligations under the licence. The authority is not concerned with the applicant's general financial position but only with it as is relevant to the licence. To assess the position the authority will look at the applicant's business plan to see how operations will be financed, but it should not try to assess the viability of the scheme. If necessary guarantees or an audit certificate can be sought by the authority but the W.M.P. warns of the limited effectiveness of these.[68]

18.30 Specific risk cover may also be necessary; for example to cope with failure of containment resulting in pollution of a watercourse. This type of provision may be required as a condition of the licence. It can be secured by insurance, self-insurance or an overdraft facility.[69] In addition cover will be necessary to secure funds in the post-closure phase of a site. This can be achieved by a bond to secure restoration of the site under planning conditions together with an escrow account, independently held trust fund or in-house funding.[70] Mutually funded operating companies may also provide a solution here if acceptable to the authority.[71]

18.31 An applicant can, instead of making actual financial provision, rely on demonstrating that he intends, and is able, to make such provision. While this is an option it is unlikely that an authority will consider an applicant a fit and proper person for these purposes unless it is clear that the applicant can show that he will be capable of funding not only during the site's

---

[65] S.I. 1996 No. 916.
[66] W.M.P. No. 4, para. 3.52.
[67] W.M.P. No. 4, paras 3.70–3.72.
[68] *ibid.* paras 3.81–3.91.
[69] *ibid.* paras 3.91–3.102.
[70] *ibid.* paras 3.103–3.113.
[71] *ibid.* paras 3.114–3.117.

operational life but also in the post-closure phase. Thus while this option may be suitable for some non-landfill operations it is probably not appropriate for landfills.[72]

3. PROTECTION OF GROUNDWATER

18.32   Article 3(b) of the E.C. Directive on the Protection of Groundwater against Pollution caused by certain Dangerous Substances[73] requires Member States to prevent the introduction of substances in List I in the Annex to it to groundwater and to limit pollution by substances in List II. For this purpose before a waste disposal licence is granted there should be an investigation of the effects of the operation on groundwater. The licence can only be granted if all the technical requirements for preventing groundwater pollution by List I or II substances are observed.[74] These provisions are implemented in Great Britain by regulation 15 of the Waste Management Licensing Regulations 1994[75] and explained in Annex 7 of Department of Environment Circular 11/94. The definitions used in that regulation will, where relevant, be those of the Directive.[76]

18.33   Before any licence is issued for any disposal or tipping operation that might lead to an indirect discharge of a List I or II substance into groundwater—reaching it by percolation through the soil—or a direct discharge—reaching it without percolation—of them to it there must be a prior investigation of the matter.[77] While this only applies to disposal or tipping operations, these terms are not defined. The operation of a transfer station could be regarded as a "tipping operation" for these purposes, even though the "tip" is not a permanent one. The only exception to this requirement is if the quantity of List I or II substances contained in a discharge of another substance is so small that any risk of pollution can automatically be discounted.[78] The new rules apply to applications under the 1990 Act and any outstanding under the provisions of the 1974 Act.[79]

18.34   The investigation must include an examination of the hydrogeological conditions of the area concerned, the possible purifying powers of the soil and sub-soil and the risk of pollution and the alteration of the quality of the groundwater from the discharge.[80] For these purposes the WRA should consult the NRA—a rivers purification authority in Scotland—about the information it will want from the investigation, require the applicant for the licence to do the work and evaluate his proposed plan; if necessary preparing

---

[72] W.M.P. No. 4, paras 3.119–3.122.
[73] Dir. 80/68 [1980] L20/43.
[74] Dir. 80/68, Arts 4.1 and 5.1.
[75] S.I. 1994 No. 1056.
[76] *ibid.* reg. 15(12).
[77] *ibid.* reg. 15(1).
[78] See D.o.E. Circular 11/94, paras 7.7–7.9.
[79] 1994 Regulations, reg. 15(11).
[80] *ibid.* reg. 15(2).

a parallel plan of work as a check.[81] The investigation must also establish whether the discharge of substances into groundwater is a satisfactory solution from the point of view of the environment.[82]

The question of whether the discharge is a satisfactory solution is, to a   18.35
certain extent, answered by reference to the criteria in regulation 15(4). List I substances should only be allowed to be discharged into groundwaters that are permanently unsuitable for other uses and if they will not impede exploitation of ground resources and will be prevented from reaching other aquatic systems or harming other ecosystems. If the authority cannot be satisfied of this then the licence can only be issued if conditions are imposed to ensure that no List I substances will reach groundwaters. For List II substances a licence can be issued but must be made subject to conditions to ensure that those substances will not cause pollution of groundwaters.[83]

Conditions for indirect discharges are specified by regulation 15(6) and for   18.36
direct discharges by regulation 15(7). These include the place of disposal, tipping or discharge, the methods to be used, the essential precautions to be taken to avoid harm to the environment and nearby water resources, the maximum quantities involved, monitoring arrangements and the technical precautions required to prevent List I substances entering groundwater, where relevant, or List II substances from polluting it.

No licence may be issued for tipping or disposal operations involving   18.37
List I or II substances until the authority is satisfied as to groundwater monitoring arrangements.[84] The licence must only be granted for a limited period.[85] Authorities are advised to set time limits that are commercially realistic.[86] Further the licence must be reviewed every four years.[87]

4. LICENCE CONDITIONS

A waste management licence will be granted on such terms and subject to   18.38
such conditions as appear to the waste regulation authority to be appropriate.[88] In Northern Ireland conditions can be imposed under article 8(a) of the 1978 Order.[89] However, this general power to impose conditions is limited to those necessary to fulfil the purposes of the legislation authorising the grant of licences. Thus under the 1974 Act conditions could only be set to ensure that water was not polluted, there was no danger to public health and no serious detriment to public amenity. In *Attorney-General's Reference (No.2 of 1988)*[90] a condition that sought to prohibit nuisances of all kinds

[81] W.M.P. No. 4, para. 5.10.
[82] 1994 Regulations, reg. 15(2).
[83] *ibid.* reg. 15(5).
[84] *ibid.* reg. 15(3).
[85] *ibid.* reg. 15(8).
[86] D.o.E. Circular 11/94, para. 7.18.
[87] 1994 Regulations, reg. 15(9).
[88] E.P.A. 1990, s.35(3).
[89] S.I. 1978 No. 1049 (N.I. 19).
[90] [1989] 3 W.L.R. 397.

was struck down. However, given the duty to ensure the objectives of Article 4 of the Directive are met such a condition would now be valid, but is not advised.[91] But W.M.P. No. 4, para. 1.6 also states that in assessing pollution, WRAs should have regard to the wider environment. They should, for example, consider the impact of emissions on global climate change as well as on local air, water, soil, flora and fauna.

18.39   Licence conditions must reconcile two objectives. The operator should have the maximum possible flexibility to operate the site in a cost effective manner. The WRA should retain the means of imposing controls that the public interest and the legislation require.[92] While conditions need to be site specific, they should reflect the guidance given in the Waste Management Papers, particularly W.M.P. No. 4. The W.M.P.s have greater authority than a government circular as they constitute the guidance to which WRAs must have regard under section 35(8) of the Environmental Protection Act 1990.

18.40   The licence must contain conditions to control the types and quantities of waste, the technical requirements, security precautions, the disposal site and the treatment method.[93] It must also deal with any other matters set out in W.M.P. No. 4. However, any condition set must be necessary, enforceable, unambiguous and comprehensive.[94] However, no condition can be imposed the sole purpose of which is to regulate health and safety.[95]

18.41   Unlike the Control of Pollution Act 1974, the Environmental Protection Act 1990 does not set out examples of the types of condition that may be imposed. Rather it provides that conditions may relate to the activities which the licence authorises and to the precautions to be taken and works to be carried out in connection with, or in consequence of, those activities. Requirements can be imposed in the licence which are to be complied with before the activities which the licence authorises have begun. These will involve preparatory works and infrastructure as set out in paragraphs 4.3 to 4.12 of W.M.P. No. 4. In addition a WRA will not usually grant a licence until it has all the working plan details. Although the working plan is not itself a licence condition, conditions may make reference to elements of the plan.[96]

18.42   Further, and in contrast to the 1974 Act, requirements can also be imposed that must be complied with after the authorised activities have ceased.[97] Thus, for example, monitoring of leachates and landfill gases could be required to continue, with associated record keeping, until a certificate of completion is issued. In addition a leachate recirculation system could be required to be kept running until that time.[98]

---

[91] W.M.P. No. 4, para. 1.31.
[92] W.M.P. No. 4, para. 2.7.
[93] 1994 Regulations, Sched. 4, para. 6.
[94] W.M.P. No. 4, para. 2.4.
[95] 1994 Regulations, reg. 13.
[96] W.M.P. No. 4, paras 2.15–2.18.
[97] E.P.A. 1990, s.35(3).
[98] W.M.P. No. 4, para. 4.47.

The Environmental Protection Act 1990 extends the powers to make 18.43 conditions beyond those imposed under the 1974 Act. The 1994 Regulations, with their emphasis on the objectives of the E.C. Waste Directive take this extension further. In particular, under the Act, where waste other than controlled waste is to be treated, kept or disposed of at a licensed site, the licence can include conditions concerning that waste.[99] This type of condition is not dealt with in the Waste Management Paper and could cause enforcement difficulties. Section 33(1)(b) of the Act makes it an offence to keep, etc., *controlled* waste—except under and in accordance with a waste management licence. If non-controlled wastes are brought on site but not dealt with in accordance with a site licence it could be argued that section 33(1)(b) cannot apply to them. However, if there are controlled wastes on the site and conditions as to non-controlled wastes are breached a court may say that the operator is not keeping those controlled wastes in accordance with the licence.[99a]

In addition the Secretary of State may make regulations to provide what 18.44 conditions are or are not to be included in a licence. These regulations may set conditions for different types of facility or be based on the type of waste handled.[1] This power will be used to implement specific requirements of E.C. Directives or other international obligations. Thus regulation 14 of the 1994 Regulations is concerned with the implementation of the "Waste Oils" Directive (Dir. 75/439) and regulation 15 with that of the "Groundwater" Directive (Dir. 80/68). More regulations under this provision are likely to implement the landfill directive when finally agreed. Conditions can also be imposed here to clarify the boundaries between waste management licensing and other regulatory regimes—hence regulations 13 and 16.

For specific licence applications the Secretary of State may give the 18.45 relevant Agency directions as to the terms and conditions that are or are not to be included in the licence and the Agency must comply with such a direction.[2] However, it is intended that this power will only be used in exceptional circumstances.

Operators are required to keep a large number of records. Paragraph 14 of 18.46 Schedule 4 to the 1994 Regulations requires establishments or undertakings carrying out the disposal or recovery of controlled waste to keep a record of the quantity, nature, origin and, where relevant, the destination, frequency of collection, mode of transport and treatment method of any waste which is disposed of or recovered. That information must be made available, on request to any of the relevant authorities. In addition W.M.P. No. 4 suggests that licence conditions should also require environmental monitoring records to be kept and that the licensee should keep a site diary. This would record significant events, with their dates, and other matters such as plant maintenance, emergencies, problems with waste received, and actions taken, and the weather. The diary should be kept in a form which can be audited.[3]

[99] E.P.A. 1990, s.35(5).
[99a] But see *London Waste Regulation Authority v. Drinkwater Sabey Ltd* [1997] Env. L.R. 137.
[1] E.P.A. 1990, s.35(6).
[2] *ibid.* s.35(7).
[3] W.M.P. No. 4, paras 4.43–4.45.

18.47   It may be a condition that entries be made in a record as to the observance of a particular licence condition. If such an entry is not made that fact will be admissible as evidence that the relevant condition has not been observed.[4] Anyone who intentionally makes a false entry in such a record will be guilty of an offence and liable on summary conviction to a fine not exceeding the statutory maximum or on indictment to a fine or up to two years' imprisonment or both.[5]

18.48   W.M.P. No. 4 also sets out the type of regime that will be required for particular facilities. In particular scrapyards should have what will be called a "Metal Recycling Licence" to distinguish them from other types of site. Special conditions that will apply to this type of operation include far less record keeping, longer hours and measures to avoid ground contamination. The licensing of Metal Recycling Sites is dealt with in W.M.P. No. 4A.

18.49   A licence holder who is aggrieved by a condition specified in a licence may appeal against it to the Secretary of State.[6] The question to be decided on appeal is whether the condition is reasonable, bearing in mind the objectives of waste management licensing. The Secretary of State has stated that he considers that conditions attached to waste disposal licences should also reflect the nature and scale of operation on, and the circumstances of, a particular site, should afford appropriate protection to local amenities from operations on the site and not impose an unreasonable burden on the operator.[7]

5. CONDITIONS FOR OFF-SITE WORKS

18.50   Waste disposal operations may have effects beyond the site on which they take place. For example, gas monitoring may be necessary on adjoining lands. Under section 35(4) of the Environmental Protection Act 1990 a condition may be set requiring the licence holder to carry out works or do other things even though he is not entitled to do them. Thus he could be required to site a borehole on land belonging to X and regularly monitor it. The section goes on to provide that any person whose consent would be required shall grant, or join in granting, the holder of the licence such rights in relation to the land as will enable the licence holder to comply with the condition. Thus X and lessor of the land on which the borehole is to be sited would have to give the holder a right of access to the land and allow him to dig the hole and maintain it thereafter. The D.o.E. consider that this power will mainly be needed on the modification of existing licences. However, they add that it is not desirable to rely on this power more than is strictly necessary; and, in particular, WRAs and applicants for licences should make every effort to avoid the need for such conditions in new licences.[8]

[4] E.P.A. 1990, s.35(7A) as added by E.A. 1995, Sched. 22, para. 66(2).
[5] *ibid.* s.35(7B) & 35(7C) as added *ibid.*
[6] *ibid.* s.43(1)(b).
[7] Appeal Dec. LW/APP/HE/195 Shakespeare Road, Herne Hill, November 29, 1985.
[8] D.o.E. Circular 11/94, paras 4.25.

Where such a condition is likely to be necessary there must be **18.51** consultation with the owner, lessee or occupier of the relevant land before the licence is issued.[9] Consultation is by way of a notice served on these persons, setting out the proposed condition, indicating the nature of the works or other matters that will be necessary to comply with it and specifying the date by which, and the way in which, any representations about the condition should be made to the authority.[10] Where the authority issues the licence it must consider any such representations made about the condition; although it need not do so if they were not made in time.[11]

Where such a condition has been imposed on a licence and relevant rights **18.52** have been granted to enable the holder to comply with it, any person who has granted or joined in granting those rights will be entitled to compensation from the licence holder.[12]

6. APPEALS

Provisions as to when an appeal may be made and the effect of making an **18.53** appeal in a particular case are set out in the sections of this chapter and the next dealing with specific aspects of the licensing regime. This section is concerned with procedure on appeal. In Northern Ireland the right of appeal is contained in article 12 of the Pollution Control and Local Government (N.I.) Order 1978[13] in similar terms to that under the 1974 Act.

Appeals under section 43 of the Environmental Protection Act 1990 must **18.54** be made within six months of either the date of the decision appealed against or the date on which the authority will be deemed to have rejected the relevant application.[14] However, the Secretary of State may extend the time allowed for making an appeal in a particular case.[15] Before doing so he will want to know the reasons for the delay.[16]

Notice of appeal must be given as required by regulation 6 of the 1994 **18.55** Regulations. This requires that the notice must be in writing and must be accompanied by a statement of the grounds of appeal.[17] A form for appeals is available from the relevant government department.[18] In addition the Secretary of State should be sent, with the notice, copies of certain documents. For an appeal against rejection of an application for a licence or modification of conditions or for its transfer or surrender these should include the application and any plans, drawings, particulars or other

---

[9] E.P.A. 1990, s.36A(1)–36A(3) & 36A(8) as added by E.A. 1995, Sched. 22, para. 69.
[10] *ibid.* s.36A(4).
[11] *ibid.* s.36A(6) & 36A(7).
[12] *ibid.* s.35A(1) & 35A(2) as added by E.A. 1995, Sched. 22, para. 67.
[13] S.I. 1978 No. 1049, (N.I. 19).
[14] S.I. 1994 No. 1056, reg. 7(1)(a).
[15] *ibid.* reg. 7(2).
[16] D.o.E. Circular 11/94, para. 10.20.
[17] S.I. 1994 No. 1056, reg. 6(1) & 6(2)(a).
[18] D.o.E. Circular 11/94, para. 10.8.

documents submitted in support of the application. For an appeal in relation to an existing licence, including one that has been suspended or revoked, a copy of the disposal licence is required. In all appeals the appellant should send the Secretary of State copies of any other relevant consent, determination or notice given by the waste regulation authority, any relevant planning permission or established use certificate, etc., and other correspondence or document that is relevant to the appeal. Further, the documentation should include a statement as to whether the appellant wants the appeal dealt with by a hearing or through written representations.[19] The appellant should also send the waste regulation authority a copy of the notice and copies of all the documents he has submitted to the Secretary of State.[20] The Secretary of State may deal with the appeal himself or delegate the matter under section 114 of the Environment Act 1995.[21]

18.56   On receipt of an appeal to be dealt with by written representations the Secretary of State will request the waste regulation authority to send him, within six weeks of his request, to respond to the grounds of appeal. A copy of the WRA's representations must be sent to the appellant. After this there may be a further stage of comments on each others' view by both parties and provision of further information needed for the determination of the appeal. Ideally this stage is expected to take about 28 days. This will be followed by a site inspection. Usually both sides will be able to be present at the inspection, but if one side does not appear the inspection may not be deferred.[22]

18.57   The decision as to whether there should be a local inquiry is not just a matter for the Secretary of State. If the appellant wants one he is entitled to request it in his notice of appeal. The WRA can also demand a hearing.[23] The costs of holding the inquiry may be recovered from the parties, while the inspector may make an award of costs between the parties.[24] Unless the person holding the inquiry has also been appointed to determine the appeal he must report his findings to the Secretary of State together with his recommendations or reasons for not making any recommendations.[25]

18.58   Following the close of written representations or an inquiry the Secretary of State or appointed person will issue his determination. This must be sent to the appellant in writing and, if an inquiry was held on behalf of the Secretary of State, must be accompanied by the inspector's report.[26] Any documents sent to the appellant must also be sent to the WRA. All the documents served on or sent to the authority during the appeal by virtue of the Regulations should be entered on its public register.[27]

---

[19] S.I. 1994 No. 1056, reg. 6(2).
[20] *ibid.* reg. 6(3).
[21] E.P.A. 1990, s.43(2A) as added by E.A. 1995, Sched. 22, para. 77.
[22] D.o.E. Circular 11/94, para. 10.23.
[23] E.P.A. 1990, s.43(3)(c).
[24] Local Government Act 1972, s.250(4) & 250(5) & E.A. 1995, s.53(2).
[25] S.I. 1994 No. 1056, reg. 8.
[26] *ibid.* reg. 9(1) & 9(2).
[27] *ibid.* reg. 10(1)(e).

If the result is to the effect that the authority's decision is to be altered it **18.59** is the duty of the authority concerned to give effect to that determination.[28] If the appellant is aggrieved by the decision he will be able to appeal against it to the High Court on the basis that it was wrong in law or that the inspector took into account matters that he should not have done or failed to take account of matters that he should have done through an application for judIcial review.

7. PROVISION OF FALSE INFORMATION

It is an offence under section 44 of the Environmental Protection Act **18.60** 1990[29] for any person who, in an application for a licence, an application for the modification of conditions or in an application for the transfer or surrender of a licence and in purported compliance with any request for information to make any statement that he knows to be false in a material particular or to recklessly make such a statement. A statement can be false for these purposes if it leaves out a material particular.[30] A person will be liable on summary conviction for this offence to a fine not exceeding the statutory maximum or on indictment to up to two years imprisonment or a fine or both.

8. FINANCIAL PROVISIONS

There were no provisions concerning fees and charges for waste disposal **18.61** licences issued under the Control of Pollution Act 1974. Finance for the exercise of functions under the 1974 Act was raised through central or local taxation.[31] However, under the 1990 Act and now the Environment Act 1995 the estimated costs of the waste regulatory system are to be met through fees and charges for licences.[32]

Sections 41 and 42 of the 1995 Act sets out the charging regime. This is **18.62** detailed in a scheme made by the Secretary of State, with Treasury approval, that prescribes the fees to be paid in respect of applications for licences or other relevant applications in respect of them and the charges payable in respect of the subsistence of a licence.[33] The current scheme is the Waste Management Licensing (Fees and Charges) Scheme 1997 which is available from the relevant government department. The scheme may be varied from time to time. In particular a scheme may provide for different fees or charges to be payable according to the type of licence—*i.e.* transfer station or mobile

---

[28] E.P.A. 1990, s.43(3).
[29] As substituted by E.A. 1995, Sched. 19, para. 4.
[30] *R. v. Lord Kyslant* [1932] 1 K.B. 442.
[31] D.o.E. Circular 55/76, para. 6.
[32] E.A. 1995, s.42(3).
[33] *ibid.* ss.41(1), (2) & (9) and 42(7).

plant—and the type and amounts of waste dealt with under it.[34] It will also set out how and when any fees or charges are to be paid and may contain any incidental, supplementary and transitional provisions as the Secretary of State considers necessary. Different schemes may be made and revised for different areas.[35]

18.63   A fee will be payable on the application for a licence. A further fee is levied on applications for the modifications of licence conditions, to surrender the licence or to transfer it.[36] The sections of the 1990 Act that are concerned with these applications state that they should be "accompanied by the prescribed fee payable under section 41."[37] If the fee does not accompany the application then it will not have been properly made and the waste regulation authority will be entitled to refuse to process it until the fee is paid.

18.64   The scheme also prescribes a subsistence charge, which is an amount that is payable in each financial year during the course of the licence. The amount is varied on a daily basis when a licence is granted during the year. Details of the annual charge must be served on the licence holder before March 1, in relation to each financial year for which it is to be levied.[38] No account is to be taken for the purposes of the charge of any period of suspension, or of any modification, revocation, surrender or transfer of the licence which takes effect after the first day of the financial year.[39] If the holder does not pay this subsistence charge the relevant Agency will be able to revoke or suspend his licence to carry on the authorised activities, although the holder will still be subject to the conditions of the licence.[40]

18.65   Where it is proposed to revoke or suspend a licence for non-payment, the Agency must follow the procedure set out in the Environmental Licences (Suspension and Revocation) Regulations 1996.[41] These require it first to serve the holder with a notice demanding payment within twenty eight days of the service of the notice. The notice should also set out the effects of suspension or revocation. Only after the 28 day period has elapsed may the Agency go on to revoke or suspend the licence A notice of suspension or revocation must be served on the licence holder. It should set out the reason for the suspension or revocation and the date and time it will take effect. If the licence is only suspended it should set out the circumstances in which the suspension will be lifted.

9. TAX RELIEFS

18.66   In *Rolfe (Inspector of Taxes) v. Wimpey Waste Management*[42] it was held that the costs of acquiring a landfill site and expenses in preparation and restoration were not deductible in computing a company's trading profits for

[34] E.A. 1995, s.41(4).
[35] *ibid*. s.41(7)(a).
[36] *ibid*. s.41(2)(a), (c) and (e).
[37] *e.g. ibid*. s.39(3).
[38] "Charges Scheme" para. 3(13) & 3(14).
[39] *ibid*. para. 3(11).
[40] E.A. 1995, s.41(6).
[41] S.I. 1996 No. 508.
[42] (1989) S.T.C. 454.

corporation tax purposes. In order to reverse, in part, this decision the government has enacted section 78 of the Finance Act 1990 which adds sections 91A and 91B to the Taxes Act 1988. The purpose of this provision is to provide a basis for tax relief for expenditure on preparing and making good landfill sites for waste disposal.

Section 91A deals with restoration payments. A "site restoration" **18.67** payment is one made in connection with the restoration of all or part of a site in order to comply with the conditions of a site licence, planning agreement or planning permission to use the site for waste disposal activities.[43] Waste disposal activities are the collection, treatment, conversion and final disposal of waste. Payments made after April 6, 1989 may be allowed as a deduction in computing the profits or gains of the business for income or corporation tax purposes for relevant period of account. But a payment that has already been deducted for a prior period of account or one that represents capital expenditure which has or may be the subject of an allowance under the Capital Allowance Act 1990 or other legislation on capital allowances cannot be deducted under section 91A.

A site preparation payment deduction may be claimed under section 91B. **18.68** Site preparation expenditure is expenditure on preparing a waste disposal site—a site used or to be used for the disposal of waste materials by their deposit on it—for that deposit and may include expenditure on earthworks.[44] A claim may be made where a person, at any time, incurs site preparation expenditure in relation to a waste disposal site, if he holds, at the time *he* first deposits waste materials on the site, a site licence. The claim must be made in the form directed by the Inland Revenue Commissioners. The allowed deduction for a period of account will be determined by a formula of,

$$(A - B) \times \frac{C}{C + D}$$

Here A is the site preparation expenditure during or before the accounting period (unless it has already been allowed or has been nor may be given a capital allowance) and B is site preparation expenditure that has already been allowed in respect of the site. C is the volume of waste materials deposited on the site during the period and D the remaining capacity, in volume, of the site. Thus the allowance is related to the proportion of the site capacity that has been filled with waste during the period. Allowance will not be made under this section for site preparation expenditure incurred before April 6, 1989.

[43] Taxes Act 1988, s.91A(4)–91A(7).
[44] *ibid.* s.91B(11)(a) & (11)(b).

19. SUPERVISION OF WASTE MANAGEMENT FACILITIES

The Environment Agencies—waste regulation authorities—have the powers of entry and inspection set out in paragraphs 2.22–2.36. These powers are supplemented by those concerning the supervision of waste installations in the Environmental Protection Act 1990. In exercising their supervisory powers, authorities may vary, suspend or revoke licences and approve their transfer. Further, registers of waste management licences under the Act will enable the public to exercise a measure of supervision themselves.

1. POWERS OF SUPERVISION

19.01   Waste regulation authorities have a duty to supervise licensed activities. This duty is imposed under section 42 of the Environmental Protection Act 1990. Supervisory duties are to be exercised to ensure that the licensed activities do not cause an environmental or health hazard or become seriously detrimental to the amenities of the area affected by them. This duty is extended by Schedule 4 to the 1994 Regulations which requires the authority to exercise its functions with the "relevant objectives" in particular those of Article 4 of the waste directive.[1] In addition the authority must ensure that licence conditions are complied with.[2] These duties only relate to licences that are in force.

19.02   The manner in which the Agency should carry out these duties is set out in "The Environment Agency, Code of Enforcement Practice".[3] This sets out the principles under which the Agency will operate. In particular before formal action is taken the Agency should serve a "minded to" notice and give the licensee a right for its view to be heard before acting. If immediate action is taken then the licensee should be provided with a written statement explaining why the action had to be immediate and the consequences of failing to take action. The Code is made under section 5 of and Schedule 1

[1] S.I. 1994 No. 1956, Sched. 4, para. 2.
[2] E.P.A. 1990, s.42(1).
[3] May 1996.

to the Deregulation and Contracting Out Act 1994. Failure to observe its terms could make a particular enforcement action unlawful.

Any establishment or undertaking that carries out the recovery or disposal 19.03 of controlled waste—and others involved with the transport and collection of waste and waste dealers—must be subject to periodic inspection by the waste regulation authority by virtue of paragraph 13 of Schedule 4 to the 1994 Regulations. The words in paragraph 13 are actually "competent authorities" however the authorities listed in Table 5 to the Regulations do not have supervisory powers under Part II of the 1990 Act. Normally therefore other such authorities like planning authorities will only inspect for the purpose of their functions under the relevant control regime. Inspection for the sake of it is discouraged.[4] Inspectors of the waste regulation authority should comply with the guidance on frequency of inspection in paragraph B9 of W.M.P. No. 4. These inspections will be carried out by an authorised officer of the authority who will have the powers of entry and other rights set out in section 108 of the Environment Act 1995. These powers have been extended by paragraphs 13(2) and (2A) of Schedule 4 to the 1994 Regulations to inspectors of other competent authorities.[5]

In an emergency an authority may carry out works on a licensed site or 19.04 equipment in the performance of its supervisory duties.[6] These works will be carried out by an authorised officer of the authority who will have the powers of entry set out in the Acts. Any expenditure incurred by the authority may be recovered from the holder of the licence or, as the case may be, its former holder. However, the holder or former holder may escape liability if he shows that there was no emergency requiring any work. In addition he will not have to pay the costs of unnecessary works.[7]

If a waste regulation authority considers that a licence condition is not 19.05 being complied with, or is likely not to be complied with, then, without prejudice to any other proceedings it may institute, it may serve the licence holder with a notice requiring compliance.[8] The notice should state the authority's opinion, specify the matters that constitute the breach or anticipated breach, the steps that are necessary to remedy or prevent the situation and the period within which those steps must be taken.[9] There is no right of appeal against the service of such a notice, and, although judicial review might lie here, a court could say that the issue of a notice is merely the expression of the authority's opinion.[10]

If the steps required by the notice have not been taken in the specified 19.06 period, the authority will have power to partially revoke the licence, to

---

[4] D.o.E. Circular 11/94, paras. 1.86 & 1.87.
[5] As amended by The Environment Act 1995 (Consequential Amendments) Regulations 1996 (S.I. 1996 No. 593), Sched. 2, para. 10(5)(d).
[6] E.P.A. 1990, s.42(3).
[7] *ibid.* s.42(4) as amended by E.A. 1995, Sched. 22, para. 76(3)
[8] *ibid.* s.42(5)(a) & (7) as amended *ibid.* para. 76(4).
[9] *ibid.* s.42(5)(a) as substituted *ibid.* para. 76(5).
[10] *R. v. London Waste Regulation Authority, ex p. Specialist Waste Management, The Times,* November 1, 1988.

revoke it entirely or to wholly or partially suspend its operation.[11] In either case the licence holder will be able to appeal against the revocation or suspension. On such an appeal the question will be whether the service of the compliance notice was reasonable in the circumstances and whether it was subsequently justified in taking the further steps to ensure compliance.[12] A revocation of a licence is unlikely in these circumstances because of the problems the authority would be faced with in dealing with the site. It is more likely that there will be suspension so as to prohibit the activities carried on under the licence while it is in force.[13] An appeal against such a decision will have no effect on the situation[14]; so that the site could not generate income for some time, although the obligations and requirements under the licence to prevent pollution, etc., will continue.[15]

19.07   Where revocation is being considered it should be remembered that it is a discretionary power. It should be distinguished from the power to prosecute. If a site does not cause significant environmental problems prosecution should be tried before the authority can justify the harsh step of revocation.[16] Where a licence is revoked or suspended under the 1990 Act the provisions of section 38(5) and (12) or (8) to (12) of that Act will apply as appropriate to the revocation or suspension.[17] Under the Environmental Protection Act the Secretary of State may give an authority directions as to the exercise of these supervisory powers and it will be the duty of the authority to comply with his directions.[18]

19.08   The situation may arise where a waste regulation authority considers that revocation or suspension of the licence, whether entirely or in part, would be an ineffectual remedy against a person who has failed to take the steps required by a compliance notice—because the activities would not cease or the operator would disregard the part suspension. In such a case the authority can apply for an injunction or interdict in the High Court, or in Scotland in any court of competent jurisdiction, to secure compliance with its notice.[19] It need not prosecute first, but it will have to show that the defendant's unlawful activities will continue unless and until effectively restrained by law and that nothing short of an injunction will be effective to restrain them.[20] Mandatory, as well as restraining, injunctions can be granted by the court.[21]

19.09   Reports produced by an authority in the exercise of its duties under section 42, including details of any correspondence with the relevant agency under section 42(2), any remedial or preventive action taken by the

[11] E.P.A. 1990, s.42(5)(b) (as amended by E.A. 1995, Sched. 22, para. 76(6)).
[12] See Appeal Decision at Local Government Review, April 8, 1989, p. 277.
[13] E.P.A. 1990, ss.38)8) & 42(7).
[14] ibid. s.43(5).
[15] ibid. ss.38(9) & 42(7).
[16] See Appeal Decision: Oldfields, Cradley Heath, March 9, 1995 Ref: LEQ/5/4/200.
[17] E.P.A. 1990, s.42(7) as amended by E.A. 1995, Sched. 22, para. 76(8).
[18] ibid. s.42(8).
[19] ibid. s.42(6A) as added by E.A. 1995, Sched. 22, para. 76(7).
[20] City of London v. Bovis Construction Ltd [1989] J.P.L. 263.
[21] Croydon London Borough Council v. Gladden [1994] 1 P.L.R. 30.

authority and notices issued by the authority under section 42(5) should be entered on the authority's public register.[22]

2. VARIATION OF LICENCES

Under section 37(1)(a) of the Environmental protection Act 1990[23] an **19.10** authority may, while a waste management licence is in force, serve a notice modifying the conditions of a licence (a variation notice) on the licence holder of its own initiative. Such modifications must appear desirable to the authority and the new conditions must be unlikely to require unreasonable expenditure by the licence holder. These powers may be restricted by regulations made under section 35(6) of the Act, although no such restrictions have been provided in the 1994 Regulations. The Secretary of State will also be able to direct an authority to issue a variation notice by virtue of section 37(3) of the 1990 Act.

The authority must serve a variation notice if it considers it necessary to **19.11** do so to ensure that the activities authorised by the licence do not cause pollution of the environment or harm to human health or do not become seriously detrimental to the amenities of the locality affected by those activities.[24] The modifications made by the notice must not exceed those necessary for those purposes. In addition such a notice may be required by regulations made under section 35(6) of the Act.[25] Regulation 15(10) of the 1994 Regulations requires a review of all licences that might lead to direct or indirect discharges of substances listed in the E.C. Groundwater Directive into underground waters. A notice will be an alternative to revocation of the licence.

The licence holder may appeal to the Secretary of State against a variation **19.12** notice under section 43(1)(c) of the Act. While an appeal is pending the variation will normally be ineffective until it is dismissed or withdrawn.[26] However, if the authority include in the notice a statement that as it was served to protect the environment or human health it should take effect immediately, it will do so.[27] The holder may then apply to the Secretary of State for a determination as to whether the authority acted unreasonably in making such a statement. If he decides that it did then, if the appeal is still pending, the new conditions will be suspended until determination and in any event the licence holder will be able to recover compensation from the authority in respect of any loss suffered as a result of the statement. Any dispute as to entitlement to, or the amount of, such compensation will be determined by arbitration or, in Scotland, by a single arbiter.[28]

---

[22] S.I. 1994 No. 1056, reg. 10(1)(g).
[23] Art. 9 of the N.I. legislation S.I. 1978 No. 1049 (N.I. 19).
[24] E.P.A. 1990, s.37(2)(a) and see 1994 Regulations, Sched. 4, paras 2(1) & 4(1).
[25] *ibid.* s.37(2)(b).
[26] *ibid.* s.43(4).
[27] *ibid.* s.43(6).
[28] *ibid.* s.43(7).

19.13   A licence holder may also apply for a modification of his licence but must pay the appropriate application fee.[29] The authority have two months, or an agreed longer period, to reach a decision on the application. If it fails to do so by the end of that time it will be deemed to have rejected it.[30] The holder may appeal against a refusal to modify under section 43(l)(a) of the Act.

19.14   Normally a proposal to issue a modification notice must go through the same consultation procedure as that for the grant of a licence. However, consultation may be postponed if the authority considers that it is dealing with an emergency. Further, an authority is not required to consult a body about a modification that it considers will not affect that other body.[31] The modification notice must state the time at which the modification made by it is to take effect.[32] Particulars of applications for modification and notices of modifications made to a licence must be entered on the authority's public register.[33]

19.15   If the authority proposes to vary a licence to add a condition requiring off site works it will have to follow the procedure in section 37A of the Act[34]; although in an emergency this procedure may be postponed.[35] The W.R.A. are required to notify the owners, occupiers and lessees of land over which rights will have to be granted to enable compliance with the condition[36] and give them time to make representations about the proposed condition.[37] Before the licence is varied the authority must consider any representations that have been made in the time allowed.[38]

3. SUSPENSION OF LICENCES

19.16   A waste management licence may be wholly or partially suspended under section 38(6) of the Environmental Protection Act 1990. The first ground of suspension is that the holder of the licence is no longer a fit and proper person to hold a licence because the person managing the site is not qualified for that purpose.[39] The other main ground is that serious environmental pollution or harm to health has resulted from, or is about to be caused by, the licensed activities or by something that will affect them and that if they, or some of them, continue then the pollution or health hazard will also continue to occur.[40] The final ground is for non-payment of licence charges, details of which are set out at paragraph 19.24.

---

[29] E.P.A. 1990, s.37(1)(b) as amended by E.A. 1995, Sched. 22, para. 70(1).
[30] ibid. s.37(6).
[31] ibid. s.37(1).
[32] ibid. s.37(4).
[33] S.I. 1994 No. 1056, reg. 10(1)(b) & (c).
[34] As added by E.A. 1995, Sched. 22, para. 71.
[35] E.P.A. 1990, s.37A(9).
[36] ibid. s.37A(3) & (4).
[37] ibid. s.37A(5).
[38] ibid. s.37A(7) & (8).
[39] ibid. ss.38(6)(a) and 74.
[40] ibid. s.38(6)(b) & (c).

The waste regulation authority may make the suspension of its own 19.17
volition or on the direction of the Secretary of State.[41] Notice in writing
must be given to the licence holder of the suspension. The notice should
state the time at which it comes into effect and when it will end or the event
that will lift the suspension.[42]

A licence that is suspended by such a notice will, during the suspension, 19.18
no longer authorise the holder to carry out the licensed activities, or such of
them as are specified in the notice.[43] Thus, if any more waste is received at
the site or is treated or otherwise dealt with there, the holder will be guilty
of an offence under section 33(6) of the Act. An appeal lies against a
suspension notice,[44] but the appeal will not have any effect on the operation
of the suspension.[45]

When serving a suspension notice a W.R.A. can require the licence holder 19.19
to take such measures to deal with or avert the pollution or harm as it
considers necessary.[46] Such requirements can include provision for off-site
works. If such a requirement is imposed, anyone whose consent would be
required to enable the licence holder to comply with it must grant him the
necessary rights.[47] Where such a requirement is contemplated the authority
must follow the procedure set out in section 36A(2) to (8) as modified[48];
although compliance with this procedure may be postponed in an emergency.[49] Anyone granting rights under this provision will be entitled to
compensation from the licence holder under section 35A of the Act.

Failure to comply with the requirements of a suspension notice without a 19.20
reasonable excuse is an offence. If the requirements are concerned with
"ordinary" waste the offence is punishable on summary conviction by a fine
not exceeding the statutory maximum or on conviction on indictment for up
to two years imprisonment or a fine or both.[50] If the waste involved is special
waste the penalties increase to up to six months imprisonment or a fine of
up to the statutory maximum or both in the magistrates court or on
indictment to up to five years imprisonment or a fine or both.[51] Alternatively, if the authority considers that prosecution would be ineffective it
can apply to the High Court or, in Scotland, any court of competent
jurisdiction, for an injunction or interdict.[52]

---

[41] E.P.A. 1990, s.38(7).
[42] *ibid.* s.38(12).
[43] *ibid.* s.38(8).
[44] *ibid.* s.43(1)(d).
[45] *ibid.* s.43(5).
[46] *ibid.* s.38(9).
[47] *ibid.* s.38(9A) as added by E.A. 1995, Sched. 22, para. 72.
[48] *ibid.* s.38(9B) and see para. 17.
[49] *ibid.* s.38(9C).
[50] *ibid.* s.38(10).
[51] *ibid.* s.38(11).
[52] *ibid.* s.38(13) as added by E.A. 1995, Sched. 22, para. 72(2).

4. REVOCATION OF LICENCES

19.21   Two types of revocation are provided for in section 38 of the Environmental Protection Act 1990; partial and total revocation. Both powers of revocation may be exercised if the waste regulation authority either considers that the licence holder has ceased to become a fit and proper person to hold a licence because he has been convicted of a relevant offence,[53] or if it considers that the continuation of the activities authorised by the licence would cause pollution of the environment or harm to human health or would be seriously detrimental to the amenities of the locality affected and that the situation cannot be remedied by modifying the conditions of the licence.[54] In addition an authority may partially revoke a licence if it considers that the manager of the site is not qualified to conduct the authorised activities,[55] or if fees payable in respect of the licence are outstanding.[56] Powers of revocation are also contained in section 42(6) of the Act to ensure compliance with conditions. A revocation notice must state the date from which the licence is revoked.[57]

19.22   A partial revocation is made under section 38(3). This will remove authorisation to carry on the activities carried out under it or such of them as the authority specifies in the revocation notice. Thus if a licence allows treatment of waste and disposal a revocation notice could ban treatment but allow disposal to continue. A partial revocation will not affect the duties imposed by conditions of the licence so that, for example, leachate would still have to be dealt with as required by the licence during the period of revocation.[58]

19.23   Alternatively a notice may revoke the licence entirely.[59] On the service of such a notice the licence ceases to have effect.[60] However, if the site is likely to cause environmental pollution or harm to health from landfill gas or from leachate a local authority will be able to exercise the powers of Part IIA of the Act in respect of it. Further, powers under section 215 to 219 of the Town and Country Planning Act 1990 [ss.179–181, 1997 Act] will enable a local planning authority to require the proper maintenance of the site.

19.24   A licence holder, or former holder, has a right of appeal to the Secretary of State against a revocation notice.[61] While the appeal is pending the revocation will normally have no effect unless the provisions, set out in paragraph 19.12 above, apply.

19.25   Where revocation is for non-payment of charges the relevant Agency must follow the procedure set out in The Environmental Licences (Suspen-

---

[53] E.P.A. 1990, ss.38(1)(a) & 74(3)(a).
[54] *ibid.* s.38(1)(b) & (c).
[55] *ibid.* s.38(2).
[56] E.A. 1995, s.41(6).
[57] E.P.A. 1990, s.38(12).
[58] *ibid.* s.38(5).
[59] *ibid.* s.38(4).
[60] *ibid.* s.35(11).
[61] *ibid.* s.43(1)(e).

sion and Revocation) Regulations 1996.[62] This requires that first a final demand be made for payment within 28 days after service of the demand notice.[63] The notice should point out that the licence may be suspended or revoked if no payment is made and the effect of revocation or suspension.[64] During that 28 day period no further action can be taken.[65] Once the period for payment has expired the Agency may then serve a further notice suspending or revoking the licence. Such a notice should give reasons for the Agency's action, the date and time at which it takes effect and, if it is suspended, the circumstances in which the suspension will be lifted.[66]

5. TRANSFER OF LICENCES

A power of transfer is given to the waste regulation authority under section 35(10) of the Environmental Protection Act 1990. A licence may be transferred in accordance with section 40 of the Act even if it is partly revoked or is suspended.[67] **19.26**

To transfer the licence the holder and the proposed transferee will have to make a joint application to the authority.[68] The application must be made in the form required by regulations and contain the required information. It must be accompanied by the fee payable under section 41(2)(c) of the 1995 Act and the existing licence.[69] The Department of the Environment is of the view that this provision does not allow for an actual form to be specified in the Regulations. Thus Schedule 2 of the Waste Management Licensing Regulations 1994[70] merely sets out the information required in respect of a transfer application. This includes details about the proposed transferee; in particular those as to previous convictions, technical competence and financial provisions to be made in respect of his obligations under the licence so that the authority can assess whether the transferee is a fit and proper person to hold a licence. A copy of the application must be entered on the authority's public register.[71] **19.27**

The only ground for rejecting such an application is that the proposed transferee is not a fit and proper person to hold a licence. The authority must satisfy itself that the transferee is a fit and proper person before effecting the transfer.[72] A decision must be reached on the application within two months of receipt, or any longer period agreed in writing. If the **19.28**

---

[62] S.I. 1996 No. 508.
[63] *ibid.* reg. 3(a).
[64] *ibid.* reg. 4.
[65] *ibid.* reg. 3(b).
[66] *ibid.* reg. 5.
[67] E.P.A. 1990, s.40(1).
[68] *ibid.* s.40(2).
[69] *ibid.* s.40(3).
[70] S.I. 1994, No. 1056.
[71] *ibid.* reg. 10(1)(b).
[72] E.P.A. 1990, ss.40(4) & 74.

authority do not either effect the transfer or inform the applicants that it will not do so within that time then it will be deemed to have rejected it.[73] The proposed transferee may appeal against the rejection of the application.[74] A transfer will be effected, either initially or as a result of an appeal, by endorsing the licence with the name and other particulars of the transferee. The transfer will take effect from a date agreed with the applicants and specified in the licence.[75]

6. REGISTERS

19.29   Under section 6(4) of the Control of Pollution Act 1974[76] waste disposal authorities established registers containing a copy of all waste disposal licences that were issued by them. These registers must have been kept at their principal offices and been open to public inspection, free of charge, at all reasonable times. In addition the public should have been able to obtain copies of entries in the register on payment of a reasonable sum. Provision as to the content of registers in Scotland was made by regulation 8 of the Control of Pollution (Licensing of Waste Disposal) (Scotland) Regulations 1977.[77] These old registers may still be of relevance in certain circumstances; although the requirement to maintain them extends only to licences "for the time being in force".[78]

19.30   The Environmental Protection Act 1990 considerably extended the amount of information to be held on public registers and also provides for certain information to be kept confidential. In addition registers will be kept not only by the waste regulation authorities (the relevant Agency) but also by waste collection authorities.

19.31   Section 64 of the 1990 Act requires each waste regulation authority to maintain a register containing prescribed particulars of all current or recently current licences. These particulars will be prescribed in regulations to be made by the respective Secretaries of State.[79] For these purposes a licence is recently current for twelve months after it has ceased to have effect.[80] The register must also contain details of current or recently current applications for licences—applications relating to current or recently current licences or, where an application was rejected or is deemed to have been rejected, for twelve months after rejection.[81]

19.32   In addition to these matters the register will also have to contain details of applications for the modification of a licence, modification notices, notices

---

[73] E.P.A. 1990, s.40(6).
[74] *ibid.* s.43(1)(g).
[75] *ibid.* s.40(5).
[76] Art. 8(4) of the N.I. legislation: S.I. 1978 No. 1049 (N.I. 19).
[77] S.I. 1977 No. 2006.
[78] C.o.P.A. 1974, s.6(4) as amended E.A. 1995, Sched. 22, para. 20(4).
[79] E.P.A. 1990, s.64(8).
[80] *ibid.* s.64(1)(a) & (3) and see S.I. 1994 No. 1056, reg. 10(4).
[81] *ibid.* s.64(1)(b) & (3).

issued by the authority suspending or revoking a licence or imposing requirements on the licence holder and notices requiring compliance with conditions under section 42(5). When an appeal is made under section 43 of the Act in respect of a licence details of the appeal must also be entered. Convictions, whether under the 1990 Act or otherwise, of licence holders must be contained on the register as must a record of any action the authority has taken under section 42 of the Act. When a site is closed a copy of the certificate of completion will be registered and any subsequent action taken by the authority under section 61 of the Act endorsed. In Scotland copies of resolutions authorising a disposal authority to deal with waste will also be entered. Finally, the register must contain any directions given to the authority under Part II of the 1990 Act, any other matter relating to the treatment, keeping or disposal of waste in the authority's area or any pollution caused by waste as may be prescribed in the regulations and any other document or information that is required to be kept in the register under any provision of the Act.[82]

Regulation 10(1)of the Waste Management Licensing Regulations 1994 sets out an expanded list of particulars to be entered in public registers. That list, as amended, is set out at the Annex to this chapter. In addition where an inspector exercises powers under section 69(3) of the 1990 Act, or now section 108(4) of the 1995 Act, a record should be entered in the register showing when the power was exercised, indicating what information was obtained and what action was taken as a consequence on that occasion.[83] **19.33**

Although inspectors' reports of actions taken should be entered on the register, this does not require it to contain information relating to, or to anything which is the subject matter of any actual or prospective criminal proceedings before those proceedings are finally disposed of.[84] **19.34**

Section 65 excludes information affecting national security from being placed in the register. Material, or material of a certain type, is excluded if and so long as, in the opinion of the Secretary of State, its inclusion would be contrary to the interests of national security.[85] No definition is given of "national security" but it is submitted that exclusion could only be justified on these grounds on the basis of a threat to national safety, diplomatic relations or to some state secret of high importance; although the Secretary of State's certificate that he considers disclosure of the relevant information to be contrary to the interests of national security will have considerable weight.[86] **19.35**

In order to exclude information relating to national security from the register, the Secretary of State may give authorities directions specifying the information, or type of information, that is to be excluded or that is to be referred to him for a decision on whether it should be excluded. Information **19.36**

---

[82] E.P.A. 1990, s.64(1).
[83] S.I. 1994 No. 1056, reg. 10(2)(a) & (aa) (as added by S.I. 1996 No. 593).
[84] *ibid.* reg. 11(1).
[85] *ibid.* s.65(1).
[86] *Williams v. The Home Office* [1981] 1 All E.R. 1151.

that is referred to him may not be entered in the register until he determines that it should be.[87] Where an authority excludes information on security grounds pursuant to a direction the Secretary of State should be notified.[88] If someone considers that information that may be entered in the register could have security consequences he may notify the Secretary of State of this, specifying the information and its apparent nature. If he does so he must also notify the relevant authority of his action. The authority must then withhold it from the register until the Secretary of State decides that it can be entered in it.[89]

19.37   A person who considers that information is commercially confidential in relation to him may apply under section 66 of the Act to have it excluded from the register. For these purposes information is commercially confidential in relation to anyone if it would prejudice, to an unreasonable degree, his commercial interests.[90] In assessing this a balance will have to be struck between the public interest in disclosure of environmental information and the right of individuals to protect their business from harm. There are four elements to be discerned in identifying what is confidential information. First, the person seeking confidentiality must believe that the release of the information would be injurious to him or of advantage to his rivals. Second, he must believe that the information is confidential or secret. Third, these beliefs must be reasonable and, fourth, the information must be judged in the light of the usage and practices of the particular industry concerned. In general this provision will be used to protect trade secrets such as the operation of a waste treatment process, but trade secrets can also include information that would lead to knowledge about costs or pricing.[91] However, even if someone considers that the information is commercially confidential, the guiding principle is that information should be freely available to the public. To establish confidentiality he will have to show that the disclosure of the information would negate or significantly diminish the commercial advantage he has over a competitor.[92]

19.38   No information relating to the affairs of any individual or business can be included in the register without the consent of that individual or person carrying on the business if and so long as the information is, in relation to him, commercially confidential. However, information is only confidential if the authority maintaining the register determine it to be so or the Secretary of State does so on appeal.[93] The phrase "if and so long as" implies a right to review the classification of information if it appears to have lost its confidentiality.

19.39   Even if information is determined to be commercially confidential, the Secretary of State may make a direction under section 66(7) that it should be

[87] E.P.A. 1990, s.65(2).
[88] *ibid.* s.65(3).
[89] *ibid.* s.65(4).
[90] *ibid.* s.66(11).
[91] *Thomas Marshall Ltd v. Guinle* [1978] 3 W.L.R. 116.
[92] Environmental Protection Act 1990, Part I Secretary of State's Guidance—Introduction to Part I of the Act. (GGI (91), para. 56.)
[93] E.P.A. 1990, s.66(1).

entered on the register.[94] Such directions may relate to specified information or descriptions of information. Alternatively the Secretary of State may direct that information that should not be on the register should be removed from it.[95]

Where information is furnished to an authority for the purposes of an application for a licence, or for the modification of one, or for complying with any condition of one or for complying with a requirement to provide information under section 71(2), the person furnishing it may request the authority to exclude it from the register on the grounds that it is commercially confidential to himself or another person. On receipt of such a request the authority must determine whether the information is or is not commercially confidential.[96] It must decide the matter within 14 days from the date of the request. If it fails to do so it will be deemed to have classified the information as confidential.[97] The Secretary of State may, by order, substitute a longer or shorter period for the time within which an authority must make its determination.[98] **19.40**

Alternatively, where it appears to an authority that any information, other than that furnished under section 66(2), that it has obtained under the provisions of Part II of the 1990 Act might be commercially confidential it must give the person to whom, or to whose business, it relates notice that it will be entered in the register unless excluded on the grounds of confidentiality. Such a person must be given a reasonable opportunity to object to the information being entered on the register and to make representations justifying his view that it is commercially confidential. Where such representations are made the authority must consider them before deciding whether or not the information is commercially confidential.[99] The duty here is to notify someone about information which it appears to the authority might be confidential. Breach of the duty may give rise to an action against an authority for any damages suffered by the person to whom the duty was owed,[1] even if the damage only takes the form of economic loss.[2] However, the word "appears" gives an authority wide discretion, so that as long as it has approached the matter properly it is unlikely to be liable. **19.41**

In all cases where it is determined that information is commercially confidential a statement must be entered on the register indicating the existence of information of the type that has been excluded.[3] The amount of information excluded from the register should be kept to the minimum necessary to safeguard the commercial confidentiality.[4] If the excluded **19.42**

---

[94] E.P.A. 1990, s.66(1)(b) and see S.I. 1994 No. 1056, reg. 10(1)(i).
[95] *ibid.* s.64(2A) as added by E.A. 1995, Sched. 22, para. 82(2).
[96] *ibid.* s.66(2).
[97] *ibid.* s.66(3).
[98] *ibid.* s.66(10).
[99] *ibid.* s.66(4).
[1] *Meade v. Haringey London Borough Council* [1979] 1 W.L.R. 637.
[2] *Brewers Bros. v. Canada* [1991] 80 D.L.R. (4th) 321 (CAN).
[3] E.P.A. 1990, s.64(2).
[4] GGI (91) para. 58.

information shows whether or not there has been compliance with a licence condition, a statement should be made in the register, based on that information, as to whether or not that condition has been complied with.[5]

19.43   If an authority determines that information is not commercially confidential it must delay entering it on the register for 21 days after it has notified its decision to the person concerned. He may appeal to the Secretary of State against that determination within that 21 day period. Where an appeal is made the relevant information may not be entered on the register until seven days after the appeal is finally determined or withdrawn.[6] These appeals will be dealt with in the manner set out in section 43(2) and regulations made under section 43(8); although such appeals must be held in private and can be delegated or referred under the provisions of section 114 of the Environment Act 1995.[7]

19.44   Where information is excluded from the register it will be treated as ceasing to be commercially confidential after four years from the date of the decision under which it was excluded. If the person who furnished it to the authority wants it kept confidential he must make a new application to the authority for its continued exclusion on the grounds that it is still commercially confidential. The authority will then make a fresh decision on the matter.[8] The wording of this provision may cause some difficulty as the right to re-apply lies with "the person who furnished" the information. If this was an individual who has transferred his licence to someone else the new holder may not be able to make such an application. There will be a right to appeal against a decision by the authority that the information is no longer confidential.[9]

19.45   Waste collection authorities in England or Wales must maintain a register, containing the particulars prescribed in regulation 10(3) of the Waste Management Licensing Regulations 1994, of information in the regulation authority's register that relates to the treatment, keeping or disposal of controlled waste in their area.[10] The information required is of current, or recently current, waste management licences, notices of modification, revocation or suspension under sections 37 and 38 of the Act and certificates of completion. It will be the duty of the Agency to provide a collection authority with the information necessary to enable it to maintain a register.[11]

19.46   Registers maintained under section 64 may be kept in any form,[12] including on a computer. The maintaining authorities must ensure that they are open to public inspection at their principal offices at all reasonable hours.

---

[5] S.I. 1994 No. 1056, reg. 10(2)(b).
[6] E.P.A. 1990, s.66(5) as amended by E.A. 1995, Sched. 22, para. 83(1) and S.I. 1994 No. 1056, reg. 7(1)(b).
[7] *ibid.* s.66(6) as substituted by E.A. 1995, Sched. 22, para. 83(2).
[8] *ibid.* s.66(8).
[9] *ibid.* s.66(9).
[10] *ibid.* s.64(4) as amended by E.A. 1995, Sched. 22, para. 82(3).
[11] *ibid.* s.64(5) as substituted by E.A. 1995, Sched. 22, para. 82(4).
[12] *ibid.* s.64(7).

The Secretary of State may prescribe the places at which such registers are to be made publicly available. Inspection of the register must be free but if copies of entries are requested—a request that must be complied with—a reasonable charge may be made for complymg with the request.[13]

The Secretary of State may issue a direction to require an authority to **19.47** remove an entry in the register that should not be there.[14] Otherwise the information on them should be retained indefinitely. There are three exceptions to this rule. First, where only "current or recently current" information is to be registered,[15] when it is no longer "recently current" it can presumably be removed. In addition monitoring information required to be included under regulation 10(1)(h) can be removed four years after being entered and any information that is superseded by later information may be removed four years after the entry of that later information.[16]

7. OTHER SOURCES OF INFORMATION

Registers under the Environmental Protection Act 1990 are the main way **19.48** in which the government will implement its duty under the E.C. Directive on Freedom of Access to Information on the Environment 90/313 which was adopted on June 7, 1990.[17] The aim of the directive is to give the public access to information on the environment held by authorities like the Environment Agency or SEPA. Information relating to the environment means, for these purposes, any available information, in whatever form, on the state of water, air, etc., and on activities likely to affect water, etc., or on activities or measures designed to protect them including administrative measures and environmental management programmes.

Under the Directive anyone may ask to see such information and, if he **19.49** considers his request has been unreasonably refused or inadequately complied with he must be able to have recourse to a court. Information may be kept secret if it affects the confidentiality of the proceedings of authorities, public security, matters that are *sub judice*, commercial or industrial confidentiality including trade secrets, personal information, material supplied by an informer, etc., or material the disclosure of which might cause further environmental harm. However, if confidential aspects can be separated from the other information then the edited version should be supplied. Documents in the course of preparation and internal communications need not be supplied. The Directive has been implemented in Great Britain by the Environmental Information Regulations 1992.[18] In Northern Ireland implementation is by way of the Environmental Information Regulations (Northern Ireland) 1993.[19]

---

[13] E.P.A. 1990, s.64(6) as amended by E.A. 1995, Sched. 22, para. 82(5).
[14] *ibid.* s.64(2A) as added by E.A. 1995, Sched. 22, para. 82(2).
[15] See S.I. 1994 No. 1056, reg. 10(4) and para. 18.
[16] S.I. 1994 No. 1056, reg. 11(2).
[17] [1990] O.J. L158/56.
[18] S.I. 1992 No. 3240.
[19] S.I. 1993 No. 45.

ANNEX

CONTENTS OF REGISTERS—WASTE MANAGEMENT LICENSING REGULATIONS 1994, REG. 10(1)

(a) current or recently current waste management licences ("licences") granted by the authority and any associated working plans;

(b) current or recently current applications to the authority for licences, or for the transfer or modification of licences, including details of—

    (i) documents submitted by applicants containing supporting information;

    (ii) written representations considered by the authority under section 36(4)(b), (6)(b) or (7)(b) or 37(5) of the 1990 Act;

    (iii) decisions of the Secretary of State under section 36(5), or, in Scotland, section 36(6), of the 1990 Act;

    (iv) notices by the authority rejecting applications;

    (v) emergencies resulting in the postponement of references under section 37(5)(a) of the 1990 Act;

(c) notices issued by the authority under section 37 of the 1990 Act effecting the modification of licences;

(d) notices issued by the authority under section 38 of the 1990 Act effecting the revocation or suspension of licences or imposing requirements on the holders of licences;

(e) notices of appeal under section 43 of the 1990 Act relating to decisions of the authority and other documents relating to such appeals served on or sent to the authority under regulation 6(3) or (4) or 9(3);

(f) convictions of holders of licences granted by the authority for any offence under Part II of the 1990 Act (whether or not in relation to a licence) including the name of the offender, the date of conviction, the penalty imposed and the name of the Court;

(g) reports produced by the authority in discharge of any functions under section 42 of the 1990 Act, including details of—

    (i) any correspondence with the National Rivers Authority or river purification authority as a result of section 42(2) of the 1990 Act;

    (ii) remedial or preventive action taken by the authority under section 42(3) of the 1990 Act;

    (iii) notices issued by the authority under section 42(5) of the 1990 Act;

(h) any monitoring information relating to the carrying on of any activity under a licence granted by the authority which was obtained by the authority as a result of its own monitoring or was furnished to the authority in writing by virtue of any condition of the licence or section 71(2) of the 1990 Act;

(i) directions given by the Secretary of State to the authority under section 35(7), 37(3), 38(7), 42(8), 50(9), 54(11) or (15), 58 or 66(7) of the 1990 Act;

(j) any summary prepared by the authority of the amount of special waste produced or disposed of in their area;

(k) registers and records provided to the authority under regulation 13(5) or 14(1) of the Control of Pollution (Special Waste) Regulations 1980, or regulation 15(5) or 16(1) of the Special Waste Regulations 1996;

(l) applications to the authority under section 39 of the 1990 Act for the surrender of licences, including details of—

  (i) documents submitted by applicants containing supporting information and evidence;

  (ii) information and evidence obtained under section 39(4) of the 1990 Act;

  (iii) written representations considered by the authority under section 39(7)(b) or (8)(b) of the 1990 Act;

  (iv) decisions by the Secretary of State under section 39(7) or (8) of the 1990 Act; and

  (v) notices of determination and certificates of completion issued under section 39(9) of the 1990 Act;

(m) written reports under section 70(3) of the 1990 Act by inspectors appointed by the authority;

(n) in Scotland, resolutions made by the authority under section 54 of the 1990 Act, including details of—

  (i) proposals made in relation to land in the area of the authority by a waste disposal authority under section 54(4) of the 1990 Act;

  (ii) statements made and written representations considered by the authority under section 54(4) of the 1990 Act;

  (iii) requests made to, and disagreements with, the authority which are referred to the Secretary of State under section 54(7) of the 1990 Act and his decisions on such references;

  (iv) emergencies resulting in the postponement of references under section 54(4) of the 1990 Act.

20. CLOSURE AND AFTERCARE OF LICENSED SITES

20.01 The holder of a licence under the Control of Pollution Act 1974 could cancel it by delivering it to the issuing authority and notifying them that he no longer required it.[1] This right no longer exists now that the waste licensing provisions of the Environmental Protection Act 1990 have entered into force, even in respect of licences existing at that time.[2] However, for mobile plant licences, the situation is relatively unchanged.

20.02 Section 39 of the 1990 Act enables any licence to be surrendered by its holder, but, for a site licence, the surrender will only be valid if it is accepted by the authority. The surrender procedure in respect of a site licence is set out in section 39(3) to (11). It is initiated by the holder making an application to the relevant Agency for the surrender of the licence to be accepted.

20.03 When an Agency receives a surrender application it must inspect the land concerned and may require the licence holder to furnish it with more information or evidence.[3] The purpose of this is to enable the Agency to determine whether the condition of the land, as a result of its having been used for the treatment, keeping or disposal of waste (whether or not in pursuance of the licence) is such that it is likely or unlikely that it will cause environmental pollution or harm to human health.[4] This is known as the "completion condition". For landfills advice on the way in which this condition can be satisfied is set out in Waste Management Paper No. 26A.

20.04 Completion is defined as that point at which the landfill has stabilised physically, chemically and biologically to such a degree that the undisturbed contents of the site are unlikely to cause pollution of the environment or harm to human health.[5] The relevant factors here are therefore:

    (a) the quantity and quality of leachate present,

    (b) the flow and concentration of gas,

    (c) the potential for polluting leachate or gas to be generated in future,

---

[1] C.o.P.A. 1974, s.8(4); Art. 10(4) S.I. 1978 No. 1049 (N.I. 19).

[2] E.P.A. 1990, s.77(1) & (2).

[3] *ibid.* s.39(4).

[4] *ibid.* s.39(5).

[5] W.M.P. No. 26A, para. 2.1.

(d) the potential for leachate or gas to reach sensitive targets,

(e) the possibility of physical instability of the waste or retaining structures, and

(f) the presence of particular problem wastes which could present a hazard in the future.

Chapter 4 of Waste Management Paper No. 26A discusses the criteria **20.05** against which to assess these factors in determining whether or not the completion condition has been met. No general criteria are laid down due to the diversity of sites. But, while the relevant Agency will set the criteria to be met by a particular site, these must take into account the guidance in Chapter 4 and must also be reasonable. As far as gas is concerned this should be assessed as satisfactory when it reaches the levels set out in Waste Management Paper No. 27.

It may take some time from the ending of disposal operations on the site **20.06** to get to the position when the site is considered complete. Waste Management Paper No. 26A states that if monitoring results show with 90 per cent confidence or more that appropriate criteria are exceeded, or if an increase in gas flows or a decrease in leachate quality becomes apparent, then any necessary control measures should be established and a suitable period (typically five years) allowed to elapse before recommencing completion monitoring.[6]

A site should only be certified as complete when all the criteria have been **20.07** met and there is no significant increase in gas flows or decrease in leachate quality (allowing for seasonal variations) over a reliable monitoring period. In cases where historical leachate and gas data indicate that biodegradation has not occurred due to lack of moisture, or a site investigation has demonstrated extensive presence of desiccated wastes, then a completion certificate should not be issued except when it is unlikely that the wastes will become active in the future.[7]

The surrender of licences for sites other than landfills are dealt with in **20.08** Chapter 7 of Waste Management Paper No. 4. Before a surrender application is accepted for such sites they should be physically stable, largely free of contamination by wastes, clear of deposited residues, etc., of wastes or residues from treatment processes and free of continuing discharges that require active site management. However, contamination due to causes other than the licensed activity should not be taken into account here.[8] Special advice as to certificates of completion for scrapyards is contained in Appendix D to Waste Management Paper No. 4.

While there is a right of appeal under section 43(l)(f) of the 1990 Act **20.09** against a decision by an authority not to accept the surrender of a licence it might be better for both sides if there was some form of dispute resolution in a case where the parties disagree as to whether the completion condition has

---

[6] W.M.P. No. 26A, para. 4.16.

[7] *ibid.* para. 4.19.

[8] W.M.P. No. 4, para. 7.8.

been met. While there is no statutory provision for this, if the parties agree to accept the decision of an independent expert, and then one does not, the Secretary of State is unlikely to be very sympathetic to that party in any appeal. This, it is suggested, should be done before any application to surrender the licence is made.

20.10   If both sides agree that the completion criteria for the site are met an application to surrender the licence should be made. The application must be made on a form provided by the Agency and contain the information prescribed by Schedule 1 to the Waste Management Licensing Regulations 1994[9] and be accompanied by such evidence about the site as provided in that Schedule. The Schedule requires general information to be provided about any site including a map or plan of it and a description of the activities carried out on it. For landfill or lagoon sites, paragraph 5 of the Schedule requires the provision of information about engineering works done on the site, geological, hydrological and hydrogeological information, monitoring data on surface and groundwater and on landfill gas as well as to the stability of the site and details of special wastes deposited at the site. For other sites the applicant will have to provide details of contaminants that may be present as a result of the operations carried out and a report of sampling done to determine the level of contaminants there.[10] However, if the waste regulation authority already has these details they need not be set out in the application.[11] Further, where the site has changed hands the information as to activities on the site and the details required by paragraphs 5 or 6(a) need only be supplied as far as they are known by the applicant or, if a company or partnership, by an officer of the company or partner.[12]

20.11   In Waste Management Paper No. 26A this information and evidence is called the Completion Report. Chapter 5 sets out how it is considered that the report should be made. In particular it suggests that where the licensee says that information about certain matters has already been provided the authority should only be satisfied of this if there is an unambiguous reference in respect of it and an explanation as to its relevance. Paragraph 7 of Schedule 1 to the Regulations enables the operator to supply any other information which the applicant wishes the waste regulation authority to take into account. Waste Management Paper No. 26A suggests that an authority, either under this head, or as part of its powers to demand further evidence, should be supplied with any relevant Environmental Statement, the planning consent and any modifications and the monitoring data in a form that demonstrates completion. However, the planning material should already be in the hands of the authority.

20.12   The holder will also have to pay the fee required under section 41 of the Environment Act 1995 for the application.[13] The fee structure is set out in the Waste Management Licensing (Fees and Charges) Scheme 1997 and the

---

[9] S.I. 1994 No. 1056, reg. 2(1) & (2).
[10] Schedule 1, para. 6.
[11] 1994 Regulations, reg. 2(3).
[12] *ibid.* reg. 2(4).
[13] E.P.A. 1990, s.39(3) as amended by E.A. 1995, Sched. 22, para. 73(2).

fee here is known as the "surrender fee". A copy of the application and details of the completion report must be entered on the register maintained by the waste regulation authority under section 64 of the 1990 Act.[14]

If the authority is satisfied that the completion condition has been met it **20.13** may accept the surrender.[15] Before doing so however it must consult the appropriate planning authority,[16] and consider any representations they make about the matter within 21 days of receipt of the application.[17] Those representations must also be entered on the authority's public register.[18]

Where a surrender is accepted the authority must issue the applicant with **20.14** a certificate of completion together with its notice of acceptance. The certificate will state that the relevant Agency is satisfied that the condition of the site is unlikely to cause environmental pollution or harm to human health. On the issue of the certificate the licence will cease to have effect.[19] A copy of the certificate must be entered on the Agency's public register.[20]

If the authority consider that the condition of the land is such that it is **20.15** likely to cause pollution or to be a health hazard then it must reject the application for surrender.[21] If it fails to determine the application within three months of receiving it, or such longer period as the parties may agree in writing, it will be deemed to have rejected it.[22] The holder will have the right to appeal to the Secretary of State against the rejection, or deemed rejection, of the application.[23]

SITE RESTORATION AND AFTERCARE

Conditions attached to a planning permission can require restoration of a **20.16** waste site after operations have ceased. For example such a condition could require that:

> "Before the commencement of development a detailed scheme for the progressive filling, restoration and landscaping of the site shall be submitted to and approved by the planning authority; such scheme shall be implemented unless approval to any variation is given in writing by the planning authority."

Other conditions in the same permission may relate to the stripping and storage of the topsoil on the site for restoration purposes, the way in which the filled area should be covered and its preparation for after-use. Finally the permission may require that:

---

[14] 1994 Regulations, reg. 10(1)(l).
[15] E.P.A. 1990, s.39(6).
[16] As defined in s.39(12).
[17] *ibid.* s.39(7) as amended by E.A. 1995, Sched. 22, para. 73(3) and ss.36(10) & 39(11) .
[18] 1994 Regulations, reg. 10(1)(l)(iii).
[19] E.P.A. 1990, ss.35(11) & 39(9).
[20] 1994 Regulations, reg. 10(1)(l)(v).
[21] E.P.A. 1990, s.39(6).
[22] *ibid.* s.39(10).
[23] *ibid.* s.43(1)(f).

"a scheme for the proposed cultivation, fertilising, planting, draining and aftercare of the restored land, detailing such steps as may be necessary to bring the land to the required standard for use for forestry and game shall be submitted to the (county) planning authority for approval not later than one year prior to the completion of the restoration."[24]

Other planning permissions may state these requirements in different ways, for example requiring an agricultural end use, but the principles are the same.

20.17   The first condition set out in paragraph 20.16 is a "restoration condition" which requires that after the depositing of waste has ceased the site shall be restored by the use of all or any of the following, namely subsoil, topsoil and soil making material.[25] This type of condition, or a scheme made under it, may deal with the origin of the soil, the depth of soil placed, its stone content, its acidity and the amount of clay or organic matter it should contain. No specific power is needed to impose such a condition as it relates fairly and reasonably to the permitted development. If the after-use of the site is to be either agricultural or forestry then MAFF or the Forestry Authority should be consulted.[26]

20.18   Powers to impose "aftercare conditions" on permissions involving the depositing of refuse or waste materials are granted by section 72(5)[27] of, and Part I of Schedule 5 to, the Town and Country Planning Act 1990 [Schedule 3, 1997 Act]. Under paragraph 2 of Schedule 5 [1997, Sched. 3] where a permission is granted for such development and is subject to a restoration condition, a further condition may be imposed requiring that such steps shall be taken as may be necessary to bring the land to the required standard for either agriculture, forestry or amenity use as specified in the condition. If the condition will require an agricultural or forestry use then, in England, the Ministry for Agriculture, Fisheries and Food or the Forestry Commission should be consulted.[28]

20.19   Landfill Restoration is also the subject of Waste Management Paper No. 26E.

CONTAMINATED LAND

20.20   Part IIA of the Environmental Protection Act 1990[29] is concerned with the remediation of contaminated land. The definition of "contaminated land" for these purposes is set out in section 78(A) of the Act and in guidance issued by the Secretary of State. At the time of writing this book the final version of that guidance had not been issued.

20.21   This book is not the place to deal with contaminated land in any depth. Reference should rather be made to the Encyclopaedia of Environmental Law

[24] Planning Appeal, *Re: Patterson's Pit, Colney Heath*, Ref: T/APP/M1900/A/89/132266/P3.
[25] T.&C.P.A. 1990, Sched. 5, paras. 2(1)(b) & 2(2)(a) [1997, Sched. 3, para. 2(2)].
[27] As amended by Planning and Compensation Act 1991, s.21 and Sched. 1, para. 2.
[28] T.&C.P.A. 1990, Sched. 5, para. 4.
[29] As added by the Environment Act 1995, s.57.

and to *Contaminated Land* by Tromans and Turrell-Clarke. It is expected that the contaminated land provisions of the 1990 Act will enter into force in early 1998.

21. LIABILITY TO LANDFILL TAX

21.01   Part III of the Finance Act 1996 introduced the landfill tax. It is under the care and management of the Commissioners of Customs and Excise.[1] This chapter considers when the tax is payable and in what amounts. The mechanism by which the tax is collected is outside the scope of this work. It is mainly set out in Schedule 5 to the Act and in the Landfill Tax Regulations 1996.[2] The tax is charged on a "taxable disposal" made on or after October 1, 1996.[3] Guidance to the tax is set out in *Landfill Tax Information Notes: 3/96: Tax Liability*.

"TAXABLE DISPOSAL"

21.02   For there to be a "taxable disposal" there must be a disposal of material as waste.[4] The definition of a disposal as "waste" for these purposes is contained in section 64 of the Act. It is different from the definition of "waste" in the Environmental Protection Act 1990 and thus no reliance should be placed on that Act to interpret the phrase "a disposal of material as waste" in the 1996 Act. Under section 64(1) a disposal of material is a disposal of it as waste if the person making the disposal does so with the intention of discarding the material. However, and this is the distinction between the 1990 Act and the 1996 Act, the fact that the person making the disposal or any other person could benefit from or make use of the material is irrelevant.[5] Where the person actually making the disposal does so on behalf of another, either at his request or in pursuance of a contract or otherwise, the disposal will be treated as being made by the person who discarded the material.[6] Thus a site operator is unable to say that he has not discarded anything and thus is not liable to pay the tax. "Material" here, means material of all kinds, including objects, substances and products of all kinds.[7]

---

[1] Finance Act 1996, s.39.
[2] S.I. 1996 No. 1527.
[3] Finance Act 1996, s.40(1) & (2)(d).
[4] *ibid.* s.40(2)(a).
[5] *ibid.* s.64(2).
[6] *ibid.* s.64(3) & (4).
[7] *ibid.* s.70(1).

To be a "taxable disposal" the material must be disposed of by way of 21.03 landfill.[8] Under section 65(1) there is a disposal by way of landfill if the material is deposited on the surface of land—including land covered by water above the low water mark—or on a structure set into the surface or it is deposited under the surface. It does not matter for these purposes whether or not the material is put into a container before it is deposited.[9] Given the use of the word "disposal" here, there could not be a "taxable disposal" where material is deposited at a transfer station. If the material is deposited under the surface of land there is a taxable disposal whether it is covered with earth—or similar matter such as sand or rocks—after deposit or it is put in a cavity such as a mine or cavern.[10] If it is put on the surface of land with a view to being covered with earth, the taxable disposal is made on deposit, not when the material is covered.[11] The Treasury may, by order, amend subsections (1) to (4) of section 65 to vary the interpretation of disposal by way of landfill.[12]

A taxable disposal must be made at a landfill site.[13] For this purpose a 21.04 disposal is made at a landfill site if the land on or under which it is made constitutes or falls within land which is a landfill site at the time of the disposal.[14] A "landfill" is either one licensed under Part II of the Environmental Protection Act 1990, or, in Scotland, is managed by a waste disposal authority pursuant to a resolution under section 54 of that Act or, in Northern Ireland is licensed or run under the Pollution Control and Local Government (Northern Ireland) Order 1978 or legislation in the province corresponding to section 35 of the Environmental Protection Act 1990.[15] It should be noted that the site must be licensed. A disposal made under one of the exemptions in the Waste Management Licensing Regulations 1994 would not therefore be a "taxable disposal."

EXEMPTIONS

Most dredging materials are exempted from the tax. There are three 21.05 exemptions here. The first is for materials removed from the bed of waters— whether natural or artificial—that are rivers, canals or watercourses or docks or harbours.[16] The second exemption is for dredgings from the bed of waters forming the approaches to a natural or artificial harbour that were removed for the purposes of navigation.[17] Finally, naturally occurring mineral material that is removed from the sea in the course of commercial aggregate extraction operations on the seabed will also be exempt from the tax.[18]

---

[8] Finance Act 1996, s.40(2)(b).
[9] *ibid.* s.65(2).
[10] *ibid.* s.65(3) & (8).
[11] *ibid.* s.65(4).
[12] *ibid.* s.65(5) & (6).
[13] *ibid.* s.40(2)(c).
[14] *ibid.* s.40(3).
[15] *ibid.* s.66.
[16] *ibid.* s.43(1) & (2).
[17] *ibid.* s.43(3).
[18] *ibid.* s.43(4).

21.06   Mining and quarrying waste material may come under the exemption provided by section 44. The material must result from commercial mining operations—whether deep or open-cast—or from commercial quarrying operations. It must also be naturally occurring material extracted from the earth in the course of such operations. The exemption is aimed at this natural material. Waste material emanating from processes separate from the mining and quarrying operation or that permanently alter the material's chemical composition, will be taxable on disposal.

21.07   Pet cemeteries are also exempt from the tax. A pet cemetery is not directly defined but for these purposes the exemption only applies if since October 1, 1996 the only disposals at the site have been material consisting entirely of the remains of dead domestic pets. At such a site the disposal of such remains is not a "taxable disposal" by virtue of section 45 of the 1996 Act.

21.08   Section 46 of the Act gives powers to vary these exemptions either by making disposals that were taxable exempt or exempting what would otherwise be taxable. This may be effected by amending the Act or in any other way the Treasury sees fit.

21.09   The Landfill Tax (Contaminated Land) Order 1996[19] has been made under section 46. This adds sections 43A and 43B to the Act. Under section 43B anyone carrying out, or intending to carry out, land reclamation[20] may apply to the Commissioners for a contaminated land certificate. This certificate will exempt the disposal of material removed from the land in question from payment of tax. The purpose of this exemption is to encourage land reclamation. Thus disposals under the certificate will no longer be exempt when the land has been restored to a condition that is suitable for development.[21] In addition any conditions attached to the certificate must be complied with.[22] If the restoration is being carried out as the result of enforcement action the exemption will not apply unless the material has been removed by a local authority or a relevant government Agency.[23]

21.10   Temporary disposals may be exempt under Part IX of the Landfill Tax Regulations 1996[24] if the materials concerned are to be recycled, incinerated or re-used otherwise than in a landfill site within a year.[25] For this exemption to apply the area of the temporary disposal must have been designated for this purpose by a customs officer and the designation must be valid at the time of the disposal.[26] A designation may contain conditions relating to the use of the material to be disposed of or such conditions may have been specified generally. If material is used in breach of those conditions the

---

[19] S.I. 1996 No. 1529.
[20] As defined in s.43B(7)–(12).
[21] Finance Act 1996, s.43A(2)(b) & (3).
[22] ibid. s.43A(2)(c).
[23] ibid. s.43A(2)(d), (4) & (5).
[24] S.I. 1996 No. 1527.
[25] ibid. reg. 38(4) & (5) and reg. 39(1)(a).
[26] ibid. reg. 38(2)(a) & (b) and see reg. 38(3).

exemption from tax will not apply.[27] In addition the exemption will only apply if the site operator has made the temporary disposal record required by regulation 39(3).[28]

LIABILITY TO PAY

The person liable to pay the tax charged on a taxable disposal is the landfill site operator.[29] This person is the person who, at the time the disposal is made, is the operator of the site at which, or under which, it is disposed of.[30] At any given time the operator is either the holder of the waste management licence in respect of the site, or, in Scotland, the waste disposal authority authorised to run the site, or in Northern Ireland, the district council running the site under the 1978 Order or a holder of a waste management licence.[31] 21.11

These site operators will be carrying out "taxable activities" for the purposes of section 69 of the Act. They do so either because they make a taxable disposal in respect of which they are liable to pay the tax or they permit someone else to make a disposal for which they, the site operator, are liable to pay tax. For these purposes they will permit disposals even if they are made without their knowledge.[32] 21.12

Those who carry out taxable activities must be registered with the Commissioners of Customs and Excise under section 47 of the Act. A person who forms the intention of carrying out such activities and is not registered must notify the Commissioners of his intentions.[33] Failure to register is not in itself a criminal offence but may give rise to civil liability under paragraph 18 of Schedule 5 to the Act. Special provision is made in section 58 for the registration of partnerships and divisions of a company, while section 59 is concerned with groups of companies. Registration, and provision for special cases, is dealt with in Part II of the Landfill Tax Regulations 1996.[34] 21.13

AMOUNT OF TAX

Usually the amount of tax payable will be £7 for each whole tonne disposed of and a proportionately reduced sum for any additional part of a tonne.[35] The weight of the material will be determined by weighing it at the 21.14

---

[27] S.I. 1996 No. 1527, reg. 38(5)(a).
[28] *ibid.* reg. 39(2).
[29] Finance Act 1996, s.41(1).
[30] *ibid.* s.41(2).
[31] *ibid.* s.67.
[32] *ibid.* s.69(2).
[33] *ibid.* s.47(2).
[34] S.I. 1996 No. 1527.
[35] Finance Act 1996, s.42(1).

time of the disposal.[36] However, the Commissioners may make rules that specify different ways of determining the weight of particular disposals.[37] Alternatively an agreement can be reached between the taxpayer and the Commissioners as to the method by which and time at which the material should be weighed and for discounting of any water that forms part of the material disposed of.[38]

21.15   "Qualifying material", listed in an order made under section 42(3) of the 1996 Act, attracts a reduced rate of £2 a tonne.[39] The general purpose of this provision is to list material that is of a kind commonly described as inactive or inert.[40] The Landfill Tax (Qualifying Material) Order 1996[41] has been made under this provision and the schedule to it sets out the materials concerned; which are to be construed in accordance with the notes to the schedule. For the lower rate to apply the conditions in column 3 to the schedule must be satisfied.[42] In addition unless the owner of the material is the landfill operator at the disposal site the transfer note concerning it must describe the material in the same way as in the schedule or contain some other accurate description.[43] In Northern Ireland, where currently there is no transfer note system, an accurate description of the material should accompany it to provide evidence that the material was "qualifying material".[44]

21.16   The Commissioners may make directions that "qualifying material" that has mixed with it a small amount of non-qualifying material must nevertheless qualify for the lower rate. The question as to whether the amount of non-qualifying material is small will be determined in accordance with the direction.[45] In addition a direction can be sought from them as to whether the lower rate applies, despite a small amount of non-qualifying material being in the waste.[46]

---

[36] S.I. 1996 No. 1527, reg. 42.
[37] *ibid.* reg. 43.
[38] *ibid.* reg. 44.
[39] Finance Act 1996, s.42(2).
[40] *ibid.* s.42(4).
[41] S.I. 1996 No. 1528.
[42] *ibid.* Art. 4.
[43] *ibid.* Arts 5 & 6.
[44] *ibid.* Art. 7.
[45] Finance Act 1996, s.63(2).
[46] *ibid.* s.63(3).

22. INCINERATION OF WASTE

Air pollution from most waste incinerators is controlled under Part I of **22.01** the Environmental Protection Act 1990. The disposal of waste is controlled under Part II of that Act. To avoid duplication of control, processes authorised under Part I will be exempt from the waste disposal regime of Part II. However, while gaseous effluents emitted to the atmosphere from the incinerator will not be waste,[1] the residues such as ash and sludge will be controlled waste and must be disposed of in accordance with Part II.[2]

E.C. WASTE INCINERATION CONTROLS

The E.C. Directive on the combating of air pollution from industrial **22.02** plants (84/360/EEC)[3] is a "framework" Directive under which "daughter" directives will be issued for the particular types of plant which fall under its controls. These are set out in Annex I to the directive and include "plants for the disposal of toxic and dangerous waste by incineration" and "plants for the treatment by incineration of other solid and liquid wastes."[4] The operation of such a plant must be authorised by the competent authority of the Member State in which it is situated.[5] Authorisations may only be issued where the authority is satisfied that all appropriate preventive measures against air pollution have been taken. These will include the application of the best available pollution control technology, unless its use would entail excessive costs.[6] Further, the plant must not cause significant air pollution, and no emission limit values to which it is subject may be exceeded.[7] Finally, in granting the authorisation, any applicable air quality standards must be taken into account.[8] Applications for authorisations and the determinations in respect of them must be made public.[9]

The directives on "the prevention of air pollution for new municipal waste **22.03** incineration plants" (89/369/EEC)[10] and on "the reduction of air pollution

---

[1] Framework Waste Directive (91/156/EEC) Art. 2.1(a).
[2] E.P.A. 1990, s.28(1).
[3] [1984] O.J. L188/20, Art. 15A added to this directive by Dir. 91/692/EEC.
[4] Dir. 84/360, Annex I, para. 5.
[5] *ibid.* Art. 3.
[6] *ibid.* Art. 4.1.
[7] *ibid.* Art. 4.2 & 4.3.
[8] *ibid.* Art. 4.4.
[9] *ibid.* Art. 9.1.
[10] [1989] O.J. L163/32.

from existing municipal waste incineration plants" (89/429/EEC)[11] are "daughter" directives to the "Industrial Plants" Directive. For the purposes of those directives a municipal waste incineration plant is defined as any technical equipment used for the treatment of municipal waste by incineration, whether or not the heat is recovered. However, plant used specifically for the incineration of sewage sludge, chemical, toxic and dangerous waste, clinical wastes from hospitals or other special wastes are excluded from the directives' control; even if they also burn municipal waste. "Municipal waste" here means domestic refuse and commercial, trade and other refuse that is similar to domestic refuse.[12] The controls imposed by the directives will cover the site of, and the entire installation comprising, the incinerator.[13]

22.04   A municipal waste incinerator that received authorisation to operate before December 1, 1990 will be "existing plant" for the purposes of Directive 89/429.[14] Such plant that has a nominal capacity of over six tonnes of waste per hour must have met the standards set by the "New Plants" Directives by December 1, 1996; although the temperature of the burn and other matters covered by Article 4 of the "existing plant" directive will continue to apply to them.[15] All other plant must, by December 1, 1995, have achieved the limit values for dust emission and the other emission standards set by Articles 3 to 5, as measured by the provisions of Article 6. Different standards are set for plants with a nominal capacity of between one to six tonnes per hour and those with a capacity of less than one tonne per hour. If those standards are exceeded the plant must be shut down and the competent authority must take steps to ensure that it is either modified to meet them or taken out of service.[16] However, these measures need not be taken where the standards are exceeded for technical reasons if the failure does not extend for more than 16 hours at any one time nor to a total of 200 hours over an operating year.[17] After December 1, 2000 all municipal waste incinerators will have to meet all the requirements of the "New Plants" Directive except those contained in Article 4. Article 4 of the "existing plants" directive will continue to apply to such plant.[18]

22.05   New plant, that is plant receiving authorisation to operate after December 1, 1990[19] must comply with the provisions of the "New Plants" Directive (89/369/EEC). Articles 3 to 5 set out the relevant emission standards they must attain and the temperatures at which the waste should be burnt,[20] while Article 6 is concerned with the measurement of those parameters. All new plant must be equipped with auxiliary burners.[21] Similar enforcement

---

[11] [1989] O.J. L203/50.
[12] Dirs 89/369 and 89/429, Art. 1.3.
[13] *ibid.* Art. 1.2.
[14] Dir. 89/429/EEC, Art. 1.5.
[15] *ibid.* Art. 2(a).
[16] *ibid.* Art. 7.1.
[17] *ibid.* Art. 7.2.
[18] *ibid.* Art. 2(b).
[19] Dir. 89/369/EEC, Arts. 1.5 & 12.1.
[20] But see Art. 10.
[21] *ibid.* Art. 7.

procedures are required as those in the "Existing Plants" Directive but the maximum time allowed for any one failure is reduced to eight hours and the cumulative total over an operating year to 96.[22]

A further daughter directive, the Directive on the Incineration of **22.06** Hazardous Waste (94/67/E.C.),[23] is concerned with incineration plant—any technical equipment used for the incineration by oxidation of hazardous wastes with or without the recovery of combustion heat generated, including pretreatment as well as pyrolysis or other thermal treatment processes in so far as their products are subsequently incinerated. This includes plant burning such wastes as regular or additional fuel for any industrial process. However, it does not cover animal carcass incinerators, incinerators for infectious clinical waste—unless other hazardous wastes are present—and municipal waste incinerators that burn such clinical waste.[24] Hazardous wastes for these purposes are those defined in Directive 91/689/ E.C.[25] but with exemptions for certain combustible liquid wastes that meet prescribed criteria, hazardous wastes resulting from offshore installations and incinerated on board, municipal wastes and sewage sludge that meets set criteria.[26]

The aim of the Directive is to provide for measures and procedures to **22.07** prevent or, if this is not possible, reduce environmental harm and harm to human health from the incineration of hazardous waste and, to that end, to set up and maintain appropriate operating conditions and emission limit values for hazardous waste incineration plants.[27] Two classes of plant are designated for this purpose; existing plant, one for which the original operating permit was granted before December 31, 1996, and new plant, one whose permit was granted on or after that date.[28] The directive goes on to lay down operating procedures for such plant and, in Article 7, emission limits for matters such as dust and certain metals. The emission of dioxins and furans must be reduced by the most progressive techniques.[29] The procedures and limits apply to new plant from December 31, 1996 and to existing plant from June 30, 1999 unless they are being prepared for closure within five years from the end of 1996.[30]

INCINERATION CONTROL IN GREAT BRITAIN

The control of atmospheric pollution in Great Britain is a function of the **22.08** Secretary of State for the Environment in England and the Secretaries of State for Scotland and Wales in their jurisdictions. The Environment

---

[22] But see Art. 8.2.
[23] [1994] O.J. L365/34.
[24] Dir. 94/67/EC, Art. 2.2.
[25] See paras 11.03–11.06.
[26] *ibid.* Art. 2.1.
[27] *ibid.* Art. 1.1.
[28] *ibid.* Arts 2.3 & 2.4.
[29] *ibid.* Art. 7.2.
[30] *ibid.* Art. 13.

Agency, or in Scotland, SEPA, are responsible for the control of particularly polluting industrial processes as prescribed in regulations while local authorities will control others. Local authorities for these purposes will usually be a unitary authority or London Borough council or a district council in England, a county or county borough council in Wales or in Scotland a constituted council. These councils will usually exercise their powers over industrial plant and for other air pollution purposes through an environmental or public health department. However, in some places these powers may be exercised by a port health authority.[31]

CONTROL OF PROCESSES UNDER THE ENVIRONMENTAL PROTECTION ACT 1990

22.09   Part I of the Environmental Protection Act 1990 established a system of integrated pollution control for certain prescribed processes. The more complex processes are controlled by the Agencies. For the purposes of this section the details of the way in which integrated pollution control works are not relevant. Part I also gives local authorities control of other industrial processes for air pollution purposes. These controls, and those of the Agencies, implement the "Air Framework" Directive in Great Britain. The provisions of Part I will apply to the Crown[32]; although the remedy for an offence is a declaration that the Crown has acted unlawfully.[33] Specific premises may be exempted from the application of this Part under a certificate issued by the Secretary of State.[34] However, these provisions will not apply to hospital incinerators as they lost their general Crown immunity under section 60 of the National Health Service and Community Care Act 1990.

CONTROLLED PROCESSES

22.10   Part 1 and Schedule 1 of the Environmental Protection Act 1990 set out the framework of industrial air pollution control. Their provisions are supplemented by regulations made under them. However, under section 7(11) the Secretary of State can give enforcing authorities guidance as to the conduct of their duties under the Act and they must have regard to such guidance. For this purpose five general guidance notes have been issued which are concerned with the scope of Part 1 and authorisations made under it. Specific guidance notes have also been issued to local authorities on individual types of process for which they have responsibility. For example the Process Guidance note PG5/2(95) gives the Secretary of State's guidance on clinical waste incineration processes of under one tonne an hour. In addition local enforcing authorities must have regard to the national air quality strategy published by the Secretary of State under section 80 of the Environment Act 1995.[35]

[31] E.P.A. 1990, s.4(12).
[32] E.P.A. 1990, s.159(1).
[33] *ibid.* s.159(2).
[34] *ibid.* s.159(4).
[35] *ibid.* s.4(4A) added by E.A. 1995, Sched. 22, para. 46(5).

The division of control of industrial processes between the Agencies and 22.11
local authorities (the enforcing authorities) is made by section 2 of the 1990
Act and Schedule 1 to the Environmental Protection (Prescribed Processes
and Substances) Regulations 1991.[36] The types of processes are grouped in
much the same way as in Annex 1 to the "Air Framework" Directive.
Schedule 1 divides types of processes in each group into either Part A or Part
B; Part A processes are the responsibility of the Agencies while those in Part
B will be dealt with by local authorities.

Chapter 5 of Schedule 1 to the "Processes" Regulations sets out which 22.12
waste disposal or recovery operations fall under central or local control. For
these purposes waste means solid, liquid or gaseous wastes. Section 5.1 of
the Chapter is concerned with incineration. The destruction by burning in an
incinerator of any waste chemicals or waste plastic arising from chemicals or
plastics manufacture comes under central control. The deliberate, as opposed
to incidental, burning of bromine, cadmium, chlorine, fluorine, iodine, lead,
mercury, nitrogen, phosphorus, sulphur or zinc will be dealt with by the
Agencies as will the incineration of any other wastes on premises where there
is plant designed to incinerate it at a rate of one tonne or more per hour.
However, if that process is related to one falling into Part B the Agencies
will not be involved. In addition where metal containers used for storing or
transporting chemicals are cleaned by burning out their contents, the process
will be centrally controlled.[37]

The Agencies will also control waste recovery processes as defined in 22.13
section 5.2 of Chapter 5. No such processes are designated for local control.
Similarly the production of fuel from waste is centrally controlled. This is
defined as the making of solid fuel from waste by any process involving the
use of heat, other than making charcoal.[38] Where waste is used as a fuel or
otherwise in a prescribed process that is not included in Chapter 5 of
Schedule 1 the disposal or treatment of that waste in the course of the
process will be added to the description of it in the relevant paragraph in
Schedule 1; regardless of how the operator of the process came by it.[39]

Local authorities will be concerned with the incinerators described in Part 22.14
B to section 5.1. These are those used for the burning of any waste,
including animal remains, unless they are related to a Part A process. An
exemption is provided for incinerators on premises where the plant is
designed to incinerate waste at a rate of not more than 50 kilogrammes per
hour; the weight being determined by reference to the weight at the time
the waste is fed into the incinerator. However, this exemption does not
extend to such incinerators that burn clinical waste,[40] sewage sludge, sewage
screenings or municipal waste. If advantage can be taken of this exemption a
waste management licence will not be required either.[41] In addition local
authorities will regulate crematoria.

[36] S.I. 1991 No. 472, as amended by S.I.'s 1991 No. 836, 1992 No. 614, 1993 Nos. 1749 &
2405 and 1994 Nos. 1271 & 1329.
[37] ibid. Sched. 1, Chap. 5, s.5.1, Part A.
[38] ibid. Sched. 1, Chap. 5, s.5.3.
[39] ibid. Sched. 2, para. 8.
[40] As defined in S.I. 1992 No. 588—see S.I. 1994 No. 1271, Sched. 1, para. 8.
[41] S.I. 1994 No. 1056, Sched. 3, para. 29.

22.15   By virtue of section 4(4) of the 1990 Act the Secretary of State may issue a direction to transfer a process in Part B from local authority control to that of one of the Agencies. Such a direction can be made in respect of a type of process or for one operated by a particular undertaking. However, a process in Part A cannot be transferred to local authority control by such a direction.

22.16   Not all the listed processes are controlled. Regulation 4 of the "Processes" regulations provide a number of exemptions where, for example the use of the process would only cause a trivial amount of pollution or it is a process carried on as a domestic activity in connection with a private dwelling. Further exceptions are set out in the rules for the interpretation of Schedule 1 provided by Schedule 2 to the regulations.[42]

22.17   Where a process is controlled some confusion may arise as to which category it belongs or how many processes are on a premises. Schedule 2 to the regulations provides a set of rules for determining this issue and states what forms part of a controlled process. In cases of difficulty the enforcing authority that is considered by the operator to be responsible for the process should be approached.

AUTHORISATION OF PROCESSES

22.18   By section 6 of the Act all processes must be authorised by the relevant enforcing authority. An application for authorisation must be made in accordance with the Environmental Protection (Applications, Appeals and Registers) Regulations 1991.[43] This must give details of, amongst other things the process, the substances to be emitted from it and the techniques to be used to prevent those substances from causing environmental harm. Information about the application must be advertised in newspapers in the locality in which the process will be carried on.[44] The applicant must also pay the prescribed fee.[45] Once made, the application will be determined by the enforcing authority under the provisions of paragraphs 2 and 5 of Schedule 1 to the 1990 Act. However, the Secretary of State may direct that certain applications should be transmitted to him for determination.[46] Schedule 3 to the "Processes" regulations lays down dates by which applications to carry on an "existing" process should have been made. As far as waste incineration processes are concerned those were not later than October 31, 1992 for Part A plants in England and Wales and September 30, 1991 for those in Part B. For Scotland the date is the same for Part A plants and July 31, 1991 for Part B.

22.19   In determining an application the enforcing authority must ensure that the objectives set out in section 7(2) of the Act can be met.[47] The principal

[42] S.I. 1994 No. 1056, Sched. 2, para. 5.
[43] S.I. 1991 No. 507.
[44] *ibid.* reg. 5.
[45] E.P.A. 1990, s.6(2) as amended by E.A. 1995, Sched. 22, para. 48.
[46] *ibid.* Sched. 1, paras. 3 & 4 and S.I. 1991 No. 507, reg. 8.
[47] *ibid.* s.7(1).

objective is that in carrying on the process the best available techniques not entailing excessive cost (BATNEEC) will be used to prevent the release of prescribed substances into the air or, if that is not possible, to reduce any release to the minimum and render it harmless. In addition BATNEEC must also be used to render harmless any other substance that might damage the environment if released from the process.[48] Substances are prescribed for this purpose in Schedule 4 to the "Processes" regulations.[49] They are; oxides of sulphur and other sulphur compounds, oxides of nitrogen and other nitrogen compounds, oxides of carbon, organic compounds and partial oxidation products, metals, metalloids and their compounds, asbestos (suspended particulate matter and fibres) glass fibres and mineral fibres, halogens and their compounds, phosphorous and its compounds and particulate matter.

Guidelines as to the meaning of BATNEEC are contained in the **22.20** "Secretary of State's Guidance—Introduction to Part I of the Act".[50] It should be noted that the United Kingdom uses the expression "Best Available Techniques" as opposed to "Best Available Technology". "Techniques" is a term that embraces both the process and how it is operated. Thus it covers the technologies used in the process and their inter-relationship as well as such matters as numbers of staff employed to operate it, their training and supervision and also the design, construction, lay-out and maintenance of buildings.[51] "Available" means procurable by the operator in question while "best" means the most effective in pollution abatement although there may be more than one best available technique for a particular process.[52] "Not Entailing Excessive Costs" has to be looked at in the context of new or existing processes. For new processes the best available techniques should be used unless the costs of applying them would be excessive in relation to the nature of the industry and the environmental protection to be achieved.[53] For processes existing before July 1987 the guidance in Article 13 of the "Air Framework" Directive is followed in that adaptation should be gradual taking into account the matters set out in that article. BATNEEC may be expressed as a performance standard.[54] Local enforcing authorities must keep up with developments in techniques and technologies in respect of processes falling under their control.[55]

The other objectives of section 7(2) must also be met. These are that any **22.21** directions given by the Secretary of State to implement international or Community obligations, and any requirements of regulations made under relevant pollution control enactments, must be complied with.[56] Under section 3(5) of the 1990 Act the Secretary of State may make an air pollution plan in respect of particular substances. Any requirements in such a plan applicable to the grant of authorisations must be complied with.[57]

---

[48] E.P.A. 1990, s.7(2)(a).
[49] S.I. 1991 No. 472.
[50] E.P.A. 1990, Part I GG1 (91) paras. 20–32.
[51] *ibid.* s.7(10) and GG1 (91) para. 23.
[52] GG1 paras. 24 & 25.
[53] *ibid.* para. 27.
[54] *ibid.* paras. 28–32.
[55] E.P.A. 1990, s.4(9) as substituted by E.A. 1995, Sched. 22, para. 46(9).
[56] E.P.A. 1990, s.7(2)(b) & (c) and see s.7(12).
[57] *ibid.* s.7(2)(d).

22.22   For processes coming under local authority control guidance on meeting these objectives is contained in the process guidance notes made under section 7(11). These will set emission limits and controls and prescribe the appropriate monitoring, sampling and measurement of emissions regime for the particular process concerned. It will also cover the way in which materials are to be handled and stored at the plant, flue gas treatment, the disposal of residues and provide requirements for chimneys. Finally it will set out standards of general operation which will include standards of staff training and action on equipment malfunction or breakdown.

22.23   The enforcing authority will ensure the objectives are met by imposing conditions on the authorisation to this end. Such conditions are imposed for environmental protection, they cannot be used to protect the health and safety of those working on the relevant premises.[58] Nor can the authorisation regulate the final disposal of any controlled waste generated to land[59]; although the enforcing authority must exercise its functions so as to ensure that the relevant objectives in Article 4 of the Waste Framework Directive are met.[60] Further, a court is likely to strike out a condition that was not relevant to pollution control or, for local authorities, air pollution control.[61] However, in addition to conditions imposed to meet objectives, an authorisation may also impose limits on the amount or composition of any substance produced by or used in the process in any period and to require advanced notification of any proposed change in the way that the process is carried on.[62] The Secretary of State may give directions to enforcing authorities as to conditions that are, or are not, to be included in authorisations, whether generally or specifically.[63]

22.24   The specific conditions imposed in an authorisation will be supplemented by the one implied into all authorisations by section 7(4). This is a general condition that the person operating the process must use BATNEEC to prevent the release of prescribed substances to any environmental medium or, where that is not practicable, to reduce release to a minimum and render them harmless and to render harmless any other substances that might be released. For Part B processes this will only apply to releases to the atmosphere.[64] The general condition will not apply where a specific condition has been imposed under section 7(1) in relation to a particular aspect of the process.[65] However, the general condition may be specifically imposed in an authorisation.

22.25   An enforcing authority, having duly considered the application may either grant the authorisation subject to the relevant conditions or refuse the application.[66] An application should not be granted if the enforcing agency

---

[58] E.P.A. 1990, s.7(1).
[59] *ibid.* s.28(1).
[60] S.I. 1994 No. 1056, Sched. 4, para. 8.
[61] *Attorney-General's Reference (No. 2 of 1988)* [1989] 3 W.L.R. 397.
[62] E.P.A. 1990, s.7(8).
[63] *ibid.* s.7(3).
[64] *ibid.* s.7(5).
[65] *ibid.* s.7(6).
[66] *ibid.* s.6(3).

considers that the applicant will not be able to carry on the process in a way that complies with the proposed conditions.[67] The Secretary of State may direct an authority to grant or refuse any particular authorisation.[68] Where an authorisation is refused by the enforcing authority the applicant may appeal against that refusal to the Secretary of State; unless the refusal implemented a direction of the Secretary of State.[69] An applicant may also appeal against any condition imposed in the authorisation.[70]

The process operator must also pay the charges due on the grant of the 22.26 authorisation and during its subsistence. A local authority's charges will be set out in a scheme made under section 8(2) of the Environmental Protection Act 1990. Those of the Agencies will be made in schemes under section 41 of the Environment Act 1995. Failure to pay the relevant charges may result in revocation of the authorisation.[71]

VARIATION AND TRANSFER OF AUTHORISATIONS

It is the duty of each enforcing authority to review each authorisation for 22.27 which it has responsibility. A review of an authorisation should be carried out at least once every four years.[72] The review will take place against the background of the enforcing authority's duty to keep up with improvements in pollution abatement measures.[73] The authority will look at the conditions of the authorisation and determine whether it should issue a variation notice under section 10 of the Environmental Protection Act 1990 in respect of them.

The authority may vary an authorisation at any time, not just on the 22.28 regular review. It must vary one if it appears to it that the conditions to which the authorisation is subject no longer meet the objectives set out in section 7(2) of the Act.[74] This will usually mean that advances in polluton control techniques result in the process not employing BATNEEC to deal with its emissions. For these purposes an authority will vary an authorisation by adding to its conditions or varying or rescinding any of them.[75] The authority will exercise its powers under section 10 of its own accord or on the direction of the Secretary of State. The Secretary may also direct the authority not to exercise its powers in any case.[76]

Where the authority determines to vary the authorisation it should serve a 22.29 variation notice on its holder specifying the variations to be made and the dates on which they are to come into effect.[77] The notice will also require the

---

[67] E.P.A. 1990, s.6(4).
[68] *ibid.* s.6(5).
[69] *ibid.* s.15(1)(a).
[70] *ibid.* s.15(1)(b).
[71] *ibid.* s.8(8).
[72] *ibid.* s.6(6).
[73] *ibid.* s.4(9) as amended by E.A. 1995, s.5(4).
[74] *ibid.* s.10(1).
[75] *ibid.* s.10(8).
[76] *ibid.* s.10(6).
[77] *ibid.* s.10(2) & (3).

holder to notify the authority within a set time of the action he proposes to take to meet its provisions and to pay any fee that may be payable as a result of the notice.[78] A variation notice may itself be varied by a further notice issued under section 10(3A). Unless the notice is withdrawn the variations set out in the final notice will take effect on the date or dates specified. A person on whom a notice is served may appeal against it to the Secretary of State unless it implements a direction of his.[79]

22.30   Where the notice will require a "substantial change" in the operation of the process a different procedure will apply. For these purposes a substantial change means a substantial change in the substances released from it, or in the amount or any other characteristic of any substance so released.[80] In General Guidance 1 more detail is provided as to what is a "substantial change".[81] In particular this is a change that has the potential to change significantly the nature of emissions to the air from the process, the introduction of new plant that would, if it stood alone, require authorisation in its own right and cases where the alterations in the process mean that it no longer resembles the process that was originally authorised. Where it is determined that the variation will result in a substantial change the procedure of advertisement and consultation laid down in paragraph 6 of Schedule 1 to the 1990 Act must be followed. Any representations received as a result of this procedure must be considered by the authority before finally determining the variations to be made. If a further variation notice is served under section 10(3A) as a result of such representation the publicity requirements in paragraph 6 will not apply to that notice.[82]

22.31   The operator of an authorised process may wish to change it in a way that would potentially alter the substances released from it or affect the amount or any other characteristic of any such substance. In such a case he would be advised to follow the procedures in section 11 of the Act or risk enforcement action. The approach of the section is that it is for the holder of the authorisation to apply for variation but he may ask the authority for a determination as to the extent to which variation is needed. If the alterations would result in a "substantial change" in the way the process is operated the proposals must be advertised and account must be taken of any representations received.[83]

22.32   Applications will be made in accordance with regulation 3 of the Environmental Protection (Applications, Appeals and Registers) Regulations 1991.[84] On receiving the application for the variation, and the appropriate fee, the authority will consider it in the light of the objectives set out in section 7(2); unless it is the subject of the Secretary of State's call-in powers under paragraphs 8 or 9 of Schedule 1 to the 1990 Act. The application

[78] E.P.A. 1990, s.10(4) as substituted by E.A. 1995, Sched. 22, para. 51(4).
[79] *ibid.* s.15(2) aa amended by E.A. 1995, Sched. 22, para. 54(2).
[80] *ibid.* s.10(7).
[81] GG1 paras 47–52.
[82] E.P.A. 1990, Sched. 1, para. 6(1A) as added by E.A. 1995, Sched. 22, para. 93(4).
[83] *ibid.* s.11(4) & Sched. I, Part II, para. 7 and sese S.I. 1991 No. 507, regs 4–7.
[84] S.I. 1991 No. 507, reg. 3 amended by S.I. 1991 No. 836.

should be determined within four months of the date it is received or an agreed longer period.[85] The authority may refuse the application or vary it by issuing a variation notice. The holder may appeal against the refusal or, if he is aggrieved by the variations, the notice, to the Secretary of State.[86]

A holder of an authorisation may transfer it under section 9 of the 1990 **22.33** Act to another person who will carry on the process in his stead. Where an authorisation is transferred the new holder must notify the enforcing authority of it in writing within 21 days of the date that the transfer took place. However, from the date of transfer, the transferee will assume full liability for the authorisation.

ENFORCEMENT

The Agencies have powers of entry and inspection under section 108 of **22.34** the Environment Act 1995. Local authorities will appoint their inspectors to deal with Part B processes. These local inspectors will also have the powers of entry and inspection provided under section 108 as they are appointed by "local enforcing authorities".[87] All enforcing authorities have power under section 19(2) of the 1990 Act to require anyone to provide them with specified information that they need to discharge their functions under the Act.

An authority may serve the holder of an authorisation with an enforce- **22.35** ment notice where it considers that he is, or is likely to be, operating his process in breach of conditions. The notice must set out the authority's opinion, specify what the contravention is or is likely to be, set out what should be done to remedy the situation and provide a time for compliance.[88] The authority may withdraw a notice it has issued.[89] An appeal can be made to the Secretary of State against such a notice; although if the Secretary has directed an authority to issue a notice to a holder there will be no appeal against such a notice.[90]

Where the authority consider that the carrying on of any authorised **22.36** process, or doing so in a particular manner, involves an imminent risk of serious environmental pollution, it must serve a prohibition notice on the holder.[91] A notice can be served here even if the process is being operated in accordance with all conditions. It must state the authority's opinion, specify the risk involved, set out the remedial action to be taken and provide a time limit for compliance and direct that the process may not continue to be operated to a specified extent until the notice is withdrawn.[92] An appeal may be made to the Secretary of State against a notice[93] or the authority may withdraw it when it is satisfied that all the remedial steps have been taken.

[85] E.P.A. 1990, Sched. 1, para. 10.
[86] *ibid.* s.15(1)(c) & (2).
[87] E.A. 1995, s.108(15).
[88] E.P.A. 1990, s.13(1) & (2).
[89] *ibid.* s.13(4) as added by E.A. 1995, Sched. 22, para. 53.
[90] *ibid.* ss.13(3) and 15(2) as amended by E.A. 1995, Sched. 22, para. 54(2).
[91] *ibid.* s.14(1).
[92] *ibid.* s.14(3).
[93] *ibid.* s.15(2).

22.37   An authority may at any time revoke an authorisation by serving a notice in writing to that effect on its holder.[94] In particular it may revoke where it has reason to believe that the relevant process has not been carried on or has been disused for a year or more. The notice will specify a date on which it will come into force. This should be at least 28 days after it was served.[95] Revocation should be a policy of last resort.[96] Anyone served with a revocation notice may appeal against it to the Secretary of State and the operation of the notice will be suspended until the appeal is finally determined.[97] Alternatively, the authority may, before it takes effect, withdraw the notice or extend the time before which it become operative.[98]

22.38   It will be an offence to operate a prescribed process without authorisation or in breach of any of its conditions or to fail to comply with an enforcement or prohibition notice. These offences are punishable on summary conviction by a fine not exceeding £20,000 or imprisonment for up to three months or both, or, on indictment, by a fine or up to two years imprisonment or both.[99] It should be noted that there is no defence of "reasonable excuse", etc., provided here. Further, in proceedings for failing to comply with the general condition imposed by section 7(4) it will be for the accused to show that there was no other BATNEEC than that used to satisfy the condition.[1] Where a condition requires records to be kept of emission levels, etc., and no entry is made about it the lack of such an entry will be evidence that the emission level or other condition has been breached.[2]

22.39   Where a person is convicted of failing to comply with the conditions of an authorisation or with an enforcement or prohibition notice a court, in addition to, or instead of, imposing any relevant punishment may make an order for him to carry out specified steps to remedy the situation within a set time.[3] If necessary the defendant may apply to the court for an extension or further extension of the time for compliance. During the operation of the order the defendant will not be liable for any offence committed as a result of non-compliance with his authorisation in relation to the matters covered by the order.[4] Further, if the offence has caused damage that can be remedied by the Agencies, the appropriate one can take reasonable steps to remedy the harm caused and recover its costs from the defendant. Before exercising these powers they must obtain the written consent of the Secretary of State or any occupier of the land affected.[5] In a case of failure to comply with an enforcement or prohibition notice the authority, if they consider that criminal action would be pointless, may take proceedings for

[94] E.P.A. 1990, s.12(1).
[95] *ibid.* s.12(3).
[96] GG1 (91) para. 88.
[97] E.P.A. 1990, s.15(1)(d) & (8).
[98] *ibid.* s.12(4).
[99] *ibid.* s.23(1)(a) & (c) and (2) as amended by E.A. 1995, Sched. 22, para. 59(3).
[1] *ibid.* s.25(1).
[2] *ibid.* s.25(2).
[3] *ibid.* s.26(1).
[4] *ibid.* s.23(2) & (3).
[5] *ibid.* s.27, as amended by E.A. 1995, Sched. 22, para. 60.

an injunction in the High Court or, in Scotland, any court of competent jurisdiction, to secure compliance with the notice.[6]

Other offences are also provided by section 23 of the Act. These include 22.40 failure to notify a transfer of an authorisation, the making of false statements or falsifying records and to fail to comply with a court order under section 26 of the Act. If any offence under Part I committed by a body corporate can be shown to have been committed with the consent or connivance or through the neglect of a director or manager of the company the relevant individual will also be guilty of it and can be punished accordingly.[7] Alternatively if someone commits an offence through the act or default of someone else that other person can be charged with, and convicted of, that offence whether or not the first person is proceeded against.[8] Proceedings for an offence cannot be brought against the Crown but the High Court or the Court of Session can, on the application of an enforcing authority, declare that any contravention by the Crown is unlawful.[9]

PUBLIC REGISTERS

Under section 20 of the Environmental Protection Act 1990 enforcing 22.41 authorities must maintain public registers in accordance with regulations 15 to 17 of the Environmental Protection (Applications, Appeals and Processes) Regulations 1991.[10] The registers will contain details of such matters as applications for authorisations, authorisations granted, notices issued in respect of them, the results of any appeals, details of convictions under the Act and particulars of monitoring information or, where that information is considered commercially confidential, a statement based on it that the process is or is not complying with relevant conditions.[11]

It is the duty of each enforcing authority to ensure that the register it 22.42 maintains is available, at all reasonable times, for public inspection free of charge and that the public can obtain copies of entries in it for a reasonable charge.[12] The register may be kept in any form, so that some authorities may use a computerised register.[13] Monitoring information should remain on the register for at least four years, while information about other matters should be retained until four years after it has been superseded by later information.[14]

Where a process operator considers that information about his process 22.43 would, if put on the register, prejudice his commercial interests to an

---

[6] E.P.A. 1990, s.24.
[7] *ibid.* s.157.
[8] *ibid.* s.158.
[9] *ibid.* s.159(2).
[10] S.I. 1991 No. 507.
[11] E.P.A. 1990, s.20(1) & S.I. 1991 No. 507, reg. 15.
[12] *ibid.* s.20(7).
[13] *ibid.* s.20(8).
[14] S.I. 1991 No. 507, reg. 17.

unreasonable degree, he may apply for it to be treated as commercially confidential under section 22 of the Act. In addition if the authority obtains information about the process that it considers might be confidential it should allow the operator time to make representations about it.[15] It will be for the operator to prove his case and a decision as to whether the information is commercially confidential is for the enforcing authority or, on appeal, the Secretary of State.[16] The protection granted by a determination that information is commercially confidential will only last for four years unless an application is made for an extension for another four years. More than one such application can be made.[17] Information will also be excluded from a register if the Secretary of State considers that it would be contrary to the interests of national security to include it.[18]

AUTHORISATIONS AND WASTE MANAGEMENT LICENCES

22.44   Under the Collection and Disposal of Waste Regulations 1988[19] a waste disposal licence was required for the use of plant or equipment to incinerate waste.[20] In addition a licence was also necessary for the use of untreated waste as fuel to produce electricity or heat.[21] This led to a potential overlap between the controls in Part 1 and Part 2 of the Environmental Protection Act 1990. To avoid dual control the Disposal of Controlled Waste (Exceptions) Regulations 1991[22] were made under the Control of Pollution Act 1974 but will be valid under section 33 of the Environmental Protection Act 1990.

22.45   The regulations provide exceptions to the requirement for activities that involve the deposit of controlled waste or the use of plant or equipment to dispose of it to be licensed. All waste disposal and recycling activities in Chapter 5 of Schedule 1 to the "Processes" regulations that are centrally controlled are exempted from licensing as long as they are subject to a current authorisation.[23] The provisions in the Exceptions Regulations are repeated in paragraphs 3 or 24 of Schedule 3 to the Waste Management Licensing Regulations 1994[24] along with other exceptions, so as allow the storage of material to be burnt at the incineration site, subject to any conditions laid down. While no licence is required the operation would have to be registered with the local enforcing authority.[25]

22.46   The Schedule to the Exceptions Regulations sets out the exemptions for processes subject to local control.[26] Thus the pulverising of bricks, tiles or

[15] E.P.A. 1990, s.22(4).
[16] *ibid.* s.22(1) & (5).
[17] *ibid.* s.22(8).
[18] *ibid.* s.21.
[19] S.I. 1988 No. 817.
[20] 1988 Regulations, Sched. 5, para. 1.
[21] *ibid.* Sched. 5, para. 8.
[22] S.I. 1991 No. 508.
[23] "Exceptions" Regulations, reg. 2(1)(a).
[24] S.I. 1994 No. 1056.
[25] *ibid.* reg. 18(10).
[26] S.I. 1991 No. 508, reg. 2(1)(b).

concrete to extract metals from mixed scrap or the depositing of glass used in making need not be licensed.[27] For the other exceptions, a distinction has to be drawn between the exemption for the burning of animal remains and the use of untreated straw, poultry litter or wood as fuel and the rest. The incineration of such remains is exempted from waste management licensing, as is the deposit of such remains at the incinerator site and other operations ancillary to the burning that form part of the process.[28] Similarly the use of straw, etc., as fuel encompasses any related activity as long as it forms part of the process.[29] The other exemptions only apply to the burning itself.

Thus the use of tyres as a fuel and any related fuel feeding is exempted 22.47 from waste disposal licensing.[30] However, the incidental storage, handling or shredding of such tyres is not included in the process[31] but these matters are exempted by the 1994 Regulations.[32] Similarly the operation of a scrap metal furnace and the loading or unloading of such a furnace, in as much as it forms part of a prescribed process does not need a licence,[33] but, the incidental storage or handling of scrap that is to be heated, other than the actual loading of the furnace, is not part of the process[34] and therefore exemptions fall under paragraphs 41, 42 or 44 of the 1994 Regulations. Finally, waste incineration as part of a process described in paragraph (a) of Part B of section 5.1 of Schedule 1 to the "Processes" regulations is exempted, under both the Exceptions Regulations and paragraph 29 of the 1994 Regulations.

OTHER CONTROLS

Emissions of smoke from any premises or any dust steam, smell or other 22.48 effluvia arising on industrial, trade or business premises that is prejudicial to health or a nuisance will be statutory nuisances for the purposes of section 79 of the Environmental Protection Act 1990.[35] These provisions could therefore apply to emissions from waste incinerators. However, if the incinerator is subject to Part I of the 1990 Act then a local authority will only be able to bring proceedinqs in respect of it with the consent of the Secretary of State.[36]

Before the entry into force of Part I of the Environmental Protection Act 22.49 1990, emissions to the atmosphere from incinerators may have been controlled by the Clean Air Acts 1956 and 1968. However, by virtue of sections 16A and 11A of those Acts,[37] those controls no longer apply once a

---

[27] S.I. 1991 No. 508, Sched. 3, paras 5 & 6 and see S.I. 1994 No. 1056, Sched. 3, para. 24.
[28] *ibid.* Sched. 3, para. 7 and see S.I. 1994 No. 1056, Sched. 3, para. 3(a) & (b).
[29] *ibid.* Sched. 3, para. 2.
[30] *ibid.* Sched. 3, para. 3.
[31] "Processess" Regulations, Sched. 2, para. 5(a).
[32] S.I. 1994 No. 1056, Sched. 3, para. 3(d) & (e).
[33] "Exceptions" Regulations, Sched. 3, para. 4 and see S.I. 1994 No. 1056, Sched. 3, para. 2.
[34] "Processess" Regulations Sched. 2, para. 5(b).
[35] E.P.A. 1990, s.79(1)(b), (d).
[36] *ibid.* s.79(10).
[37] Added by E.P.A. 1990, Sched. 15, paras 6 and 12.

process prescribed under Part I has received an authorisation or such an authorisation has been refused—whether on appeal or otherwise.

PLANNING PERMISSION FOR INCINERATORS

22.50   An incinerator is, it is submitted, "waste development" for the purposes of article 8 of the General Development Procedure Order 1995[38]—development for the purpose of disposing of refuse or waste materials and the land it is on is put to such a use. Thus any application for permission will have to be given the publicity required by that article. There will also have to be consultation with the Agency under paragraph (r) of the table to article 10 of the Order if waste will be deposited at the incinerator before it is burnt. If special waste will be incinerated there must be an environmental assessment, otherwise the planning authority may require such an assessment.[39]

22.51   The relationship between planning and pollution controls was dealt with by the Court of Appeal in *Gateshead Metropolitan Borough Council v. The Secretary of State for the Environment*[40] a case involving a clinical waste incinerator. In that case the court held that if during the planning process it becomes clear that some of the discharges were bound to be unacceptable, so that the Agency would refuse authorisation, then planning permission should also be refused. Otherwise authorisation should be left to the relevant Agency as the grant of planning permission does not prevent a refusal.

22.52   Waste incinerators are considered in paragraphs 5.16–5.19 of P.P.G. 23. There it is said that planning authorities will need to take into account visual impact, noise, storage facilities, traffic considerations and transport requirements. The impact of emissions however should only be taken into account by planning authorities to the extent that they have land use implications and are not controlled by the appropriate pollution control authority.

22.53   Planning conditions should not therefore deal with chimney heights or use of pollution abatement equipment. Nor, where storage areas require a waste management licence should they deal with site drainage, etc., however, they can validly deal with other matters. In particular these can include hours of operation, control of noise and site restoration. Yet while a restoration condition may be appropriate this could not extend to those available under section 72(5) of and Part I of Schedule 5 to the Town and Country Planning Act 1990 [Sched. 3, 1997 Act][41] as the operation of an incinerator is unlikely to the held to be the "depositing of refuse or waste materials" for the purposes of those provisions.

FIRES ON TRADE PREMISES

22.54   Under section 2(1) of the Clean Air Act 1993 it is an offence for dark smoke to be emitted from any industrial or trade premises. However, if the smoke is emitted from a chimney controlled under section 1 of the Act, or if

---

[38] S.I. 1995 No. 419.
[39] S.I. 1988 No. 1199, Sched. 1, para. 9 and Sched. 2, para. 11(c).
[40] [1995] Env. L.R. 37.
[41] As amended by Planning and Compensation Act 1991.

it is emitted in the course of a process prescribed under Part I of the Environmental Protection Act 1990[42] then this offence will not apply. Otherwise if dark smoke is emitted from the premises on any day their occupier[43] will be liable on summary conviction for the offence to a fine not exceeding £20,000. However, it will be a defence to such proceedings to show that the emission complained of was inadvertent and that all practicable steps had been taken to prevent or minimise the emission of dark smoke.[44]

For these purposes, industrial and trade premises mean any premises used **22.55** for industrial or trade purposes or premises that are not so used but on which matter is burnt in connection with any industrial or trade purpose.[45] In *Sheffield City Council v. A.D.H. Demolition Ltd*[46] it was held that a vacant demolition site constituted "premises" despite the absence of buildings or structures on it and that the burning of demolition rubbish on the site that caused dark smoke was an offence under section 2.

Smoke is defined in section 64(1) of the Clean Air Act 1993 to include **22.56** soot, ash, grit and gritty particles emitted in smoke. Dark smoke means smoke which if compared with a Ringelmann Chart[47] would appear to be as dark as or darker than shade two on the chart. However, a court can rely on other evidence to establish that the smoke emitted was "dark smoke".[48]

The Clean Air (Emission of Dark Smoke) (Exemption) Regulations 1969[49] **22.57** exempt the burning of certain matter from the operation of section 2(1) as long as the conditions applicable to the burning of the matter in question are complied with. In particular timber and other waste materials from a demolition site can be burnt as long as there is no other reasonably safe and practicable manner of disposing of it, that the burning is carried out in a way that minimises the emission of dark smoke and that the fire is under the direct and continuous supervision of the occupier or one of his employees.

Where the smoke is not "dark" it may be a statutory nuisance under **22.58** section 79(1)(b) of the Environmental Protection Act 1990.[50] Smoke is defined in the same way as in the 1993 Act. Fumes and gases are dealt with under section 79(l)(c), while dust, steam, smell or other effluvia from trade or industrial premises may be a statutory nuisance under section 79(1)(d) of the Act.

CABLE BURNING

A person who burns insulation from a cable with a view to recovering **22.59** metal from it will be guilty of an offence under section 33 of the Clean Air Act 1993 unless the burning is part of a process subject to Part I of the

---

[42] Clean Air Act 1993, s.41.
[43] See *ibid*. s.64(2).
[44] *ibid*. s.2(4).
[45] *ibid*. s.2(6).
[46] (1984) 82 L.G.R. 117.
[47] British Standard BS 2742:1969.
[48] Clean Air Act 1993, s.3(2).
[49] S.I. 1969 No. 1263.
[50] See also E.P.A. 1990, s.79(3).

Environmental Protection Act 1990. Contravention of this provision will render a person liable on summary conviction to a fine not exceeding level 5 on the standard scale.

23. DISPOSAL AT SEA

INTERNATIONAL CONTROLS

The London Convention on the Prevention of Marine Pollution by 23.01
Dumping of Wastes and other Matter 1972 (the London Convention)
provides a global mechanism to regulate dumping at sea, while the Oslo
Convention for the Prevention of Marine Pollution by Dumping from Ships
and Aircraft 1972 (the Oslo Convention) is concerned with the seas around
Europe. In addition the declarations of the Conferences on the Protection of
the North Sea of 1987 and 1990 have also dealt with the disposal of waste
into that sea. The Oslo Convention will be replaced by the Convention for
the Protection of the Marine Environment of the North-East Atlantic 1992,
when this enters into force.

The London Convention[1] requires all parties to take all practicable steps 23.02
to prevent the pollution of the sea by the dumping of waste and other
matter that is liable to create hazards to human health, harm living resources
and marine life, to damage amenities or to interfere with other legitimate
uses of the sea. Annex I to the Convention comprises a list of substances that
may not be disposed of to the sea at all, while Annex II lists wastes that may
be dumped under a permit issued by a contracting party. Other wastes that
are not listed may be dumped under a general permit. Permits are only to be
issued after all the factors in Annex III have been taken into consideration.
Each party must take steps to prevent unlawful dumping in their territory or
by vessels under their jurisdiction.[2] Operations on the high seas may only be
dealt with by the flag state of the vessel concerned.

The Oslo Convention[3] is cast in similar terms to the London Convention 23.03
although there are differences in the substances listed in Annexes I and II.
Dumping will be allowed of substances that are not prohibited under Annex
I by special permit if they fall into Annex II or by authorisation. Before a
permit or authorisation is granted the composition of the substance to be
dumped must be ascertained[4] and the environmental and other criteria in

---

[1] Cmnd. 6486 as amended by Cmnd. 7656 (Incineration of Waste at Sea) and Cmnd. 8555
(Additions to Annexes I and II).
[2] *ibid.* Art. 7.
[3] Cmnd. 6228.
[4] Oslo Convention, Art. 12.

Annex III taken into account. Overall supervision of the way in which the Convention is implemented is conducted by the Oslo Commission. Every party must send the Commission records concerning substances dumped under permits or authorisations.

23.04   In September 1992 a Convention for the Protection of the Marine Environment of the North-East Atlantic was agreed by the parties to the Oslo Convention and some other states. This new Convention, when it enters into force, will replace the Oslo Convention and the Paris Convention for the Prevention of Marine Pollution from Land-Based Sources to provide a unified pollution control regime for the North-East Atlantic.

23.05   The Declarations following the Second and Third North Sea Conferences have included sections of dumping at sea. Under the second conference declaration (1987) it was accepted that dumping of polluting materials into the North Sea should be ended at the earliest practical date and that from January 1989 no material should be disposed of in the sea unless there are no practical alternatives on land and the material will not harm the marine environment—the Prior Justification Procedure. Dumping of industrial waste, other than inert materials of natural origin, such as colliery spoil or substances proven harmless (through the Prior Justification Procedure) should be phased out.[5] The Prior Justification Procedure is operated through the Oslo Commission and licences granted under it must be reviewed every three years.[6]

23.06   The United Kingdom gave an undertaking to the Third Conference that it would end industrial waste dumping at sea as soon as possible and in any event by the end of 1992 other than of those substances covered by paragraph 22(a) of the 1987 Declaration.[7] It also gave the Third Conference a further undertaking that it would stop the dumping of sewage sludge by, at the very latest, 1998. Wastes that states used to dump at sea by ship should not be disposed of to the North Sea by pipeline or in other seas in a way that increases pollution there.[8]

23.07   At present there is no European Community legislation directly concerned with the dumping of waste at sea. A proposed directive on the subject was issued in 1985[9] and amended following consultation with the European Parliament[10] but has not been adopted.

23.08   Any waste material dumped at sea will normally be "Directive waste" and therefore be subject to the provisions of Council Directive 75/442/EEC,[11] as amended.[12] Its deposit in the sea will be a disposal operation by virtue of paragraph 7 of Annex IIA of the directive. The directive is implemented in

[5] Second Declaration (1987) para. 22(a).
[6] Decision of the Oslo Commission, June 14, 1989.
[7] Third Declaration (1990) para. 18.
[8] Second Declaration (1987) para. 22(d).
[9] COM (85) 373; [1985] O.J. C245.
[10] COM (88) 8.
[11] [1975] O.J. L194/39.
[12] [1991] O.J. L78/32.

the United Kingdom by the Environmental Protection Act 1990 and the Waste Management Licensing Regulations 1994.[13]

LICENCES FOR DUMPING WASTE AT SEA

In the United Kingdom the deposit of substances or articles in the sea or their incineration there is regulated by Part II of the Food and Environment Protection Act 1985. Under section 5 a licence is required for deposits in the United Kingdom or United Kingdom controlled water,[14] whether in the sea or under the sea-bed, or from British vessels, hovercraft, aircraft or marine structures, or from floating containers controlled by them, anywhere in the world. **23.09**

Deposits into United Kingdom waters or United Kingdom controlled waters from a structure on land that is constructed or adapted wholly or mainly for that purpose must also be licensed.[15] The terms of this requirement emphasise the difference between the deposit of liquids and solids. There is no distinction in relation to deposits from vessels, etc. However, once the land-based sources—pipelines, etc., become involved the 1985 Act will generally only control deposits of solid matter; discharges of liquid into the sea being regulated under Part III of the Water Resources Act 1991. But if any such discharge is licensed under the 1985 Act then no offence will be committed under the 1991 Act.[16] **23.10**

In addition a licence is required by a person loading a vessel, aircraft, hovercraft, marine structure or floating container in the United Kingdom or United Kingdom waters (or a vehicle in the United Kingdom) with substances or articles to be deposited anywhere in the sea or under the sea-bed.[17] These provisions also regulate the scuttling of vessels at sea. Incineration licences are required under section 6 for the incineration of substances or articles on a vessel or marine structure in the United Kingdom, or United Kingdom controlled waters or on any British vessel or structure anywhere at sea. **23.11**

Exemptions from licensing requirements are provided by the Deposits in the Sea (Exemptions) Order 1985.[18] The effect of these exemptions has been reduced by articles 4 and 5 of the Order which were added by regulation 21 of the Waste Management Regulations 1994.[19] Under those articles the exemptions only apply in the case of an establishment or undertaking involved in the recovery or disposal of waste if it is carrying out its own **23.12**

---

[13] S.I. 1994 No. 1056.
[14] Food and Environment Protection Act 1985, ss.5(a) and 24(1) as amended by E.P.A. 1990, s.146
[15] *ibid.* s.5(a)(iii).
[16] W.R.A. 1991, s.88(1)(d).
[17] Food and Environmental Act 1995, s.5(f) & (g).
[18] S.I. 1985 No. 1699 as amended by S.I. 194 No. 1056, reg. 21.
[19] S.I. 1994 No. 1056.

waste disposal at the place where the waste was produced or is recovering waste by means of an operation listed in Part IV of Schedule 4 to the Regulations. Further, the disposal or recovery operation must be consistent with the aim of avoiding environmental harm or nuisance and must be registered with the appropriate licensing authority.

23.13　A licence will be issued by the Minister—the Minister for Agriculture, Fisheries and Food or the Secretary of State for Scotland or Northern Ireland—responsible for fisheries at the place where the licensed operation would take place or commence.[20] The Ministers are licensing authorities for the purposes of the Act and also "competent authorities" for the purposes of the implementation of the amended waste directive.[21]

23.14　An application for a licence will be made to the appropriate Minister who may require further information or tests to be carried out in order to decide whether it should be issued.[22] A fee will be charged for processing the application or making any necessary tests.[23] It will be an offence to knowingly or recklessly supply false information in respect of an application or to intentionally fail to disclose any material particular.[24] In determining whether to grant a licence the Minister must have regard to the need to protect the marine environment, its ecosystem and human health and to prevent interference with legitimate uses of the seas.[25] He must have regard to the requirements of article 4 of the "Framework Waste" Directive[26] and should also consider practical alternatives to sea disposal for the articles or substances concerned.[27]

23.15　In granting a licence the Minister must attach conditions to it for the protection of the marine environment and ecosystem and human health and to prevent interference with legitimate uses of the sea as he considers necessary or expedient and may add other conditions that he considers appropriate.[28] A licence must also cover the type and quantities of waste involved, the technical requirements, the security precautions to be taken, the disposal site and any treatment methods.[29] In addition a condition may require further consent to be obtained before specified operations are carried out or, in a licence authorising loading operations, that the deposit, etc., should take place at a specified site.[30] Automatic recording equipment may also be required so as to ensure operations take place at specified sites. The record produced by such equipment will be evidence of the matters appearing in it.[31]

---

[20] Food and Environment Protection Act 1985, s.24(1).
[21] S.I. 1994 No. 1056, Sched. 4, para. 2.
[22] Food and Environment Protection Act 1985, s.8(5).
[23] ibid. s.8(7) & 8(8).
[24] ibid. ss.9(2) and 21(3) & (4).
[25] ibid. s.8(1).
[26] S.I. 1994 No. 1056, Sched. 4, para. 4.
[27] F.E.P.A. 1985, s.8(2).
[28] ibid. s.8(3).
[29] S.I. 1994 No. 1056, Sched. 4, para. 6.
[30] F.E.P.A. 1985, s.8(4)(a)(i) & (b).
[31] ibid. s.8(3)(a)(ii) & (6).

If the Minister refuses to grant a licence or grants one subject to 23.16 conditions, the applicant may ask him for his reasons, and, if still dissatisfied, may appeal against that decision[32]; the procedure for which is set out in Schedule 3 to the 1985 Act. Fees may be charged for supervising the operation of the licence.[33] The Minister may vary or revoke a licence, either because there has been a breach of its conditions or because he considers he ought to do so due to a change in circumstances relating to the marine environment, its ecosystem or human health or increased scientific knowledge concerning them or for any other relevant reason.[34] The licensee may obtain the reasons for such a decision and appeal against it in accordance with Schedule 3.[35]

An establishment or undertaking responsible for a licensed operation, but 23.17 not an exempt activity, must keep a record of the quantity, nature, origin and where relevant, the destination, frequency of collection, mode of transport and treatment method of any waste which is disposed of or recovered. Such records must be made available, on request, to the licensing authority.[36]

The Ministers are under a duty, imposed by section 14 of the 1985 Act,[37] 23.18 to keep public registers of information relating to licences. Information may be excluded in the interests of national security or if the Minister considers it would prejudice someone's commercial interests to an unreasonable degree.[38] The registers must be available for free public inspection at all reasonable times and copies of entries from it may be obtained on payment of a reasonable charge.[39]

These registers should contain the particulars set out in regulations 3 to 23.19 12 of the Deposits in the Sea (Public Registers of Information) Regulations 1996.[40] These require details to be entered in respect of applications for licences to deposit, licences to scuttle and licences to incinerate, details of those licences, if granted, and details of refusals to grant. Variations and revocations of licences must also be shown, with the prescribed details. Particulars of convictions of offences under section 9 of the Act should be entered, together with information obtained or furnished under the Act in the form required by regulation 11. The Minister will also enter details of remedial operations carried out under section 10 of the Act.

Inspectors to enforce Part II of the Act are appointed by the Ministers 23.20 under section 11. Certificates of their authority must be produced on request.[41] They will have powers of entry where they have reasonable

---

[32] F.E.P.A. 1985, s.8(12) and Sched. 3, paras 1–3.
[33] *ibid.* s.8(8)(b) & (9).
[34] *ibid.* s.8(10) & (11).
[35] *ibid.* Sched. 3, para. 4.
[36] S.I. 1994 No. 1056, Sched. 4, para. 14.
[37] As substituted by s.147 of the E.P.A. 1990.
[38] F.E.P.A. 1985, s.14(2).
[39] *ibid.* s.14(5).
[40] S.I. 1996 No. 1427.
[41] F.E.P.A. 1985, Sched. 2, para. 5.

grounds to believe that any substances or articles intended to be disposed of at sea may be, or may have been, present. These powers extend to land and vehicles in the United Kingdom, foreign vessels, etc., in the United Kingdom or United Kingdom controlled waters and British registered vessels, etc., anywhere.[42] However, they are barred from entering premises only used as a dwelling unless they have a warrant to do so.[43] In addition they have the powers to require information, stop and detain vessels or order them into port, open containers and take samples set out in Schedule 2 of the Act. They also have the powers of Agency inspectors under section 108 of the Environmental Act 1995.[44]

23.21   Anyone who disposes of waste at sea without a licence, where one is needed, or in breach of its conditions or who causes or permits someone else to do so will be guilty of an offence and be liable on summary conviction to a fine of up to £50,000 or on indictment to a fine or to imprisonment for up to two years or both.[45] Company officers may be liable for acts of their company.[46] It will be a defence to such a charge to show that the alleged illegal operation was done in the interests of safety and that steps were taken to inform the relevant Minister about it; but the operation must have been reasonable and not brought about by the defendant's own conduct.[47] It will also be a defence to show that the operation was carried out beyond the United Kingdom or United Kingdom territorial waters under licence from a party to the London or Oslo Convention.[48] A further defence is provided by section 22 of the Act if the defendant can show that he took all reasonable precautions and exercised all due diligence to avoid the commission of the offence.

23.22   An establishment or undertaking which, after December 31, 1994, carries on an exempt activity without being registered by the licensing authority will commit an offence for which it will be liable on summary conviction to a fine not exceeding level 2 on the standard scale.[49]

23.23   Under paragraph 5 of Schedule 4 to the Waste Management Licensing Regulations 1994[50] the licensing authorities must prepare an offshore waste management plan. This should contain an authority's policies in relation to the recovery or disposal of waste so that the relevant objectives set out by the Framework Waste Directive[51] can be met in those waters under the control of the authority. A joint plan may be prepared by two or more authorities. The plan should, in particular, contain statements about the type, quantity and origin of waste to be recovered or disposed of, general

---

[42] F.E.P.A. 1985, s.11.(2).
[43] ibid. s.11(4) and Sched. 2, para. 7.
[44] S.I. 1994 No. 1056, Sched. 4, para. 13 as amended by S.I. 1996 No. 593, Sched. 2, para. 10(d).
[45] F.E.P.A. 1985, s.9(1) and 21(2A) as added by E.P.A. 1990, s.146(6)(b).
[46] ibid. s.21(6) & (7).
[47] ibid. s.9(3) & (4).
[48] ibid. s.9(5)–(7).
[49] S.I. 1985 No. 1699, Arts. 5(1) & (5) as added by S.I. 1994 No. 1056, reg. 21.
[50] S.I. 1994 No. 1056.
[51] See paras 2.02–2.07.

technical requirements, any special arrangements for particular wastes and suitable disposal sites or installations. Copies of the plan should be made available to the public on payment of a reasonable charge.

24. CIVIL LIABILITY FOR WASTE MANAGEMENT OPERATIONS

24.01   Civil liability arising from adverse affects caused by waste management operations is, at present, mainly based on common law tort doctrines; although European legislation has been in the pipeline for sometime. Section 73(6) of the Environmental Protection Act 1990 does impose strict liability for damage consequent on the commission of an offence under sections 33(1) or 63(2) of the Act (which are concerned with the unlawful keeping, treatment and disposal of controlled or uncontrolled waste), and Part IIA of the Environmental Protection Act 1990 is concerned with liability for contaminated land, but otherwise, for the present, the common law remedies must suffice.

A. LIABILITY FOR DAMAGE CAUSED BY A SITE

1. AT COMMON LAW

24.02   Harm may be caused to persons beyond the boundary of a waste disposal site by fires on the site, by gases or leachate produced by the landfill escaping to adjoining land or by movement of the waste itself. The type of harm may be smell from the waste, pollution of water or land by leachate, dangers caused by landfill gases or invasion of property by a tip movement. The most relevant case on these matters is the Canadian decision of *Gertsen v. Municipality of Metropolitan Toronto*[1] which concerned the escape of methane gas from a landfill site to adjoining private garages. The area had been developed before the landfill operation took place but the garages were erected after the site had been closed in 1959. Subsequently in 1965 emissions of gas led to a fire in one of the garages so that residents ceased to use them. However, on receiving reassurances that all was well, they later started using them again. Methane gas continued to escape from the site to the garages and, in 1969, caused an explosion that injured Mr Gertsen. He sued the local authority that brought the waste onto the site and the local authority that operated it. The court found that while those authorities had

---

[1] (1973) 41 D.L.R. (3d) 646.

316

statutory powers concerning waste disposal they had not used them in this case so that the proceedings were dealt with as if the authorities were ordinary persons. It held that both authorities were liable to Mr Gertsen under the rule in *Rylands v. Fletcher*, nuisance and negligence.

Rylands v. Fletcher

In England and Wales strict liability under the rule in *Rylands v. Fletcher* is **24.03** only imposed on those whose accumulations and operations are of such a nature and magnitude that it is not reasonable for nearby occupiers, or their insurers, to carry the risk of accidental escapes from control or containment. The doctrine is not easily applied as it depends on a number of elements being present. These are that the defendant is the owner and occupier of land onto which he has brought or allowed to be brought something that is likely to do mischief if it escapes so that his land can be said to be subject to a special use. The thing must then escape from the land and cause damage to the plaintiff. The rule will only apply if all these factors are present. In *Cambridge Water Co. Ltd v. Eastern Counties Leather plc*[2] it was held that forseeability of harm is an essential element of the principle. Further, the risk of harm has to have been foreseeable at the time at which the operations were carried out.

As far as waste disposal sites are concerned the main question is whether **24.04** the site should be considered as land subject to special use. In *Gertsen* it was held that, "a landfill project as a means of disposing of garbage is a non-natural user of land in a heavily populated residential district." This begs the question as to whether such a use in a rural area could be considered "natural". In *Read v. Lyons*, Lord Porter stated[3] that it,

> ". . . seems to be a question of fact . . . as to whether the particular object can be dangerous or the particular use can be non-natural, and in deciding this question I think that all the circumstances of the time and place and practice of mankind must be taken into consideration so that what might be regarded as dangerous or non-natural may vary according to those circumstances."

The particular dangers of landfill sites are landfill gas and leachates. Landfill gas can harm crops as well as pose a risk to buildings. Thus there are risks both in the town and in the country from sites. For these purposes it will not matter that the gas itself was not brought onto the site by the operator.[4] In the circumstances it is submitted that, despite modern reluctance for courts to apply the rule, it could, according to the particular case, be applied to an escape of gas or leachate from a site. However, in *Delaney v. F.S. Evans & Sons Pty Ltd*[5] it was considered that the rule might not apply to a site authorised by a site licence; although it is doubtful whether this would be followed in the United Kingdom.

---

[2] [1994] 1 All E.R. 53.
[3] [1947] A.C. 156 at 176.
[4] *Miles v. Rock Forest Granite Co.* (1918) 34 T.L.R. 500.
[5] (1985) 58 L.G.R.A. 405 (AUS) at 442–447.

Nuisance

24.05   A nuisance can either be something that physically affects a person's land or that interferes with his enjoyment of land. In *Priest v. Manchester Corporation*,[6] a case in which rain falling on a tip caused a gully to be eroded in a nearby street, it was said that,

> "The defendants may indeed use the land for tipping, but not for tipping in such a way as to cause a nuisance. It might possibly have been different if it were impossible to use the land for tipping without causing a nuisance."

Despite this caveat it is unlikely that a court would consider that the impossibility of using land without causing a nuisance would be a defence to such an action, although it might award damages instead of granting an injunction in respect of it.[7]

24.06   In *Gertsen* the plaintiff succeeded in private nuisance on the grounds that the gas caused a substantial and unreasonable interference with the use of his land. It is submitted that a United Kingdom court would follow this decision. Similarly if leachate from a site flows into a river or seeps into underground water and adversely affects the owner or occupier of land as a result the site operator will be liable.[8] Further, if the waste causes physical damage to adjoining land or chemicals in it harm neighbouring property the site owner will also be liable.[9] In such a case it will not be a defence that waste disposal is done for the benefit of the community.[10]

24.07   The defence of "statutory authority" is unlikely to apply in this situation. In *Harvey v. Truro Rural District Council*[11] it was said that a local authority's consent to something could not legalise that which was otherwise illegal. Thus planning permission for the site does not authorise any nuisance that may emanate from it.[12] However, a waste management licence under the Environmental Protection Act 1990 goes further than mere permission to establish a site in a particular location. It could be argued that it is statutory authority to operate it in accordance with the conditions of the licence.[13] Thus, a site operator might escape liability if he could show that any nuisance caused was inevitable or, to put it another way, that he has used all reasonable care and skill in the light of contemporary knowledge to prevent it.[14] However, this line of argument has been weakened by the decision in *Wheeler and Another v. J.J. Saunders*[15] where it was considered that the courts should be slow to acquiesce in the extinction of private rights without compensation as a result of administrative decisions which cannot be appealed and are difficult to challenge.

---

[6] (1915) 84 L.J.K.B. 1734.

[7] *Miller v. Jackson* [1977] Q.B. 966.

[8] *Ballard v. Tomlinson* (1885) 29 Ch.D. 115.

[9] *Maberly v. Peacock* [1946] 2 All E.R. 192.

[10] *Bamford v. Turley* (1860) 3 B. & S. 65.

[11] [1903] 2 Ch. 638.

[12] *Wheeler & Anor v. J.J. Saunders & Anor* [1996] Ch. 19.

[13] See *F.S. Evans & Sons Pty Ltd v. Delaney* (1985) 58 L.G.R.A. 405 at 445/446 and *Edwards v. Blue Mountains City Council* (1961) 78 Weekly Notes 864 (AUS).

[14] *Manchester Corporation v. Farnworth* [1930] A.C. 171, but see W.M.P. 26, para. 3.98.

[15] [1996] Ch. 19.

Normally an occupier's liability for a nuisance will cease when he gives up 24.08 possession of the land. However, if he has actively created a nuisance he will remain liable even after vacating.[16] This was confirmed in *Fennel v. Robson Excavations Pty*[17] where it was said that liability in nuisance did not depend on occupation. In the American case of *Tadier v. Montgomery County*[18] a local authority used a landfill site for 12 years under a lease that expired in 1962. Subsequently the property changed hands several times until, in 1980, an explosion caused by the build up of methane gas on the site occurred. The court considered the position where an operator has created a state of events on land which he no longer owns and approved earlier decisions in which it had been held that a non-owner or occupier could be liable for the creation of a dangerous condition or nuisance. Thus, if this line of cases is followed in United Kingdom courts, a site operator may be held liable for nuisances caused by a site he has vacated; even if he has been issued with a certificate of completion by the relevant Agency.

Negligence

Where a plaintiff seeks to establish negligence he must first show that the 24.09 defendant owed him a duty of care. In *Caparo Industries plc v. Dickman*[19] It was held that to show this he must show forseeability of damage, proximity of relationship between him and the defendant and that it is reasonable for a court to impose such a duty. As far as proximity is concerned this will be dealt with on the facts, looking at previous cases, to determine whether the particular damage suffered was the kind of damage that the defendant was under a duty to prevent and whether there are circumstances from which the court can pragmatically conclude that a duty of care exists.

In *Gertsen*[20] the defendant site operator was held to owe a duty of care to 24.10 occupiers of property adjacent to the site. In *Delaney v. F.S. Evans & Sons Pty Ltd*[21] an operator was liable to someone five kilometers away from the site whose property was damaged by the spread of a fire that had been lit on it. The fact that a site is operated under a waste management licence will not affect the duty of care if the negligence arises from a breach of licence conditions, although it may be otherwise if the damage occurred in the course of the reasonable conduct of lawful operations.[22]

Once a duty of care is shown to exist it must then be established that the 24.11 defendant has failed to exercise proper standards of care. Operations on a site or in the collection of waste may be regulated by a licence or be dealt with by a code of practice. Codes issued under section 34(7) of the 1990 Act concerning the duty of care of waste managers will be admissable in any proceedings and, if relevant, should be taken into account in determining the

---

[16] *Roswell v. Prior* (1701) 12 Mod. Rep. 635.
[17] (1977) 2 NSWLR 486 (AUS).
[18] (1985) 487 A(2d) 658 (US).
[19] [1990] 1 All E.R. 568, H.L.
[20] *Gertsen v. Municipality of Metropolitan Toronto* [1973] 41 D.L.R. 646.
[21] (1985) Aust. Tort Reports, para. 80–714.
[22] *Casley-Smith v. Evans & Sons Pty Ltd* (1989) Aust. Tort Reports para. 80–227 at 68, 373.

issue.[23] The principles a court will use in looking at a code will follow those of cases concerned with the Highway Code. A breach of that Code,

> "Creates no presumption of negligence calling for an explanation, still less a presumption of negligence making a real contribution to causing an accident or injury. The breach is just one of the circumstances on which one party is entitled to rely in establishing the negligence of the other and its contribution to causing the accident or injury . . . it must be considered with all the other circumstances . . . It must not be elevated into a breach of statutory duty which gives a right of action to anyone who can prove that his injury resulted from it."[24]

Breaches of licence conditions would be treated in the same way.[25]

24.12   In *Gertsen* the site operator was found to have failed to exercise proper standards of care because it allowed methane gas to escape from the site onto adjoining properties and failed to take adequate precautions to prevent the escape. It was also liable for failing to inspect the site properly and to warn adjoining property owners of the potential dangers of methane gas.

24.13   These cases only apply where damage has actually occurred. A court is unlikely to grant relief against prospective losses or compensate a plaintiff for works done in anticipation of damage, such as work to make a house safe against the risk of landfill gas from an adjoining site. In *Midland Bank v. Bardgrove Property Services Ltd*[26] such a claim was classified as economic loss caused by the need to carry out preventive works. It could not form part of a claim for wrongful interference with the plaintiff's land. Nor are damages recoverable as reasonable mitigation of prospective losses or as parasitic damages.

2. STATUTORY LIABILITY

24.14   Under section 73(6) of the Environmental Protection Act 1990 a person who deposits waste on land, or who causes or knowingly permits it to be deposited there in circumstances in which he would commit an offence under section 33(1) or 63(2) of the Act will be liable for any damage caused by that waste. "Damage" here includes not only damage to property but also death or injury to anyone, any disease that may be contracted or any physical or mental disability that may result from the defendant's actions.[27] He need not have actually been prosecuted or found guilty of such an offence. However, if he can show that one of the defences provided in section 33(7) of the Act would apply to his case he will escape liability.[28]

24.15   He will also avoid liability if the damage caused was wholly due to the fault of the person who suffered it.[29] Fault here means negligence, breach of

---

[23] E.P.A. 1990, s.34(10).
[24] Stephenson L.J.: *Powell v. Phillips* [1972] 3 All E.R. 864 at 868.
[25] *The Raithwaite Hall* (1874) 30 L.T. 233, Sir R. Phillimore.
[26] (1990) 24 Con.L.R. 98.
[27] E.P.A. 1990, s.73(8).
[28] *ibid.* s.73(7).
[29] *ibid.* s.73(6).

statutory duty or other act or omission which gives rise to liability in tort. Thus while a plaintiff may be able to recover damages under section 73(6) even if he was partially to blame, his damages may be reduced under the provisions of the Law Reform (Contributory Negligence) Act 1945.[30] A person who voluntarily accepts the risk of suffering damage as a result of the unlawful deposit will have no recourse under this provision.[31]

Part IIA of the Environmental Protection Act 1990 imposes liability for 24.16 the costs of remediation works to contaminated land. Remediation works required to clean up such land may extend to land adjacent to the site causing the contamination.

Liability for economic loss

Often the existence of a waste disposal site or former site will cause no 24.17 physical damage to property or any health risk but will depreciate the value of nearby properties. This may be particularly a problem where people have bought houses and now find that permission has been given for a landfill site in their area. Any depreciation in value will be considered by the courts as "economic loss". It is not physical loss but rather a financial one.

Normally there will be no recovery of such loss under the principle in 24.18 *Rylands v. Fletcher* or in nuisance. Both causes of action require some damage to have been suffered. Whether the fact that someone is living on land that is slightly contaminated, and thus does not fall into the provisions of Part IIA of the Environmental Protection Act 1990 nor forms any health risk, can be said to have suffered interference with the enjoyment of their property is open to question. However, if because of the site they are not allowed to develop their land in the way they would wish or are not granted permits for operations they wish to carry out, this might be sufficient to ground a cause of action in nuisance.[32] While in *Smith v. Jeffrey*[33] it was held that causing a nuisance that diminishes the value of the plaintiff's property was actionable, this decision must be viewed with caution in the light of more recent cases on economic loss.

The position in negligence was clarified by the House of Lords in *D. & F.* 24.19 *Estates Ltd v. Church Commissioners for England*[34] where it was held that a builder, in the absence of any contractual duty or of a special relationship of proximity introducing the *Hedley Byrne* principle of reliance, owes no duty of care in tort in respect of the quality of his work. To hold that there was such a duty of care to anyone acquiring an interest in the property would be to impose on the builder the obligations of an indefinitely transmissible warranty of quality. This decision was affirmed by the House in *Murphy v. Brentwood District Council*[35] where it was held that while there was a duty to

[30] E.P.A. 1990, s.73(9)(c).
[31] *ibid.* s.72(6).
[32] *Exxon Corporation v. Yarema* (1986) 516 A2d 990 (US).
[33] (1886) 2 T.L.R. 480.
[34] [1989] A.C. 177.
[35] [1990] 3 W.L.R. 414.

guard against defects that cause injury or damage, that duty only applies to latent defects. Once the defect is discovered before any such injury or damage, it is no longer latent, and the expense incurred in putting it right is pure economic loss. If there is to be recovery for such damage it should be imposed by statute as in the Defective Premises Act 1972. This principle would certainly extend to houses built on contaminated land and also to those on sites near to a former landfill. Further the "no liability for economic loss rule" would apply to the diminution in value of houses next to operating sites.

24.20 However, in nuisance, if physical damage is caused, or there is interference with the reasonable enjoyment of property, and it can be shown that this has an adverse effect on the value of the property, damages may be recoverable. In *Hunter v. Canary Wharf*[36] it was considered that all nuisances were interferences with the use and enjoyment of land. Where physical damage is caused the measure of damages is diminution in market value. Where there is loss of amenity the measure is the value of that loss that can be measured—for example by loss in notional rental value. Consequential losses may also be recoverable. Further, an innocent buyer of a house on contaminated land or near a landfill might be able to recover from the vendor despite the rule of *caveat emptor*.[37]

B. LIABILITY FOR INJURY ON SITE

24.21 Statutory provision for the protection of employees and others on waste disposal sites or transfer stations is made by Part I of the Health and Safety at Work Act 1974. Of particular relevance may be the Control of Major Accident Hazards Regulations[38] and the Control of Substances Hazardous to Health Regulations 1988.[39] Under the latter Regulations an assessment should have been made of the health risks created by work involving substances prescribed by Schedule 1 to those Regulations.

24.22 The person responsible for premises to which the 1974 Act applies owes his employees a duty under section 2 to ensure that it is safe for them to work on it. He must also give them proper instruction on site safety. In addition he owes inspectors of an Agency or other people who go on the premises in the course of their employment a duty to ensure that they will not be exposed to risks to their health and safety.[40] This duty extends to giving them information or instruction about site safety.[41] Under section 4 of the 1974 Act the person responsible for premises must also ensure, as far as reasonably practicable, that the site is safe and without risk to health. However, this only requires him to take precautions against risks that are reasonably foreseeable.[42]

[36] April 24, 1997, H.L.
[37] *Heighington v. The Queen* (1988) 41 D.L.R. (4th) 208.
[38] S.I. 1984 No. 1902 as amended by S.I. 1988 No. 1462.
[39] S.I. 1988 No. 1657.
[40] Health and Safety at Work Act 1974, s.3.
[41] *Carmichael v. Rosehall Engineering Works* [1983] I.R.L.R. 480.
[42] *Austin Rover Group Ltd v. H.M. Inspector of Factories* [1989] W.L.R. 520.

Under the Occupiers Liability Act 1957 the occupier of premises owes 24.23
people that he invites onto his land a duty to take such care that is
reasonable in the circumstances to see that they will be reasonably safe in
using it for the purposes for which they were invited.[43] This duty is also
owed to anyone, such as a local authority officer, entering in the exercise of a
right conferred by law.[44] However, the occupier is entitled to assume that
those entering to carry out certain duties will appreciate, and guard against,
any risks ordinarily inherent in performing them.[45] Section 5 of the 1957 Act
is concerned with visitors entering premises under contract.

Trespassers are also owed a limited duty under the 1984 Occupiers 24.24
Liability Act. If the occupier of premises has reason to believe that a danger,
such as unsafe surfaces or landfill gas, exists on them and that the risk from
it is one against which he may reasonably be expected to give people some
protection he will owe a duty to take care for the safety of trespassers whom
he has reasonable grounds to suppose to be coming into the vicinity of the
danger.[46] This duty may be discharged by taking reasonable steps to warn of
the danger or to discourage persons from taking the risk[47]; for example by
maintaining adequate fencing and erecting warning signs.[48]

SCOTTISH LAW

In Scottish law harm caused by a waste disposal site will be remedied 24.25
under the law of delict. In general the same principles as to nuisance[49] and
strict liability under *Rylands v. Fletcher* apply in Scotland as they do in
England and Wales, however, in the latter instance liability must be based
on some fault on the part of the defender.[50] The liability of a site operator
for injury to those on the site will be determined under the Occupiers
Liability (Scotland) Act 1960 or Part I of the Health and Safety at Work Act
1974.[51]

---

[43] Occupiers Liability Act 1957, s.2(2).
[44] *ibid.* s.2(6).
[45] *ibid.* s.2(3) and *Roles v. Nathan* [1963] 1 W.L.R. 117.
[46] Occupiers Liability Act 1984, s.1(3) & (4).
[47] *ibid.* s.1(5).
[48] *White v. St Albans City Council, The Times*, March 12, 1990.
[49] *Watt v. Jamieson* [1954] S.C. 56.
[50] Per Lord Fraser: *RHM Bakeries (Scotland) Ltd v. Strathclyde Regional Council* (1985) S.L.T.
214 at 218.
[51] Health and Safety at work Act 1974, s.84(1).

INDEX